Bacteriophage Applications in Biotechnology

Bacteriophage Applications in Biotechnology

Editor: Jayde Mason

Callisto Reference
www.callistoreference.com

Callisto Reference,
118-35 Queens Blvd., Suite 400,
Forest Hills, NY 11375, USA

Visit us on the World Wide Web at:
www.callistoreference.com

ISBN: 978-1-64116-794-9 (Hardback)

Cataloging-in-Publication Data

Bacteriophage applications in biotechnology / edited by Jayde Mason.
p. cm.
Includes bibliographical references and index.
ISBN 978-1-64116-794-9
1. Bacteriophages. 2. Biotechnology. 3. Lysogeny. 4. Transduction. 5. Viruses. I. Mason, Jayde.
QR342 .B33 2023
579.26--dc23

Table of Contents

Preface

This book aims to highlight the current researches and provides a platform to further the scope of innovations in this area. This book is a product of the combined efforts of many researchers and scientists, after going through thorough studies and analysis from different parts of the world. The objective of this book is to provide the readers with the latest information of the field.

A phage or bacteriophage is a virus that infects and replicates within bacteria and archaea. The diversity and abundance of phages in the biosphere makes them prime model organisms for studying the principles of evolution and ecology. Some major areas of research related to phages include the study of phages as promoters of bacterial evolution, novel therapeutic agents, and sources of genetic and diagnostic tools. There are various biotechnological applications of phages and phage-derived proteins in industries such as health, food science and safety, and agriculture. A major therapeutic application of phage is phage therapy, which is seen as a potential therapy for treating pathogenic bacterial infections. In the food industry, phages are extensively used for securing food products and preventing food spoilage caused by bacteria. This book outlines the applications of bacteriophages in the field of biotechnology. It presents researches and studies performed by experts across the globe. The book will prove to be immensely beneficial to students and researchers in this field.

I would like to express my sincere thanks to the authors for their dedicated efforts in the completion of this book. I acknowledge the efforts of the publisher for providing constant support. Lastly, I would like to thank my family for their support in all academic endeavors.

Editor

1

Genomic and Proteomic Characterization of Bacteriophage BH1 Spontaneously Released from Probiotic *Lactobacillus rhamnosus* Pen

Piotr Jarocki [1,*], Elwira Komoń-Janczara [1], Marcin Podleśny [2], Oleksandr Kholiavskyi [1], Monika Pytka [1] and Monika Kordowska-Wiater [1]

1 Department of Biotechnology, Microbiology and Human Nutrition, University of Life Sciences in Lublin, 8 Skromna St., 20-704 Lublin, Poland; elwira.komon.janczara@up.lublin.pl (E.K.-J.); okuplublin@wp.pl (O.K.); mdpuplublin@wp.pl (M.P.); kwmuplublin@wp.pl (M.K.-W.)

2 Process and Development Department, Al. Tysiąclecia Państwa Polskiego 13, Grupa Azoty Zakłady Azotowe "Puławy" S.A, 24-110 Puławy, Poland; mpuplublin@wp.pl

* Correspondence: piotr.jarocki@up.lublin.pl

Abstract: *Lactobacillus rhamnosus* Pen is a human endogenous strain used for the production of probiotic formula, which is effective in the prevention of antibiotic-associated diarrhoea. Our study showed that this probiotic strain releases bacteriophage BH1 without the addition of any inducing agent. Our research revealed that phage BH1 has a circular genome with a length of 40721 nt and a GC content of 44.8%. The genome of phage BH1 possesses 57 open reading frames which could be divided into functional modules associated with DNA packaging, morphogenesis, lysis, integration, genetic switch, and replication. In spite of similarity in morphology and genomic organization, comparative analysis revealed substantial genetic diversity and mosaic genomic architecture among phages described for the *Lactobacillus casei* group. Additionally, qPCR and ddPCR analysis confirmed earlier microscopic observations indicating that *L. rhamnosus* Pen liberates bacteriophage particles during growth. This occurs spontaneously, and is not a result of external inducing factors. For samples collected after 4 and 24 h of *L. rhamnosus* Pen culture, the number of *att*B and *att*P copies increased 2.5 and 12 times, respectively. This phenomenon, by introducing resistance to other phages or enhancing the biofilm-forming capabilities, may increase the survivability of microorganisms in their natural ecological niche. Conversely, spontaneous phage induction may be an important virulence factor for bacteria, posing a potential threat for the human host.

Keywords: *L. rhamnosus*; probiotics; bacteriophage; spontaneous prophage induction

1. Introduction

Human intestines form an ecosystem in which many species of bacteria reside. The presence of some species stabilizes the digestive system, in part by preventing various pathological conditions. Such strains of microbes with health-promoting characteristics are collectively called probiotics. This group of bacteria includes various strains of the genus *Lactobacillus*, which are used intensively in the dairy and pharmaceutical industries due to their health-promoting properties, which are increasingly supported by research [1,2]. One example of commercial health-promoting bacteria is *Lactobacillus rhamnosus* Pen, which is a component of a pharmaceutical used to alleviate diarrhoea associated with antibiotic therapy [3]. It is commonly believed that probiotic bacteria stabilize intestinal microflora, inhibit the growth of pathogenic microbes, eliminate or reduce symptoms of lactose intolerance, prevent or alleviate bacterial and viral diarrhoea, and normalize intestinal motility disorders. It has also been shown that enrichment of the human diet with probiotic bacteria may stimulate the immune system and positively influence the regulation of blood cholesterol levels [4,5].

However, it remains important to conduct further studies on the safety of using bacteria as dietary supplements, as there are also reports of negative features of some microorganisms considered to be probiotics [6,7]. Additionally, prophage sequences are widespread in genomes of *Lactobacillus* bacteria [8,9]. In general, the presence of sequences encoding bacteriophage proteins in the genome was considered a negative phenomenon. Often, their presence is connected with a high probability of induction and release of bacteriophage particles through cell lysis. Unsurprisingly, such processes result in large losses in the food industry, where bacteria of the genus *Lactobacillus* are very widely applied [10,11]. Interestingly, scientific studies indicate that the presence of prophage sequences in bacterial genomes not only increases the genetic variability, but may also have a positive effect on the bacterial host. Therefore, prophage sequences cannot be treated only as a potential threat to bacterial cells, which would undergo lysis at the time of activation of prophages and their transition to the lytic cycle. The presence of prophages in bacterial genomes requires further analysis, taking into account their positive effects, such as increasing the viability of bacterial cells and enabling them to gain an environmental advantage in a given ecological niche [12,13].

The analysis of genome sequences publicly available in bioinformatic databases indicates that, despite high affinity, the bacteria belonging to the *L. casei* group (*L. casei*, *L. paracasei*, and *L. rhamnosus*) possess substantially diversified sequences of phage origin in their genomes. Furthermore, results of studies also conducted by our research group suggest that bacteriophages associated with the above species of intestinal bacteria are characterized by various forms of existence in genomes of these microorganisms. A good example in this case may be the lytic phage Lc-Nu [14] or bacteriophage Lrm1, described by Durmaz et al. [15] which, due to the lack of lytic activity against different strains of *L. rhamnosus*, was classified as defective. We also demonstrated that BH1 phage has a diverse nature. The analysed probiotic strain — *L. rhamnosus* Pen-released high amounts of this bacteriophage during culture growth, and interestingly, this phenomenon did not result from the application of an external inducing agent [16]. There is a potential physiological importance of such a co-existence of the phage and the bacterium; some researchers suggest that the phage may bring potential benefits for the bacterial host, and could play an important role in bacterial virulence [17].

In this study, we carried out genomic and proteomic characterization of bacteriophage BH1 isolated from probiotic *Lactobacillus rhamnosus* Pen. Additionally, the process of spontaneous prophage induction (SPI) during bacterial growth was monitored using qPCR and droplet digital PCR method. Finally, the potential physiological significance of SPI for bacteria and their host was analysed and briefly discussed.

2. Materials and Methods

2.1. Bacterial Strains and Culture Conditions

Lactobacillus rhamnosus Pen obtained from Biomed Serum and Vaccine Production Plant Ltd. in Lublin (Poland) was routinely cultured anaerobically in a Man-Rogosa-Sharpe broth (Difco) at 37 °C. For prophage induction, 1 mL of an overnight culture was added to 100 mL of MRS with 10 mM $CaCl_2$ and then when OD_{600} reached 0.2, mitomycin C was supplemented at a final concentration of 0.5 μg/mL. Next, the culture was incubated at 37 °C for 18 h, and afterwards lysate was centrifuged (5000× *g*, 10 min, 4 °C) and filtered using 0.45 pore diameter membranes. Phage particles were precipitated by the addition of 10% (*w*/*v*) PEG 8000 and 0.5 M NaCl (overnight, at 4 °C). After centrifugation (10,000× *g*, 1 h, 4 °C), the phages were resuspended in a 1 ml SM buffer. The residual PEG was removed by extraction with chloroform [18].

2.2. Electron Microscopy

Phage particles were fixed on formvar carbon-coated copper grids and stained negatively with 1% (w/v) phosphotungstic acid. Then, bacteriophages were imaged on LEO 912AB transmission electron microscope (LEO Electron Microscopy Inc., Thornwood, NY, USA).

2.3. Protein Identification

For electrophoresis (SDS-PAGE, 2DE), the protein samples were prepared with a 2D Clean-Up kit (GE Healthcare, Chicago, IL, USA) and separated on 7 cm linear IPG strips (pH 3–10) using Protean IEF (Bio-Rad, Hercules, CA, USA) and mini Protean Tetra cell (Bio-Rad) with 4%–20% precast polyacrylamide gels. Selected protein bands were excised from the gel and analysed by liquid chromatography coupled to the mass spectrometer in the Laboratory of Mass Spectrometry, Institute of Biochemistry and Biophysics, Polish Academy of Sciences (Warsaw, Poland). Samples were concentrated and desalted on a RP-C18 precolumn (Waters, Milford, MA, USA), and further peptide separation was achieved on a nano-ultra performance liquid chromatography (UPLC) RP-C18 column (Waters, BEH130 C18 column, 75 μm i.d., 250 mm long) of a nanoACQUITY UPLC system, using a 45 min linear acetonitrile gradient. Column outlet was directly coupled to the electrospray ionization (ESI) ion source of the Orbitrap Velos type mass spectrometer (Thermo, Waltham, MA, USA), working in the regime of data dependent MS to MS/MS switch with HCD type peptide fragmentation. An electrospray voltage of 1.5 kV was used. Raw data files were pre-processed with the Mascot Distiller software (version 2.4.2.0, MatrixScience, Boston, MA, USA). The obtained peptide masses and fragmentation spectra were matched to the National Center Biotechnology Information (NCBI) nonredundant database, with a bacteria filter using the Mascot search engine (Mascot Daemon v. 2.4.0, Mascot Server v. 2.4.1, MatrixScience, Boston, MA, USA). The following search parameters were applied: Enzyme specificity was set to trypsin, peptide mass tolerance to ± 20 ppm, and fragment mass tolerance to ± 0.1 Da. The protein mass was left as unrestricted, and mass values as monoisotopic, with one missed cleavage being allowed. Alkylation of cysteine by carbamidomethylation as fixed, oxidation of methionine was set as a variable modification.

Protein identification was performed using the Mascot search engine (MatrixScience, Boston, MA, USA), with the probability-based algorithm. The expected value threshold of 0.05 was used for analysis, which means that all peptide identifications had less than 1 in 20 chance of being a random match.

2.4. Genomic Analysis

Phage DNA was prepared using Virus Mini AX transfect kit (A&A Biotechnology, Gdansk, Poland) according to the manufacturer's instructions. Genome sequencing was carried out at the Genomed using an Illumina MiSeq 250 bp paired-end run with a 300 bp insert library (Illumina, San Diego, CA, USA). The raw reads were trimmed and de novo assembled using CLC Genomics Workbench (Qiagen, Hilden, Germany). The assembly was verified by Sanger sequencing. The complete sequence was annotated by using Prokka tool v. 1.12 (Victorian Bioinformatics Consortium, Melbourne, Australia) and the identified ORFs were analysed by BLASTp (NCBI, Bethesda, MD, USA) and InterProScan (EMBL-EBI, Hinxton, UK). The following programs were used to analyse the obtained BH1 sequence: Easyfig version 2.2.3, PhagTerm version 1.0.12, Signal IP version 5.0, MegaBlast (NCBI, Bethesda, MD, USA) and CLC Sequence Viewer 8 (Qiagen, Hilden, Germany). The genome sequence of BH1 has been deposited in the GenBank database under accession number MH983004.1.

2.5. Analysis of Prophage Induction—Sample Preparation

In an experiment aiming to examine the BH1 phage induction using mitomycin C, *L. rhamnosus* Pen was cultured in a MRS medium (vol. 10 mL) at 37 °C, until OD_{600} reached the value of 0.5. The bacteria were then centrifuged (7000× *g*, 5 min) and suspended in fresh MRS medium. Subsequently, a specific amount of inducer was added to the culture (final concentration of mitomycin was 0.25, 0.5, 1 and

2 μg/mL). The control sample consisted of cultures not treated with an inducer. After 6 h of induction, OD_{600} was measured and 2 mL of each culture was collected. The samples were centrifuged for 5 min at 7000× *g*, after which the cell deposit was frozen at 20 °C. The experiment was performed in three biological replicates.

In the case of spontaneous prophage induction (SPI), 50 mL of MRS medium was inoculated with an initial culture at a ratio of 1:100. Then, every 4 h (after 4, 8, 12, 16, 20, and 24 h, respectively), the optical density was measured and samples for DNA isolation were collected. In this case, the control samples were those collected after 8 h. Samples were prepared as described above. The experiment was conducted in five biological replicates.

After thawing, bacterial deposits were used to isolate total DNA with the genomic mini AX bacteria+ kit from A&A Biotechnology. The isolation was performed in accordance with the maker's recommendations, and the concentration of nucleic acids was estimated using the Nanodrop 2000c spectrophotometer (Thermo, Waltham, MA, USA). The obtained DNA samples were subsequently used for quantitative analysis of *att*P, *att*B, *att*L, and *att*R sequences by qPCR and droplet digital PCR methods.

2.6. *Quantitative PCR and Droplet Digital PCR*

The quantitative analysis of the excision of the prophage from *L. rhamnosus* Pen genome required designing four pairs of primers flanking the integration loci of the phage BH1 (*att*P, *att*B, *att*L, and *att*R) and one pair of primers (Lrh) that generated the reference product (total bacteria count) (Table 1) (Figure 3). Real-time PCRs were performed in 15 μL with the use of a SG qPCR master mix (Eurx, Gdansk, Poland) reagents on the CFX apparatus from Biorad (Hercules, CA, USA). In the reaction, primers with a final concentration of 0.3 μM and appropriately diluted DNA solutions were used. qPCR was conducted under conditions recommended by the reagent manufacturer (predenaturation at 95 °C, 3 min, denaturation at 94 °C, 15 s; annealing at 60 °C, 30 s; elongation at 72 °C, 30 s). At the end of the reaction, to confirm the specificity of amplification, the melting curves of the obtained PCR products were analysed. The results were analysed with the use of CFX Manager v. 3.1 software (Biorad, Hercules, CA, USA), taking into account the yield of reaction [19].

Table 1. PCR primers used in qPCR and droplet digital (dd)PCR reactions.

Primer Set	Amplified Fragment	Primer Sequences (5′–3′)	Position	GenBank Accession Number
1	*att*B	AGGTCTCAATTTGGTCTCAAAA	2237443–2237464	CP020464.1
		ACGGGTTTTGCCTGTATCAA	2278301–2278320	CP020464.1
2	*att*R	ATTTTGTGCACCGGAGGAA	21564–21582 2278166–2278184	MH983004.1 CP020464.1
		ACGGGTTTTGCCTGTATCAA	2278301–2278320	CP020464.1
3	*att*L	AGGTCTCAATTTGGTCTCAAAA	2237443–2237464	CP020464.1
		AAATATGCACAAAATGTGGCTTT	21408–21430 2237597–2237619	MH983004.1 CP020464.1
4	*att*P	TGTGCACCGGAGGAAAAA	21561–21578 2278170–2278187	MH983004.1 CP020464.1
		AAATATGCACAAAATGTGGCTTT	21408–21430 2237597–2237619	MH983004.1 CP020464.1
5	Lrh	TTGACAAGGGACTCAAGGAT	2208780–2208799	CP020464.1
		TATGATAGCCGGAATCAGCA	2208910–2208929	CP020464.1

Reactions for droplet digital (ddPCR) were prepared in 20 μL volumes with 10 μL QX200 ddPCR EvaGreen Supermix, 2 μL of 1 μM forward and reverse primers, and 1 μL of 10 or 100× water-diluted DNA. As a no template control (NTC) 1 μL of water was used. The 20 μL droplet digital PCR reaction mixture was then loaded into the disposable droplet generator cartridge (DG8 cartridge, Biorad). 70 μL of droplet generation oil was loaded into the oil well for each sample. The cartridge was placed into the droplet generator (QX100 Droplet Generator, Biorad). The generated droplets were transferred to 96-well reaction plate (ddPCR plater 96-well, Semi-Skirted, Biorad). The plate was heat-sealed with a sealer (PX1 PCR Plate Sealer, Biorad). Then, the plate was placed on a thermal cycler (T100 Thermal Cycler, Biorad) and amplified to the endpoint. Thermal-cycling conditions were 95 °C × 10 min (1 cycle), 95 °C × 30 s (ramp rate 2 °C/s), and 55 °C × 60 s (ramp rate 2 °C/s) (40 cycles), 98 °C × 10 min (1 cycle), and a 12 °C hold. Following PCR amplification, the 96-well PCR plate was loaded on a droplet reader (QX200 Droplet Reader, Biorad), which reads the droplets from each well of the plate. Analysis of the ddPCR data was performed with the QuantaSoft analysis software version 1.7.4.0917. Results with droplets number >10,000 were analysed.

2.7. Statistical Analysis

The data from qPCR and ddPCR were analysed by the Microsoft Excel 2016 (Microsoft Corporation, Redmond, WA, USA) and the Statistica version 13.3 (StatSoft, Tulsa, OK, USA) using ANOVA procedure for analysis of variance. $p < 0.05$ was considered statistically significant.

3. Results and Discussion

3.1. Morphological and Proteomic Characterization of Bacteriophage BH1

The spontaneous releasing of bacteriophage particles by *L. rhamnosus* Pen has been observed in previous work. This phenomenon did not require any inducing agent, and did not significantly affect bacterial growth. Additionally, similar to phage Lrm1, the propagating host for BH1 was not identified in the standard experimental conditions [15,16]. In this work, we carried out a thorough characterization of BH1 bacteriophage. Phage particles were obtained using standard procedures, in which mitomycin C-induced cultures were concentrated using polyethylene glycol. The final samples were used to take photographs with a transmission electron microscope (Figure 1). Our observations confirmed that these methods successfully isolated complete phage particles. The measurements showed that the BH1 bacteriophage particles had tails approximately 287 nm long and isometric heads approximately 58 nm in diameter. As in the case of phage Lrm1, the morphology indicated that the bacteriophage BH1 can be classified in the *Siphoviridae* family [15,20].

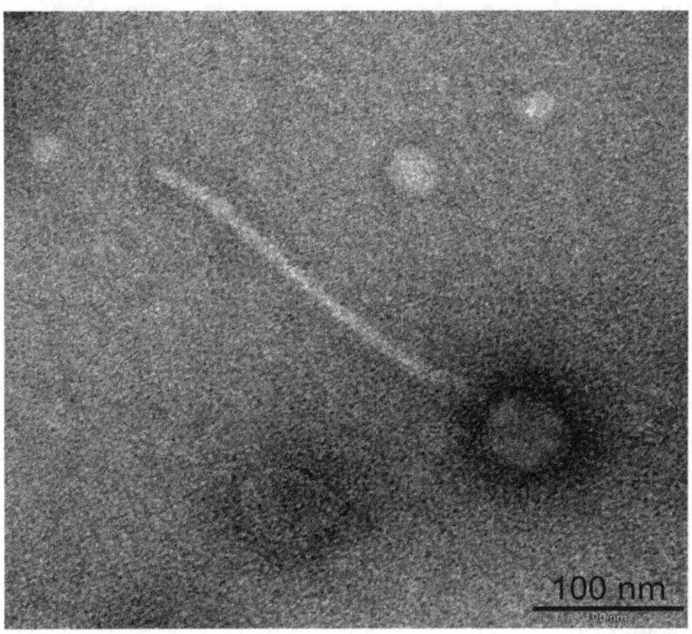

Figure 1. Transmission electron microscope micrograph of bacteriophage BH1 released by the *L. rhamnosus* strain Pen.

Next, phage samples were concentrated and prepared for protein electrophoretic examination (SDS-PAGE and 2DE—Figure 2). The resultant electrophoretic profiles of the analysed protein sample showed numerous protein bands, which were excised from the gels. For protein identification, the samples were digested with trypsin and then peptide mixtures were analysed by nano-HPLC coupled to an ESI-orbitrap spectrometer.

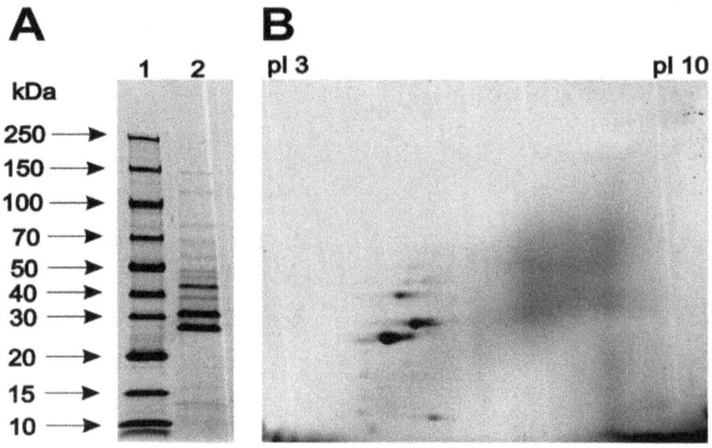

Figure 2. Electrophoretic separation of bacteriophage-derived proteins. Analysis was conducted using SDS-PAGE (**A**) and two-dimensional electrophoresis procedure (**B**).

Among the detected proteins were the following phage-derived structural polypeptides: Tail proteins, phage tail tape measure protein, capsid protein, portal protein, and head-tail adapter protein. Apart from typical structural proteins, peptides derived from proteins performing other functions were also identified, such as DNA-packaging protein, holin, transcriptional regulator, and peptidase U35 (Table 2). As well as proteins derived from phage, bacterial proteins were also detected. These proteins could represent impurities in the samples, as these contained bacteriophages, as well as bacterial host secreted proteins. However, in some cases, identified bacterial proteins may play an important role in the composition of mature phage particles. For example, the proteins GroEL and GroES, which belong to the chaperonin family, are essential for the correct assembly of bacteriophage tails and heads [21].

Table 2. Identification of bacteriophage-derived proteins using LC–MS/MS procedure.

Protein Description	Mascot Score [a]	No. of Unique Peptides Matched [b]	Protein Coverage [% aa]	Protein Molecular Mass	Orf
phage tail tape measure protein [Lactobacillus rhamnosus]	1978	28	19.4	173159	14
phage tail protein [Lactobacillus rhamnosus]	2851	35	37.4	107736	16
phage tail protein [Lactobacillus rhamnosus]	1375	18	32.4	71309	15
phage portal protein [Lactobacillus rhamnosus]	1214	16	47.7	46346	4
major capsid protein [Lactobacillus rhamnosus]	2371	25	78	43998	6
hypothetical protein [Lactobacillus rhamnosus]	192	2	49.4	7776	7
phage tail protein [Lactobacillus rhamnosus]	1345	16	75.1	22056	12
phage tail protein [Lactobacillus rhamnosus]	755	11	95.3	14653	11

Table 2. *Cont.*

Protein Description	Mascot Score [a]	No. of Unique Peptides Matched [b]	Protein Coverage [% aa]	Protein Molecular Mass	Orf
phage head-tail adapter protein [Lactobacillus rhamnosus]	388	7	58.7	12598	9
DNA-packaging protein [Lactobacillus rhamnosus]	439	7	47.9	13581	8
hypothetical protein [Lactobacillus rhamnosus]	1239	17	72.7	17915	36
phage holin [Lactobacillus rhamnosus]	120	2	6.6	14373	20
peptidase U35 [Lactobacillus rhamnosus]	37	1	5.7	23706	5
hypothetical protein [Lactobacillus rhamnosus]	234	4	31.2	14024	10
transcriptional regulator [Lactobacillus rhamnosus]	49	1	8.4	9473	29
hypothetical protein [Lactobacillus rhamnosus]	43	1	21.7	6948	46
hypothetical protein [Lactobacillus rhamnosus]	252	4	36.4	7358	31

[a] Protein scores derived from individual ion scores of Mascot-identified tryptic peptides as a nonprobabilistic basis for ranking protein hits. [b] Total number of tryptic peptides which the Mascot program assigned to a database protein. To compute this number, multiple matches to lactobacillar peptides with the same primary sequence but representing different charge or modification states were counted as one.

3.2. Bacteriophage BH1 Genome Analysis

Next, the bacteriophage particles were used for DNA isolation, and complete genomic sequence of BH1 was determined using an Illumina MiSeq system (a total of 147,084 reads with mean read length of 236.37 bp resulting in 848.73-fold average coverage of analysed genome). The obtained sequence was verified using Sanger sequencing. The final assembled consensus sequence revealed a circular, double stranded DNA genome with a size of 40,721 bp. The total length of the BH1 sequence was similar to previously described temperate phage Lrm1 isolated from industrial *L. rhamnosus* M1 (39,989 bp) [15]. Interestingly, the complete genome of the virulent phage Lc-Nu of probiotic *L. rhamnosus* strain Lc 1/3 was ~4.5 kb shorter. This difference was due to the lack of some lysogenic regions similar to integrase coding gene or parts of sequences such as *att*P and CI-like lytic cycle repressor, which are associated with virulence of Lc-Nu bacteriophage [14]. The G + C content of BH1 genome was 44.8%; very close to previously described bacteriophages for the *Lactobacillus casei* group: Lc-Nu (44.3%), ϕAT3 (44.6%), A2 (44.9%), Lca1 (44.8%), and Lrm1 (45.5%). The G + C content also corresponded to the overall GC-content determined for chromosomal DNA of bacterial host of BH1—*L. rhamnosus* Pen (46.8%), and for other previously reported *L. rhamnosus* strains (average value of 46.7%) [22–24].

The analysis of raw data from NGS sequencing using PhageTerm showed the presence of—5′-CGATCGACCT-3′ cos sequences similar to those previously determined for bacteriophages produced by other strains belong to the *L. casei* group. However, two mismatches within the cos sequence of BH1 were also detected [15]. The alignment of *att*P sequences showed that the BH1 phage attachment site is localized between ORF22 and ORF23, adjacent to integrase coding gene. The attachment site of BH1 consists of a 90-nucleotide sequence with 15-nt core (Figure 3), showing 87.5% similarity with phage Lrm1, and 96% similarity with Lc-Nu bacteriophage (with only 58% sequence coverage), seems to be characteristic of this phage.

```
attB   TAGGTCTCAATTTGGTCTCAAAATTTTTACAACATCCCAAAAGAAAATAAAAATCCCAAA
attR   CACATTTTGTGCACCGGAGGAAAAAAACAACCGAAAATAAAAGGAAACAAAAATCCCGAA
attL   TAGGTCTCAATTTGGTCTCAAAATTTTTACAACATCCCAAAAGAAAATAAAAATCCCAAA
attP   CACATTTTGTGCACCGGAGGAAAAAAACAACCGAAAATAAAAGGAAACAAAAATCCCGAA
                                              ***** *** ********* **

attB   ACACTATCATATCAGCGTTTTGGGAAGCTATGGAAAGCAACTAGAGACAAGAAAACGGAG
attR   ATGCCTTTATATCAGCGTTTTGGGAAGCTATGGAAAGCAACTAGAGACAATAAAACGGAG
attL   ACACTATCATATCAGCGTTTTGGGAAGCTATGGAAAGCAACTAGAGACAAGAAAACGGAG
attP   ATGCCTTTATATCAGCGTTTTGGGAAGCTATGGAAAGCAACTAGAGACAATAAAACGGAG
       *  *  * ****************************************** *********

attB   AGTAAGGGATTCGAACCCTTGATACAGGCAAAACCCGTATACATGGTTTCCAACCATGCT
attR   AGTAAGGGATTCGAACCCTTGATACAGGCAAAACCCGTATACATGGTTTCCAACCATGCT
attL   AGTAAGTGCGAAAACAAACGTATATACAATAGTATAAAGCCACATTTTGTGCATATTTTG
attP   AGTAAGTGCGAAAACAAACGTATATACAATAGTATAAAGCCACATTTTGTGCATATTTTG
       ****** *
```

Figure 3. Alignment of phage attachment sites derived from phage BH1 and *L. rhamnosus* Pen genome. The genome sequence of *L. rhamnosus* was marked in red; the phage sequence/prophage sequence in blue; the core sequences are underlined; the motifs used to design the primers (Table 1) were highlighted in italics.

In general, genome sequence of ϕBH1 showed the highest similarity with prophage sequences detected in genomes of several *L. rhamnosus* strains: CLS17, Lrh10, DS3, DS9, DS12, DS18, ASCC 3029, R0011, B9, ARJD, IBL027, Lr032, and ATCC 21052 (Dataset S2). A lower level of homology was revealed with two *L. rhamnosus* phages—Lrm1 (NC_011104.1, identity 94%, query coverage 59%) and Lc-Nu (NC_007501.1, identity 91%, query coverage 28%) (Figure 4) and other bacteriophages and prophages described for the *L. casei* group (Dataset S1, S3, S4). These findings, in conjunction with other reports, suggest a high diversity of *L. rhamnosus* phage genomic sequences, highlighting that sequences of a bacteriophage origin may be a very interesting source of information regarding bacteria phylogenetics [9,25]. Moreover, our results also confirm earlier works by Brandt and Allatosava [26] and Zago et al. [27], demonstrating that phage sequences have a high potential for the identification of bacteria of the genus *Lactobacillus*, even at the strain level. Interestingly, in genomes of many *L. rhamnosus* strains, short, approximately 30 nucleotide sequences were detected, which were either identical or very similar to the sequences observed in the BH1 genome. These sequences were located within the CRISPR (clustered regularly interspaced short palindromic repeat) modules, indicated by both flanking sequences and the proximity of genes encoding Cas endonuclease and type II-A CRISPR-associated Csn2 proteins. The presence of phage sequences within the CRISPR modules likely makes these bacterial strains resistant to infection caused by the BH1 bacteriophage. Previous studies have shown that this phenomenon is a highly specific, sequence-dependent defense system against phage infections [25,28].

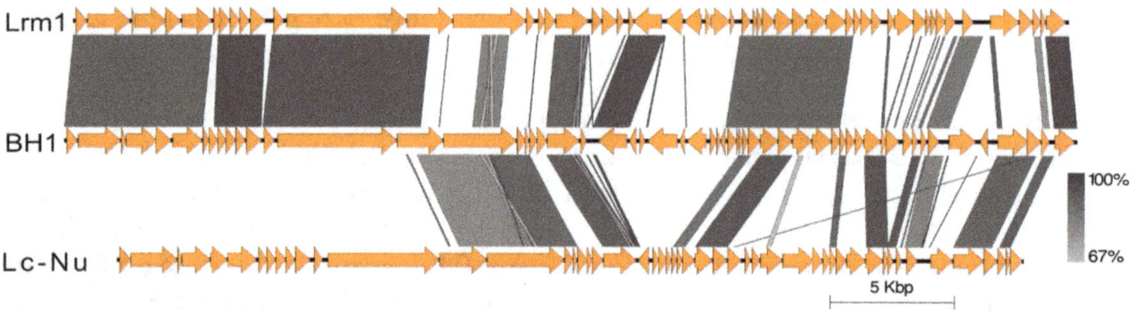

Figure 4. Visualization of genome maps alignment of three *Lactobacillus rhamnosus* bacteriophages—BH1, Lrm1, and Lc-Nu. Putative open frames are marked with arrows. Lines represent the level of homology between analyzed phage sequences (BLAST identity is presented in gray scale).

3.3. Genome Organization of Bacteriophage BH1

Phage BH1 has a genome organization typical for other *Lactobacillus* bacteriophages with the following modules: Packaging, structural proteins, lysis, integration, genetic switch, and replication. Fifty-seven open reading frames (fifty genes on the positive strain and seven genes on the negative strand) were predicted, which harboured nearly 91.4% of the whole sequence determined. The ORF's GC-content ranged from 35.2% (ORF24) to 50.8% (ORF7). Regarding the start codon, forty-four ORFs start with ATG, eight with GTG, and five genes had the TTG start codon. Among those revealed, most ORFs were preceded by atypical Shine-Delgarno sequence, complementary to the 3′ end of the 16S rRNA gene of *Lactobacillus* species (3′-UCCUCCAA-5′) [29]. The location of predicted ORFs and their putative functions are presented in Table 3.

Comparative studies revealed that ORF1 and ORF2 correspond to the terminase small and large subunits, respectively. Presumably, these two terminase subunits are responsible for specific DNA binding (HTH-domain), and cutting concatemeric DNA into genome lengths. Therefore, they most likely represent the bacteriophage DNA packaging module [30]. Similar to phage Lrm1 and *L. casei* phage A2, ORF57 encodes HNH endonuclease, which may be also involved in the DNA packaging system. Since Garcia et al. [31] showed that this HNH protein can be also classified as a phage terminase (small subunit), we speculated that the predicted protein products of BH1 ORF1 and ORF57 likely have redundant activity.

As in the case of phage Lrm1, ORF3 encodes a 63-amino-acid putative integral membrane protein, possessing two transmembrane-spanning domains. The next three ORFs (ORF4, ORF5, and ORF6) are involved in head morphogenesis. The products of these genes—portal protein, head maturation protease, and major head protein (with a CCCAAAA slippery sequence)—show an amino acid identity of about 90% with bacteriophage Lrm1 (91.9%, 94.3%, and 89.3%, respectively), and also high aa similarity with three other *L. casei* group phages—J-1, PL-1, and A2 [15,31,32]. ORFs 8 to 11 constitute a typical neck region consisting of DNA-packaging protein, phage head-tail adapter, head-tail joining protein, and phage-related head-to-tail joining protein. Interestingly, we observed an additional open reading frame (ORF7) between the head and neck module, not detected in the genome of Lrm1. This gene encodes a hypothetical protein containing putative Ig-like domain, which is found in a variety of bacterial and phage surface proteins involved in bacterial host-cell interaction [33]. The tail module of BH1 (ORF12 to ORF16) is highly related to the genome sequence generated for phage Lrm1. This cluster also bears resemblance in both gene organization and nucleotide sequences to the tail morphogenic region of J-1, PL-1, and A2 (Dataset S1). However, BH1 ORF15, encoding a phage tail component, shares high homology only with N-terminus of Lrm1 ORF14 (~310 aa). Interestingly, the second half of ORF15 exhibits high similarity with the C-terminus of the corresponding protein described for bacteriophage Lc-Nu (~385 aa). Such a mosaic structure was also observed in the next annotated gene (ORF16), which is probably associated with endopeptidase activity. Notably, another two genes (ORF17 and ORF18) with an unknown function are not present in the genome of Lrm1. However, they display a high level of nucleotide sequence identity (99.7% and 96.5%, respectively) with adequate genes of phage Lc-Nu. Additionally, ORF19 was not found either in the genome of Lrm1 and also Lc-Nu, but was nearly 100% identical with sequences determined for bacteriophage A2 and iLp84 (Dataset S1). These results suggest that BH1 has a chimeric character, and may be derived from different ancestors [34,35].

Table 3. General features of putative ORFs of bacteriophage BH1 and nucleotide sequence homology with genome sequences of phages Lrm1 and Lc-Nu.

Orf Order/Strand	Predicted Start Stop Site	GC (%)	Orf Length (aa)	Predicted Molecular Mass (kDa)	pI	Description	Representative orf (% Coverage/% nt Identity)		Biological Process/Molecular Function [a]
							Lrm1 NC_011104.1	Lc-Nu NC_007501.1	
1F	90–545	41.01	151	17.06	8.85	Terminase, small subunit	100/99.34	0/0	-
2F	567–2279	43.96	570	65.10	5.15	Terminase, large subunit	100/93.35	0/0	-
3F	2291–2482	42.19	63	6.66	9.31	Hypothetical protein	100/98.44	0/0	-
4F	2488–3741	46.65	417	46.32	5.07	Portal protein	100/81.64	0/0	-
5F	3695–4324	47.62	209	23.72	5.45	Prohead protease	100/87.78	0/0	-
6F	4366–5568	46.22	400	44.03	5.42	Major capsid protein	98.92/ 84.37	0/0	-
7F	5586–5825	50.83	79	77.63	4.53	Hypothetical protein	0/0	0/0	GO:0030246
8F	5836–6195	41.94	119	13.52	4.24	Putative DNA-packaging protein	46.67/91.67	0/0	-
9F	6185–6514	47.58	109	12.61	9.57	Head-tail adaptor	100/98.48	0/0	-
10F	6514–6900	47.80	128	14.03	5.53	Putative tail component	100/96.12	0/0	-
11F	6900–7286	45.22	128	14.66	4.39	Putative head-to-tail joining protein	100/95.61	0/0	-
12F	7320–7937	44.82	205	22.07	4.56	Major tail protein	96.12/95.79	0/0	-
13F	8036–8449	42.51	137	15.24	6.30	Hypothetical protein	100/98.31	0/0	-
14F	8572–13434	45.86	1620	173.27	9.42	Tail tape measure protein	97.27/95.48	0/0	-
15F	13435–15351	46.69	638	71.17	5.21	Tail component protein	47.99/89.67	59.36/82.86	-
16F	15352–18306	45.69	984	107.75	4.78	Prophage tail endopeptidase	36.75/82.60	55.40/79.54	-
17F	18322–18651	45.76	109	11.90	4.38	Hypothetical protein	0/0	100/99.70	-
18F	18648–18791	42.36	47	5.29	4.86	Uncharacterized protein, XkdX family	0/0	100/96.53	-
19F	18823–19122	45.33	99	11.30	6.63	Hypothetical protein	0/0	0/0	-
20F	19137–19550	49.76	137	14.38	4.82	Holin	0/0	0/0	-
21F	19561–20859	48.11	432	47.06	8.98	Glycoside hydrolase, lysin	100/90.87	100/92.33	GO:0009253 GO:0016998 GO:0003796
22F	20904–21128	45.78	74	8.18	4.97	Hypothetical protein	0/0	0/0	-
23R	21592–22719	42.55	375	43.28	9.71	Site-specific integrase	100/98.49	0/0	GO:0006310 GO:0015074 GO:0003677
24R	22827–23090	35.23	87	10.24	9.80	Hypothetical protein	0/0	0/0	-
25F	23229–23450	42.34	73	8.13	4.32	Hypothetical protein	0/0	0/0	-
26R	23560–24741	40.78	393	44.33	9.00	Restriction endonuclease, type I, HsdS	0/0	0/0	GO:0006304 GO:0003677
27R	24832–25050	47.03	72	8.07	6.53	LexA-like peptidase	0/0	0/0	-
28R	25122–25895	42.77	257	28.97	4.61	Cro/CI-type transcriptional repressor	0/0	0/0	GO:0003677
29F	26053–26304	39.29	83	9.48	10.2	XRE family transcriptional regulator	0/0	0/0	GO:0003677
30F	26301–26450	44.00	49	5.45	6.12	Hypothetical protein	0/0	0/0	-

Table 3. *Cont.*

Orf Order/Strand	Predicted Start Stop Site	GC (%)	Orf Length (aa)	Predicted Molecular Mass (kDa)	pI	Description	Representative orf (% Coverage/% nt Identity)		Biological Process/Molecular Function [a]
31R	26447–26647	37.81	66	7.30	5.33	Hypothetical protein	0/0	0/0	-
32F	26722–27078	42.58	118	13.92	9.15	DUF771 domain-containing protein	100/100	0/0	-
33F	27078–27170	47.31	30	3.36	5.96	Hypothetical protein	100/97.85	0/0	-
34F	27163–27315	45.10	50	5.69	9.52	Hypothetical protein	100/95.42	98.69/96.69	-
35F	27320–27523	45.59	67	7.51	6.00	Hypothetical protein	100/86.27	100/90.20	-
36F	27542–28027	46.30	161	17.92	8.73	Siphovirus-like Gp157 protein	100/96.91	0/0	-
37F	28028–28735	45.34	235	26.57	6.77	Nucleotide-binding protein	100/94.92	100/92.95	-
38F	28739–29296	43.37	185	20.92	5.50	DUF669 domain-containing protein	100/95.17	97.13/96.13	-
39F	29311–30108	43.86	265	31.13	9.36	Putative replication protein	41.73/85.29	0/0	-
40F	30095–30877	46.87	260	29.73	9.68	IstB-like ATP binding protein	100/86.75	0/0	GO:0005524
41F	30874–31218	43.77	114	13.09	4.51	Hypothetical protein	94.49/ 93.56	0/0	-
42F	31205–31459	50.20	84	9.42	9.75	Hypothetical protein	0/0	100/90.59	-
43F	31456–31821	43.71	121	14.39	6.92	Holliday junction resolvase RusA-like	0/0	0/0	GO:0006281 GO:0006310 GO:0000287
44F	31832–32170	46.02	112	12.44	5.38	Putative endonuclease	0/0	0/0	-
45F	32182–32883	45.44	233	27.11	5.01	SAM-dependent DNA methyltransferase	0/0	96.15/93.63	GO:0006306 GO:0003677 GO:0008170
46F	32880–33062	41.53	60	6.89	4.55	Hypothetical protein	55.19/94.06	100/94.54	-
47F	33059–33601	49.35	180	20.64	4.85	DUF1642 domain-containing protein	8.47/93.48	0/0	-
48F	33757–34137	45.41	126	14.21	4.40	Hypothetical protein	10.24/97.44	0/0	-
49F	34134–34343	40.95	69	8.34	6.73	Hypothetical protein	0/0	100/ 94.79	-
50F	34471–34692	42.34	73	8.42	9.52	HTH-transcriptional regulator	0/0	0/0	GO:0003677
51F	34762–35205	47.97	147	16.85	7.88	Transcriptional regulator, ArpU family	55.86/ 97.98	55.63/97.57	-
52F	35599–36672	35.85	357	40.65	6.33	Hypothetical protein	0/0	0/0	-
53R	36834–37133	42.00	99	10.81	9.37	Stress response protein CsbD	0/0	0/0	-
54F	37552–38769	45.07	405	46.64	6.07	GcrA cell cycle regulator	0/0	100/95.81	-
55F	38717–39286	44.91	189	21.44	9.28	HNH endonuclease	0/0	22.81/84.62	GO:0016788
56F	39290–39613	46.91	107	12.69	9.01	Ribonucleoside-diphosphate reductase	81.48/78.79	80.56/89.27	-
57F	39816–40610	45.28	264	31.14	9.05	Small terminase subunit/HNH endonuclease	100/97.99	0/0	GO:0003676 GO:0004519

[a] Biological process and molecular function of proteins was determined using InterProScan.

The lysis module consists of two genes, and resembles the structure reported for many prophages in the *L. casei* group (Dataset S2, S3, S4). ORF20 codes for holin, an enzyme which causes bacterial membrane lysis [36]. The next gene, which typically encodes lysin, is conserved between *L. rhamnosus* phages (>90% nt identity) (Dataset S1). Similar to other previously described bacteriophages of lactobacilli, this enzyme is endolysin. Endolysin exhibits 1,4-beta-*N*-acetylmuramidase activity, and by hydrolysis of the amide bond in peptidoglycan, participates in bacterial cell wall degradation [37]. Typically, lysins have two LysM domains on the C-terminus and Glyco-25 motif on N-terminus. It has also been shown that the analyzed protein contains a signal peptide. For BH1, the predicted cleavage site was identified between position 28 and 29 of the amino acid sequence (probability of 0.959).

The next coding region (ORF22) of BH1 encodes protein of unknown function, and was observed only for *L. paracasei* phage iLp84 (coverage 100%, identity 99%) and *L. casei* phage Lca1 (coverage 96%, identity 96%). Comparative analysis showed that the sequence between ORF22 and ORF23 contains the attachment region (*att*P) of 90 bp, with a putative core site 15 nucleotides in length. Based on the complete genome sequence of the bacterial host and results obtained using Sanger sequencing, the *att*B, *att*L, and *att*R regions were also determined (Figure 3). This allowed for accurate identification of the location of prophage sequence in the genome of *L. rhamnosus* Pen. BLASTn comparison of *att*B sequence generated for *L. rhamnosus* Pen with genome sequences of other *L. rhamnosus* isolates revealed that many bacterial strains possess an empty integration region, which is located between $tRNA^{ser}$ and $tRNA^{val}$ genes. Additionally, similar *att*P sequence was also observed in the genomes of phage Lrm1 and Lc-Nu [14,15]. It was typical that the core of *att*B was located within the tRNA ser gene (close to 3′ end) and presumably phage integration does not affect on tRNA integrity [29,38]. It is also noteworthy that the location of prophage BH1 is slightly different in comparison to most of the prophages that possess a similar attachment site. In the case of BH1, we observed an additional ORF encoding bacterial integrase between the typical $tRNA^{val}$ gene and *att*L sequence. Interestingly, this gene appears in six copies in the genome of *L. rhamnosus* Pen.

Open reading frame (ORF 23)—localized adjacent to *att*P—codes for integrase, an enzyme which catalyses site-specific recombination between the integration site of bacteriophage, and those located in the bacterial genomes. Among all phages described for the *L. casei* group, a similar gene for integrase was only detected for Lrm1 (98.5% identity). Analysis of the amino acid sequence of BH1 integrase revealed the presence of several distinctive protein domains, which are likely to bind DNA and cause recombination. On this basis, this enzyme can be classified as tyrosine integrases, which, unlike serine recombinases, utilize tyrosine to mediate DNA cleavage, recognize definitely longer *att*P sequences, and require host cofactors [39].

For the sequences of the next four genes, from ORF24 to ORF27, we detected no similarity to the previously described phages for *L. casei* group. Nevertheless, similar or even identical sequences have been identified in many genome sequences of *L. rhamnosus, L.casei,* and *L. paracasei*, in presumably prophage regions. The analysis of amino acid sequences of proteins encoded by ORF24 and ORF25 did not indicate the presence of characteristic protein domains. As a result, it is difficult to make conclusions about the presumable functions of these proteins. The ORF24 sequence is usually located in the immediate vicinity of the gene encoding the phage integrase. Since the orientation of these two genes is also the same, we assume that the protein encoded by this gene is important for the integration of phage DNA within the bacterial chromosome. Furthermore, an integrase sequence similar to ORF23 of the phage BH1 has been identified in 81 strains of *L. rhamnosus,* and for most of these strains (74) an accompanying gene (similar to ORF24) has also been detected. The presence of a gene similar to ORF24 has not been observed when there was no sequence encoding the integrase (Dataset S2). The amino acid sequence of the protein encoded by ORF26 has conserved domains characteristic for restriction endonuclease subunit S. Interestingly, as well as the sequences located within the genomes of phages and prophages, a similar sequence was also observed within the genomes of many bacteria, not only of the genus *Lactobacillus*. In this case, these genes were part of the type I restriction-modification system, typically composed of three subunits: Restriction (HsdR) subunit, HsdM modification (M) subunit, and HsdS recognition (S for specificity) subunit [40,41]. Preliminary analysis of the ORF27 sequence did not indicate the function of this gene. However, using the conserved domains database (NCBI), in the amino acid sequence, protein domains characteristic for peptidase S24 LexA-like proteins were identified. These are involved in the SOS response, leading to the repair of single-stranded DNA [42]. It has also been shown that these proteins can indirectly influence the level of spontaneous prophage induction [17,43,44]. The next gene, similar to ORF27, is also transcribed leftward, and the protein encoded by this ORF presumably regulates the transcription of genes when there is cellular stress associated with DNA damage, which may be related to the genetic switch of phages from a lysogenic pathway to a lytic pathway [42,45]. A similar sequence has also been observed in the genomes of J-1 phages (coverage 69%, identity 87%) and phiAT3 (coverage 100%, identity 91%). Despite its function similar to that of ORF28, that is, to regulate the gene transcription (XRE family transcriptional regulator), ORF29 was homologous to two other bacteriophages, namely C_L1 and C_L2. This is another example of the immense variability of bacteriophages described for the *L. casei* group.

Identifying the function of the genes encoded in the next portion of the bacteriophage BH1 genome (ORF30–ORF40) proved challenging. For example, the next two genes (ORF30 and 31) encode short-chain proteins with an unknown function (hypothetical protein). Interestingly, similar genes were not observed in the *L. casei* group bacteriophages described so far (Dataset S1). Illustrating the mosaic nature of the bacteriophage genome, subsequent genes were identified in sequences of different phages, e.g., Lrm1 (ORF32–ORF41), phiAT3 (ORF32–ORF40), J-1 (ORF32–ORF35), PL-1 (ORF32–ORF35), Lc-Nu (ORF: 34, 35, 37, 38), C_L2 (ORF: 33, 34, 37), A2 (ORF: 34, 37), C_L1 (ORF33, ORF34), and Lca1 (ORF40). For bacteriophage Lrm1, a region very similar to the sequence located between ORF32 and ORF39 has been described as a DNA replication module [15]. However, since in most cases there are no characteristic protein domains, the molecular function of individual genes seems to be, at most, hypothetically determinable (Table 3).

Comparative analysis has shown that a sequence very similar to ORF34 is present in genomes of many phages and prophages. Based on the similarity to *S. aureus* and *S. epidermidis*, we suggest that the product of this gene may act as a transposase, however, is likely nonfunctional [22]. Interestingly, the protein encoded by ORF36 has the domain Sipho_Gp157 (pfam05565), the presence of which is associated with bacterial resistance to bacteriophages [46]. Another gene encodes a phage-derived protein with an unknown function, often found among phage proteins of Gram-positive bacteria. These proteins usually contain characteristic P-loop motifs (G/A-X-X-G-X-G-K-T), which are located at their N-termini. The amino acid sequences of proteins encoded by ORF37–ORF40 implies that the analysed region of phage BH1 is associated with DNA replication, recombination, and repair. This is supported by the identification of protein domains characteristic for phage nucleotide-binding protein (ORF37), single-stranded DNA-binding protein (ORF38), IstB-like ATP binding protein (ORF40), DNA replication protein DnaC (ORF40), and Holliday junction DNA helicase (ORF40).

The final portion of the phage BH1 was found to be even further differentiated. A few genes are characteristic only for the phage BH1, e.g., ORF50, ORF52, and ORF53, as well as ORF45 and ORF49 (excluding the phage Lc-Nu). Functional analysis has shown that some of the examined genes can encode protein components of restriction-modification systems, such as endonucleases of ORF43, ORF44, and ORF55, etc. or methylases of ORF45. In the case of ORF50–ORF54, based on existing domains, the protein products of these genes are likely to control the expression of viral genes as part of the lysogeny/lytic growth switch. Interestingly, ORF56 encodes a protein that contains domains specific to Glutaredoxin-like protein NrdH, which may serve as a hydrogen donor for the ribonucleotide reductase during deoxyribonucleotide biosynthesis [47].

3.4. Spontaneous Phage Induction Analysis Using QPCR and Digital PCR

The final purpose of our study was the quantitative analysis of spontaneous induction of bacteriophage BH1, using the qPCR and digital PCR droplet techniques. The designed testing system included the quantitative analysis of sequences of *att*B (level of re-established *att*B sites in bacterial genome), *att*P (level of extrachromosomal phage DNA), *att*L and *att*R (amount of integrated prophages in the bacterial chromosome), in comparison to the reference gene originating from the genome of *L. rhamnosus* bacteria. This method has been previously described by Lunde et al. and is particularly useful in cases where the plate tests aimed to quantify the bacteriophage do not give positive results (lack of suitable indicator strain) [48]. Based on the genome sequence of the phage BH1 and genome of the *L. rhamnosus* Pen, five pairs of primers were designed and used in qPCR and ddPCR.

Cells of *L. rhamnosus* Pen were exposed to mitomycin C for 6 h, and DNA isolated. The results showed a significant increase in the amount of a circular phage DNA, with an increase in the concentration of mitomycin (*att*P system) (Figure 5 and Figure S1A). Even at the lowest concentration (0.25 μg/mL), the *att*P/Lrh ratio increased more than 5-fold at 6 h after the induction. For the highest concentration of mitomycin, the *att*P/Lrh ratio increased more than 300-fold compared to the control sample. The ddPCR analysis of the *att*P absolute value has shown that it increased 12-fold and, in the other case, over 400-fold (for mitomycin concentrations of 0.25 and 2 μg/mL, respectively). The frequency of induction/excision of prophage BH1 was also observed using the *att*B/Lrh ratio, which corresponded to the proportion of bacteria with the re-established chromosome (after prophage excision). For this experimental system, the *att*B/Lrh ratio increased more than 8-fold for the mitomycin concentration of 0.25, and over 230-fold for its concentration of 2 μg/mL (Figure S1B). The results obtained in the ddPCR show a 25-fold and over 850-fold increase, respectively, in comparison to the control sample that was not induced. These results and previous studies suggest that the ddPCR method, compared to qPCR, has a significantly higher sensitivity, especially in experiments aiming to detect a small number of DNA matrix copies [49,50]. As in the case of studies by Lunde et al. [48], for experiments in which the amount of integrated prophages *att*L and *att*R was measured, no significant differences in the *att*L/Lrh and *att*P/Lrh ratio were obtained for samples treated with increasing concentrations of mitomycin (data not shown). Similar results were also obtained using the ddPCR method (Figure 5). Since the absolute number of lysogenic bacterial cells is similar to the total number of bacteria in the tested samples (even under induced conditions), primer systems designed for *att*L and *att*P have proven to be unsuitable for observing the induction of prophage sequences [48].

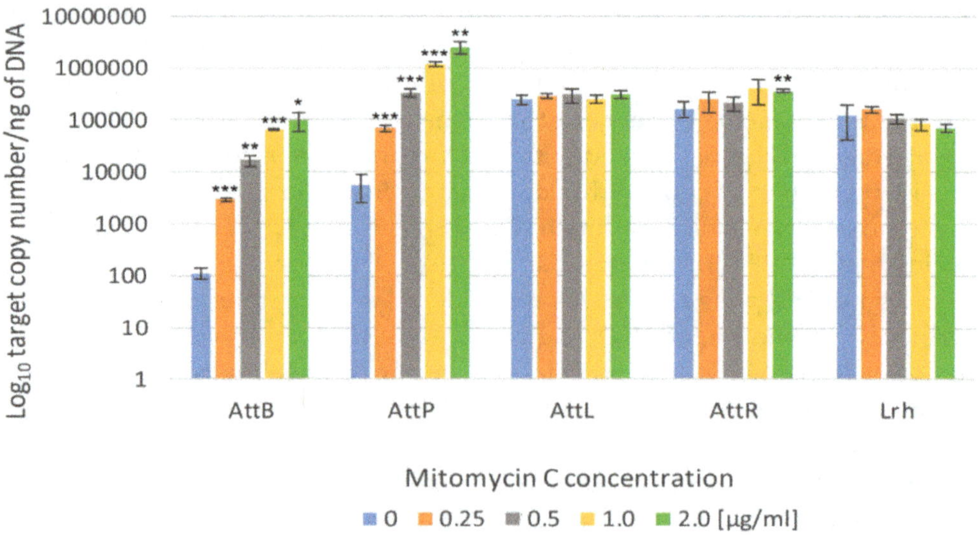

Figure 5. Droplet digital PCR analysis of phage BH1 induction in the presence of different concentrations of mitomycin C. Induction level was observed by measuring the amount of phage attachment sequences (*att*B, *att*P, *att*L, and *att*R). Additionally, the quantity of bacterial DNA (Lrh) was also monitored. The *L. rhamnosus* cultures were induced using increasing amounts of mitomycin C; 0.25 μg/mL—orange bars, 0.5 μg/mL—gray bars, 1 μg/mL—yellow bars, and 2 μg/mL—green bars. The obtained results were analyzed according to the control culture (without inducing agent—blue bars). Error bars show the standard deviations of the means. $^{*}p < 0.05$, $^{**}p < 0.01$, $^{***}p < 0.001$.

In summary, our results demonstrate that the designed primer sets for *att*P and *att*B allow the observation of the prophage BH1 induction. However, it should be stressed that the number of *att*P copies may be affected by the replication of circular phage particles, as well as by the gradual lysis and release of mature phage particles from induced cells. These processes can be monitored by analysing the changes in the *att*P/*att*B ratio. When evaluating the amount of cells that have been induced, the observation of the amount of copies of *att*B seems to be the most reliable approach. However, the lysis of cells, which occurs as a result of induction, may affect the results obtained. It is also worth noting that, especially in terms of monitoring the *att*B copy count, ddPCR has proven to be a much more sensitive method, enabling the precise detection of even a small number of copies of examined DNA molecules [49].

We also analysed the induction of the prophage BH1 during the growth of the bacterial host, expecting a significantly lower level of induction. ddPCR was used to monitor the number of copies of *att*P, *att*B, and the reference gene (bacterial chromosomal gene) during the culture of *L. rhamnosus* Pen. Measurements were taken every 4 h during a 24 h incubation at 37 °C. Analysis of the results showed a gradual increase in the number of copies of the *att*P and *att*B sequences during the bacterial culture, in the absence of any known inducer (Figure 6). In the case of *att*B, there was an almost 2.5-fold increase in the amount of the sequences resulting from the release of the prophage from the bacterial genome at the 4 and 24 h samples. For the *att*P sequence, the increase was almost 12-fold (similar results were obtained with the qPCR method—Figure S2). In both cases, a significant increase in induction was observed between 8 and 12 h of the culture, for the mid-exponential growth phase of tested bacteria. These results unambiguously confirmed previous microscopic observations of *L. rhamnosus* Pen releasing bacteriophage particles during growth. This release was spontaneous, and did not result from the use of inductive factors [16].

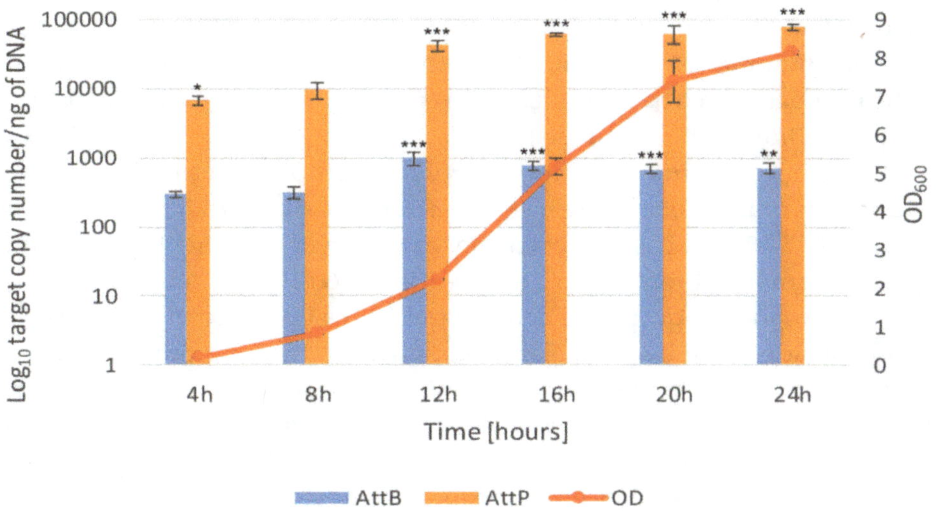

Figure 6. Analysis of spontaneous prophage induction during *L. rhamnosus* Pen growth using droplet digital PCR. SPI was monitored by measuring excision events (*att*B—blue bars) and level of extrachromosomal phage DNA (*att*P). Bacterial culture growth was observed based on OD_{600} measurements (red line). Error bars show the standard deviations of the means. * $p < 0.05$, ** $p < 0.01$, *** $p < 0.001$.

Mechanistically, the factors causing spontaneous induction, as well as the multi-faceted significance of this process, have been frequently analysed. Previous studies indicate that the phenomenon of spontaneous lysogen induction may be caused by extracellular stress related to UV radiation, reactive oxygen species (ROS), temperature, and pH changes. Intracellular factors may also cause a subpopulation of cells to start induction and release phage particles. These triggers include stalled replication forks, reactive oxygen species, and noise in gene expression [17]. From the perspective of technological processes using living microbes for the production of fermented food, spontaneous induction of prophages is unfavourable. When lysis takes place in starter strains used in fermentation processes, or for probiotic strains used in production of protective drugs, large financial losses occur. However, it should also be emphasized that many bacteriophages have a temperate nature, and for this reason, their induction does not have to lead to total lysis and the obstruction of fermentation. It is also worth noting that spontaneous phage induction is likely a natural process, which may take on a unique nature depending on the kind of bacteriophage and bacterial strain, giving unique characteristics to the products being fermented [11,15,51,52].

By analysing the spontaneous induction from the "perspective" of bacteria, it can be concluded from previous studies that this process may be important for a given strain and for the whole ecological niche [12]. Studies have shown that prophages that can be induced spontaneously have a positive effect on the general fitness of bacterial population under diverse environmental conditions [13]. It has been demonstrated that phage induction plays an important role in the formation of biofilm, superinfection exclusion, and in shaping the equilibrium of a specific bacterial community [12,17,53]. Furthermore, the spontaneous induction appears also to be important in building the genetic variability of bacteriophages and bacterial hosts. Baugher et al. have shown that SPI may lead to a horizontal gene transfer, thus giving microbes new characteristics that can give them an advantage in the competition for an ecological niche [54]. It is also worth considering the extreme variability present in the genetic material of the bacteriophages themselves [55]. Bacteriophages, prophages, and prophage-like elements are considered to be the so-called dark matter of biodiversity [34]. It appears that the reason for such an astounding variability of bacteriophages is the high virus mobility resulting from the spontaneous prophage induction, underestimated in its function. Compared to a very "spectacular" lytic cycle, only a small subpopulation of bacteria undergoes SPI, thereby releasing a relatively small number of phage particles. However, by scaling this process up to the level of an entire ecological niche, the importance of SPI in shaping biodiversity of microbial and phage communities appears to be substantial. It has also been demonstrated that many environmental factors can enhance the SPI, potentially contributing to an increase in the number of free bacteriophages in the natural environment, causing phages to have even a greater impact on the ecological niche [52,56].

Finally, SPI has an important role in bacterial virulence. Studies have demonstrated that prophage induction may indirectly contribute to the release of bacterial toxins, and increase the ability of bacteria to adhere to host cells [17,57,58]. It has also been shown that by facilitating binding to human platelets, SPI may lead to development of infective endocarditis [59,60]. It is therefore important to ask the question of whether spontaneous prophage activity promotes host virulence—not only in the case of bacteria commonly considered to be pathogenic—but also in the case of probiotic microorganisms, which are a rich reservoir of prophage sequences [25]. There are numerous cases of bacteremia and endocarditis caused by probiotic bacteria belonging to the species *Lactobacillus rhamnosus* in the literature [61–63]. An example is the *L. rhamnosus* Pen, which has many documented pro-health properties, but has also been associated with sepsis in a cardiosurgical patient [3,64]. Importantly, in relation to the resistance to many antibiotics, the treatment of infections caused by lactobacilli can be challenging [65]. Therefore, it is crucial that further research is conducted in order to identify potential determinants that may be essential for the development of bacteremia and infective endocarditis. Future research should cover not only typically pathogenic microbes, but also probiotic strains used for the production of protective medications and those used as food additives. We suggest that prophage sequences in their genomes and spontaneous prophage induction represent important factors promoting the virulence of these bacteria.

4. Conclusions

To sum up, a thorough insight into the genome phages and prophages of lactobacilli belonging to the *L. casei* group showed that these mobile elements are genetically extremely diverse. Moreover, spontaneous induction of ϕBH1 was also observed using qPCR and ddPCR, and interestingly, this phenomenon did not significantly affect the growth of bacterial culture. Despite the lack of direct evidence on the physiological significance of this process for the tested strain, it has been previously suggested that SPI may enhance biofilm formation and lead to horizontal gene transfer. On the other hand, by facilitating binding to human platelets, SPI may also play an important role in bacterial virulence. Nevertheless, further research is needed to confirm both the positive effects of SPI and the potential risks associated with prophages and their spontaneous induction. This will not only allow us to learn the biological significance of the spontaneous process of prophage induction, but can also contribute to a more conscious use of living microorganisms in the industry. Additionally, we hope that the obtained results and further research will also lead to the extension of knowledge regarding interactions between phages and bacteria, which—excluding the classical lytic cycle—seems to be still insufficient.

Supplementary Materials:
Figure S1: qPCR analysis of phage BH1 induction in the presence of different concentrations of mitomycin C. Induction was observed by measuring the relative amount of extrachromosomal phage DNA (*att*P) (**A**) and the level of reestablished *att*B sites in bacterial genome (**B**). * $p < 0.05$, ** $p < 0.01$, *** $p < 0.001$. Figure S2: Relative level of extrachromosomal phage DNA (*att*P) during growth of *L. rhamnosus* Pen monitored using Q-PCR method. * $p < 0.05$, ** $p < 0.01$, *** $p < 0.001$. Dataset S1: Comparison of coding sequences of bacteriophage BH1 with genome sequences of phages described for *Lactobacillus casei* group. Dataset S2: Comparison of coding sequences of bacteriophage BH1 with prophage sequences derived from *Lactobacillus rhamnosus* strains. Dataset S3: Comparison of coding sequences of bacteriophage BH1 with prophage sequences derived from *Lactobacillus casei* strains. Dataset S4: Comparison of coding sequences of bacteriophage BH1 with prophage sequences derived from *Lactobacillus paracasei* strains.

Author Contributions: Conceptualization, P.J.; Methodology, P.J., E.K.-J., and M.P. (Marcin Podleśny); Software, P.J. and O.K.; Investigation, P.J., E.K.-J., M.P. (Marcin Podleśny), O.K., and M.P. (Monika Pytka); Resources, P.J., E.K.-J., and M.P. (Marcin Podleśny); Data curation, P.J. and E.K-J.; Writing—original draft preparation, P.J; Writing—review and editing, E.K.-J., M.P. (Marcin Podleśny), M.P. (Monika Pytka), and M.K.-W.; Supervision, M.K.-W.; Funding acquisition, P.J. and M.P. (Monika Pytka).

References

1. Vijaya Kumar, B.; Vijayendra, S.V.N.; Reddy, O.V.S. Trends in dairy and non-dairy probiotic products - a review. *J. Food Sci. Technol.* **2015**, *52*, 6112–6124. [CrossRef]
2. Putta, S.; Yarla, N.S.; Lakkappa, D.B.; Imandi, S.B.; Malla, R.R.; Chaitanya, A.K.; Chari, B.P.V.; Saka, S.; Vechalapu, R.R.; Kamal, M.A.; et al. Probiotics: Supplements, food, pharmaceutical industry. In *Therapeutic, Probiotic, and Unconventional Foods*, 1st ed.; Grumezescu, A., Holban, A.M., Eds.; Academic Press, Elsevier: Cambridge, MA, USA, 2018; pp. 15–25.
3. Ruszczyński, M.; Radzikowski, A.; Szajewska, H. Clinical trial: Effectiveness of *Lactobacillus rhamnosus* (strains E/N, Oxy and Pen) in the prevention of antibiotic-associated diarrhoea in children. *Aliment. Pharmacol. Ther.* **2008**, *28*, 154–161. [CrossRef] [PubMed]
4. Bermudez-Brito, M.; Plaza-Diaz, J.; Munoz-Quezada, S.; Gomez-Llorente, C.; Gil, A. Probiotic Mechanisms of Action. *Ann. Nutr. Metab.* **2012**, *61*, 160–174. [CrossRef] [PubMed]
5. Plaza-Diaz, J.; Ruiz-Ojeda, F.J.; Gil-Campos, M.; Gil, A. Mechanisms of action of probiotics. In *Proceedings of the Advances in Nutrition*; Oxford University Press: Oxford, UK, 2019; Volume 10, pp. S49–S185.
6. Sanders, M.E.; Akkermans, L.M.A.; Haller, D.; Hammerman, C.; Heimbach, J.T.; Hörmannsperger, G.; Huys, G. Safety assessment of probiotics for human use. *Gut Microbes* **2010**, *1*, 164–185. [CrossRef] [PubMed]
7. Daliri, E.B.-M.; Lee, B.H.; Oh, D.H. Safety of probiotics in health and disease. In *The Role of Functional Food Security in Global Health*, 1st ed.; Watson, W.R., Singh, R., Takahashi, T., Eds.; Academic Press, Elsevier: Cambridge, MA, USA, 2019; pp. 603–622.
8. Zaburlin, D.; Mercanti, D.J.; Quiberoni, A. A fast PCR-based method for the characterization of prophage profiles in strains of the *Lactobacillus casei* group. *J. Virol. Methods* **2017**, *248*, 226–233. [CrossRef]

9. Ceapa, C.; Davids, M.; Ritari, J.; Lambert, J.; Wels, M.; Douillard, F.P.; Smokvina, T.; de Vos, W.M.; Knol, J.; Kleerebezem, M. The variable regions of *Lactobacillus rhamnosus* genomes reveal the dynamic evolution of metabolic and host-adaptation repertoires. *Genome Biol. Evol.* **2016**, *8*, 1889–1905. [CrossRef]
10. Casjens, S. Prophages and bacterial genomics: What have we learned so far? *Mol. Microbiol.* **2003**, *49*, 277–300. [CrossRef]
11. Garneau, J.E.; Moineau, S. Bacteriophages of lactic acid bacteria and their impact on milk fermentations. *Microb. Cell Fact.* **2011**, *10* Suppl. 1, S20. [CrossRef]
12. Bondy-Denomy, J.; Davidson, A.R. When a virus is not a parasite: The beneficial effects of prophages on bacterial fitness. *J. Microbiol.* **2014**, *52*, 235–242. [CrossRef]
13. Wang, X.; Kim, Y.; Ma, Q.; Hong, S.H.; Pokusaeva, K.; Sturino, J.M.; Wood, T.K. Cryptic prophages help bacteria cope with adverse environments. *Nat. Commun.* **2010**, *1*, 147. [CrossRef]
14. Tuohimaa, A.; Riipinen, K.A.; Brandt, K.; Alatossava, T. The genome of the virulent phage Lc-Nu of probiotic *Lactobacillus rhamnosus*, and comparative genomics with *Lactobacillus casei* phages. *Arch. Virol.* **2006**, *151*, 947–965. [CrossRef] [PubMed]
15. Durmaz, E.; Miller, M.J.; Azcarate-Peril, M.A.; Toon, S.P.; Klaenhammer, T.R. Genome sequence and characteristics of Lrm1, a prophage from industrial *Lactobacillus rhamnosus* strain M1. *Appl. Environ. Microbiol.* **2008**, *74*, 4601–4609. [CrossRef] [PubMed]
16. Jarocki, P.; Podleśny, M.; Pawelec, J.; Malinowska, A.; Kowalczyk, S.; Targoński, Z. Spontaneous release of bacteriophage particles by *Lactobacillus rhamnosus* Pen. *J. Microbiol. Biotechnol.* **2013**, *23*, 357–363. [CrossRef] [PubMed]
17. Nanda, A.M.; Thormann, K.; Frunzke, J. Impact of spontaneous prophage induction on the fitness of bacterial populations and host-microbe interactions. *J. Bacteriol.* **2015**, *197*, 410–419. [CrossRef] [PubMed]
18. Sambrook, J.; Russell, D. *Molecular Cloning: A Laboratory Manual*, 3rd ed.; Cold Spring Harbor Laboratory Press: New York, NY, USA, 2001.
19. Pfaffl, M.W. A new mathematical model for relative quantification in real-time RT-PCR. *Nucleic Acids Res.* **2001**, *29*, 45e. [CrossRef] [PubMed]
20. Villion, M.; Moineau, S. Bacteriophages of *Lactobacillus*. *Front. Biosci.* **2009**, *14*, 1661–1683. [CrossRef] [PubMed]
21. Grimaud, R.; Toussaint, A. Assembly of both the head and tail of bacteriophage Mu is blocked in *Escherichia coli* groEL and groES mutants. *J. Bacteriol.* **1998**, *180*, 1148–1153.
22. Mercanti, D.J.; Rousseau, G.M.; Capra, M.L.; Quiberoni, A.; Tremblay, D.M.; Labrie, S.J.; Moineau, S. Genomic diversity of phages infecting probiotic strains of *Lactobacillus paracasei*. *Appl. Environ. Microbiol.* **2016**, *82*, 95–105. [CrossRef]
23. Jarocki, P.; Podleśny, M.; Krawczyk, M.; Glibowska, A.; Pawelec, J.; Komoń-Janczara, E.; Kholiavskyi, O.; Dworniczak, M.; Targoński, Z. Complete genome sequence of *Lactobacillus rhamnosus* Pen, a probiotic component of a medicine used in prevention of antibiotic-associated diarrhoea in children. *Gut Pathog.* **2018**, *10*, 5. [CrossRef]
24. Wuyts, S.; Wittouck, S.; De Boeck, I.; Allonsius, C.N.; Pasolli, E.; Segata, N.; Lebeer, S. Large-scale phylogenomics of the *Lactobacillus casei* group highlights taxonomic inconsistencies and reveals novel clade-associated features. *mSystems* **2017**, *2*, e00061–17. [CrossRef]
25. Douillard, F.P.; Ribbera, A.; Kant, R.; Pietilä, T.E.; Järvinen, H.M.; Messing, M.; Randazzo, C.L.; Paulin, L.; Laine, P.; Ritari, J.; et al. Comparative genomic and functional analysis of 100 *Lactobacillus rhamnosus* strains and their comparison with strain GG. *PLoS Genet.* **2013**, *9*, e1003683. [CrossRef] [PubMed]
26. Brandt, K.; Alatossava, T. Specific identification of certain probiotic *Lactobacillus rhamnosus* strains with PCR primers based on phage-related sequences. *Int. J. Food Microbiol.* **2003**, *84*, 189–196. [CrossRef]
27. Zago, M.; Rossetti, L.; Reinheimer, J.; Carminati, D.; Giraffa, G. Detection and identification of *Lactobacillus helveticus* bacteriophages by PCR. *J. Dairy Res.* **2008**, *75*, 196–201. [CrossRef] [PubMed]
28. Barrangou, R.; Fremaux, C.; Deveau, H.; Richards, M.; Boyaval, P.; Moineau, S.; Romero, D.A.; Horvath, P. CRISPR provides acquired resistance against viruses in prokaryotes. *Science* **2007**, *315*, 1709–1712. [CrossRef]
29. Lo, T.C.; Shih, T.C.; Lin, C.F.; Chen, H.W.; Lin, T.H. Complete genomic sequence of the temperate bacteriophage ΦAT3 isolated from *Lactobacillus casei* ATCC 393. *Virology* **2005**, *339*, 42–55. [CrossRef]
30. Kutter, E.; Sulakvelidze, A. *Bacteriophages Biology and Applications*, 1st ed.; CRC Press: Boca Raton, FL, USA, 2004.

31. García, P.; Ladero, V.; Suárez, J.E. Analysis of the morphogenetic cluster and genome of the temperate *Lactobacillus casei* bacteriophage A2. *Arch. Virol.* **2003**, *148*, 1051–1070. [CrossRef]

32. Dieterle, M.E.; Piuri, M.; Bowman, C.; Hatfull, G.; Batthyany, C.; Turjanski, A.; Lanzarotti, E. Exposing the secrets of two well-known *Lactobacillus casei* phages, J-1 and PL-1, by genomic and structural analysis. *Appl. Environ. Microbiol.* **2014**, *80*, 7107–7121. [CrossRef]

33. Kelly, G.; Prasannan, S.; Daniell, S.; Fleming, K.; Frankel, G.; Dougan, G.; Connerton, I.; Matthews, S. Structure of the cell-adhesion fragment of intimin from enteropathogenic *Escherichia coli*. *Nat. Struct. Biol.* **1999**, *6*, 313–318.

34. Hatfull, G.F. Dark matter of the biosphere: the amazing world of bacteriophage diversity. *J. Virol.* **2015**, *89*, 8107–8110. [CrossRef]

35. Grose, J.H.; Casjens, S.R. Bacteriophage diversity. In *Reference Module in Life Sciences*; Elsevier: Amsterdam, The Netherlands, 2019.

36. Gründling, A.; Manson, M.D.; Young, R. Holins kill without warning. *Proc. Natl. Acad. Sci. USA* **2001**, *98*, 9348–9352.

37. Kashige, N.; Nakashima, Y.; Miake, F.; Watanabe, K. Cloning, sequence analysis, and expression of *Lactobacillus casei* phage PL-1 lysis genes. *Arch. Virol.* **2000**, *145*, 1521–1534. [CrossRef] [PubMed]

38. Dieterle, M.E.; Fina Martin, J.; Durán, R.; Nemirovsky, S.I.; Sanchez Rivas, C.; Bowman, C.; Russell, D.; Hatfull, G.F.; Cambillau, C.; Piuri, M. Characterization of prophages containing "evolved" Dit/Tal modules in the genome of *Lactobacillus casei* BL23. *Appl. Microbiol. Biotechnol.* **2016**, *100*, 9201–9215. [CrossRef] [PubMed]

39. Groth, A.C.; Calos, M.P. Phage integrases: biology and applications. *J. Mol. Biol.* **2004**, *335*, 667–678. [CrossRef] [PubMed]

40. Williams, R.J. Restriction endonucleases: Classification, properties, and applications. *Appl. Biochem. Biotechnol. Part B Mol. Biotechnol.* **2003**, *23*, 225–243. [CrossRef]

41. Obarska-Kosinska, A.; Taylor, J.E.; Callow, P.; Orlowski, J.; Bujnicki, J.M.; Kneale, G.G. HsdR subunit of the type I restriction-modification enzyme EcoR124I: biophysical characterisation and structural modelling. *J. Mol. Biol.* **2008**, *376*, 438–452. [CrossRef]

42. Erill, I.; Campoy, S.; Barbé, J. Aeons of distress: An evolutionary perspective on the bacterial SOS response. *FEMS Microbiol. Rev.* **2007**, *31*, 637–656. [CrossRef]

43. Fornelos, N.; Browning, D.F.; Pavlin, A.; Podlesek, Z.; Hodnik, V.; Salas, M.; Butala, M. Lytic gene expression in the temperate bacteriophage GIL01 is activated by a phage-encoded LexA homologue. *Nucleic Acids Res.* **2018**, *46*, 9432–9443. [CrossRef]

44. Nanda, A.M.; Heyer, A.; Krämer, C.; Grünberger, A.; Kohlheyer, D.; Frunzke, J. Analysis of SOS-induced spontaneous prophage induction in *Corynebacterium glutamicum* at the single-cell level. *J. Bacteriol.* **2014**, *196*, 180–188. [CrossRef]

45. Ndjonka, D.; Bell, C.E. Structure of a hyper-cleavable monomeric fragment of phage λ repressor containing the cleavage site region. *J. Mol. Biol.* **2006**, *362*, 479–489. [CrossRef]

46. Foley, S.; Lucchini, S.; Zwahlen, M.C.; Brüssow, H. A short noncoding viral DNA element showing characteristics of a replication origin confers bacteriophage resistance to *Streptococcus thermophilus*. *Virology* **1998**, *250*, 377–387. [CrossRef]

47. Gvakharia, B.O.; Hanson, E.; Koonin, E.K.; Mathews, C.K. Identification of a second functional glutaredoxin encoded by the bacteriophage T4 genome. *J. Biol. Chem.* **1996**, *271*, 15307–15310. [CrossRef] [PubMed]

48. Lunde, M.; Blatny, J.M.; Lillehaug, D.; Aastveit, A.H.; Nes, I.F. Use of real-time quantitative PCR for the analysis of φLC3 prophage stability in lactococci. *Appl. Environ. Microbiol.* **2003**, *69*, 41–48. [CrossRef] [PubMed]

49. Tang, H.; Cai, Q.; Li, H.; Hu, P. Comparison of droplet digital PCR to real-time PCR for quantification of hepatitis B virus DNA. *Biosci. Biotechnol. Biochem.* **2016**, *80*, 2159–2164. [CrossRef] [PubMed]

50. Maheshwari, Y.; Selvaraj, V.; Hajeri, S.; Yokomi, R. Application of droplet digital PCR for quantitative detection of *Spiroplasma citri* in comparison with real time PCR. *PLoS One* **2017**, *12*, e0184751. [CrossRef] [PubMed]

51. Alexeeva, S.; Guerra Martínez, J.A.; Spus, M.; Smid, E.J. Spontaneously induced prophages are abundant in a naturally evolved bacterial starter culture and deliver competitive advantage to the host. *BMC Microbiol.* **2018**, *18*, 16. [CrossRef] [PubMed]

52. Ho, C.H.; Stanton-Cook, M.; Beatson, S.A.; Bansal, N.; Turner, M.S. Stability of active prophages in industrial *Lactococcus lactis* strains in the presence of heat, acid, osmotic, oxidative and antibiotic stressors. *Int. J. Food Microbiol.* **2016**, *220*, 26–32. [CrossRef]
53. Mahony, J.; Lugli, G.A.; van Sinderen, D.; Ventura, M. Impact of gut-associated bifidobacteria and their phages on health: Two sides of the same coin? *Appl. Microbiol. Biotechnol.* **2018**, *102*, 2091–2099. [CrossRef]
54. Baugher, J.L.; Durmaz, E.; Klaenhammer, T.R. Spontaneously induced prophages in *Lactobacillus gasseri* contribute to horizontal gene transfer. *Appl. Environ. Microbiol.* **2014**, *80*, 3508–3517. [CrossRef]
55. Díaz-Muñoz, S.L.; Koskella, B. Bacteria-phage interactions in natural environments. *Adv. Appl. Microbiol.* **2014**, *89*, 135–183.
56. Lunde, M.; Aastveit, A.H.; Blatny, J.M.; Nes, I.F. Effects of diverse environmental conditions on φLC3 prophage stability in *Lactococcus lactis*. *Appl. Environ. Microbiol.* **2005**, *71*, 721–727. [CrossRef]
57. Xu, X.; McAteer, S.P.; Tree, J.J.; Shaw, D.J.; Wolfson, E.B.K.; Beatson, S.A.; Roe, A.J.; Allison, L.J.; Chase-Topping, M.E.; Mahajan, A.; et al. Lysogeny with Shiga toxin 2-encoding bacteriophages represses type III secretion in enterohemorrhagic *Escherichia coli*. *PLoS Pathog.* **2012**, *8*, e1002672. [CrossRef] [PubMed]
58. Parsek, M.R.; Singh, P.K. Bacterial biofilms: an emerging link to disease pathogenesis. *Annu. Rev. Microbiol.* **2003**, *57*, 677–701. [CrossRef] [PubMed]
59. Seo, H.S.; Xiong, Y.Q.; Mitchell, J.; Seepersaud, R.; Bayer, A.S.; Sullam, P.M. Bacteriophage lysin mediates the binding of *Streptococcus mitis* to human platelets through interaction with fibrinogen. *PLoS Pathog.* **2010**, *6*, e1001047. [CrossRef] [PubMed]
60. Mitchell, J.; Siboo, I.R.; Takamatsu, D.; Chambers, H.F.; Sullam, P.M. Mechanism of cell surface expression of the *Streptococcus mitis* platelet binding proteins PblA and PblB. *Mol. Microbiol.* **2007**, *64*, 844–857. [CrossRef] [PubMed]
61. Presterl, E.; Kneifel, W.; Mayer, H.K.; Zehetgruber, M.; Makristathis, A.; Graninger, W. Endocarditis by *Lactobacillus rhamnosus* due to yogurt ingestion? *Scand. J. Infect. Dis.* **2001**, *33*, 710–714. [PubMed]
62. Boumis, E.; Capone, A.; Galati, V.; Venditti, C.; Petrosillo, N. Probiotics and infective endocarditis in patients with hereditary hemorrhagic telangiectasia: a clinical case and a review of the literature. *BMC Infect. Dis.* **2018**, *18*, 65. [CrossRef]
63. Aaron, J.G.; Sobieszczyk, M.E.; Weiner, S.D.; Whittier, S.; Lowy, F.D. *Lactobacillus rhamnosus* endocarditis after upper endoscopy. *Open Forum Infect. Dis.* **2017**, *4*, ofx085. [CrossRef]
64. Kochan, P.; Chmielarczyk, A.; Szymaniak, L.; Brykczynski, M.; Galant, K.; Zych, A.; Pakosz, K.; Giedrys-Kalemba, S.; Lenouvel, E.; Heczko, P.B. *Lactobacillus rhamnosus* administration causes sepsis in a cardiosurgical patient-is the time right to revise probiotic safety guidelines? *Clin. Microbiol. Infect.* **2011**, *17*, 1589–1592. [CrossRef]
65. Naqvi, S.S.B.; Nagendra, V.; Hofmeyr, A. Probiotic related *Lactobacillus rhamnosus* endocarditis in a patient with liver cirrhosis. *IDCases* **2018**, *13*, e00439. [CrossRef]

2

Characterization and Therapeutic Potential of Bacteriophage-Encoded Polysaccharide Depolymerases with β Galactosidase Activity against *Klebsiella pneumoniae* K57 Capsular Type

Nikolay V. Volozhantsev [1,*], Anna M. Shpirt [2], Alexander I. Borzilov [1], Ekaterina V. Komisarova [1], Valentina M. Krasilnikova [1], Alexander S. Shashkov [2], Vladimir V. Verevkin [1] and Yuriy A. Knirel [2]

1 State Research Center for Applied Microbiology and Biotechnology, 142279 Obolensk, Moscow Region, Russia; borzilov@obolensk.org (A.I.B.); ekaterina20009@mail.ru (E.V.K.); krasv55@mail.ru (V.M.K.); lab-mdgip@mail.ru (V.V.V.)

2 N.D. Zelinsky Institute of Organic Chemistry, Russian Academy of Sciences, Leninsky Prospekt 47, 119991 Moscow, Russia; asya@ioc.ac.ru (A.M.S.); shash@ioc.ac.ru (A.S.S.); knirel@ioc.ac.ru (Y.A.K.)

* Correspondence: nikvol@obolensk.org

Abstract: Bacteriophages and phage enzymes are considered as possible alternatives to antibiotics in the treatment of infections caused by antibiotic-resistant bacteria. Due to the ability to cleave the capsular polysaccharides (CPS), one of the main virulence factors of *Klebsiella pneumoniae*, phage depolymerases, has potential in the treatment of *K. pneumoniae* infections. Here, we characterized in vivo two novel phage-encoded polysaccharide depolymerases as therapeutics against clinical isolates of *K. pneumoniae*. The depolymerases Dep_kpv79 and Dep_kpv767 encoded by Klebsiella phages KpV79 (Myoviridae; Jedunavirus) and KpV767 (Autographiviridae, Studiervirinae, Przondovirus), respectively, were identified as specific β-galactosidases that cleave the *K. pneumoniae* K57 type CPS by the hydrolytic mechanism. They were found to be highly effective at combating sepsis and hip infection caused by *K. pneumoniae* in lethal mouse models. Here, 80–100% of animals were protected against death by a single dose (e.g., 50 μg/mouse) of the enzyme injected 0.5 h after infection by *K. pneumoniae* strains of the K57 capsular type. The therapeutic effect of the depolymerases is because they strip the capsule and expose the underlying bacterium to the immune attack such as complement-mediated killing. These data provide one more confirmation that phage polysaccharide depolymerases represent a promising tool for antimicrobial therapy.

Keywords: bacteriophage; polysaccharide depolymerase; *Klebsiella pneumoniae*; antimicrobial therapy

---

1. Introduction

Antibiotic resistance of pathogenic bacteria has become an increasingly pressing clinical issue worldwide. *Klebsiella pneumoniae* is a well-known Gram-negative bacterium notorious for contributing to multidrug resistance. So far, *K. pneumoniae* has been considered as an opportunistic pathogen, primarily causing different types of healthcare-associated infections in immunocompromised patients, including pneumonia, urinary and intestinal tract infections, and wound or surgical site infections [1,2]. However, in recent years, a new type of hypervirulent *K. pneumoniae* (hvKp) which causes community-acquired invasive life-threatening infections characterized by pyogenic liver abscesses complicated with meningitis and endophthalmitis, has emerged worldwide, especially in Southeast Asia [2–4].

K. pneumoniae cells produce a number of virulence factors including fimbrial adhesins, lipopolysaccharides (O antigen), siderophore iron acquisition systems, and a polysaccharide capsule (K antigen) [2]. An essential virulence factor and a defense barrier of *K. pneumoniae* are a polysaccharide capsule (CPS) that allows a bacterial cell to survive and spread inside the host overcoming the protective mechanisms of the immune system. The capsule impairs phagocytosis and opsonophagocytosis of the *K. pneumoniae* cells by immune cells hinders the bactericidal action of antimicrobial peptides and blocks complement components, such as C3, thus preventing complement-mediated killing [2,5,6]. There is a tremendous amount of structural variability among *K. pneumoniae* capsules, and at least 134 capsular types have been identified, including 77 K-types recognized by serological typing [7] and new K-types characterized by molecular genotyping and sequencing of the capsule synthesis locus (K-locus, KL) [8,9]. Of all CPS-types *K. pneumoniae*, K1, K2, K5, K20, K54, and K57 strains overproducing capsular polysaccharides are the most virulent (hvKp) human pathogens [10]. In addition, the CPS is recognized as a primary receptor for bacteriophages that possess capsule depolymerization activities [11]. These enzymatic activities allow phage to get access to the cell surface and bind to the secondary receptor. Bacteriophage-encoded polysaccharide-degrading enzymes, known as polysaccharide (PS) depolymerases, are highly specific enzymes, typically associated with the tail structure of the virions such as tail fibers, tailspikes, base plate or tail tubular proteins [12–14]. Since depolymerases target and destroy cell structures, which are important for bacteria survival and virulence, they have been suggested as potential anti-virulence agents to control and prevent bacterial infections. Recently, there have been several reports on the successful application of PS-depolymerases in the treatment of infections caused by *Escherichia coli* [15], *Pasteurella multocida* [16], *Pseudomonas aeruginosa* [17], *Acinetobacter baumannii* [18]. It was shown that depolymerases successfully rescued mice and *Galleria mellonella* larvae infected by *K. pneumoniae*, including hvKp strains [19–22].

As previously reported, a number of PS-depolymerases specific for *K. pneumoniae* capsular type K1 [19,23], K2 [24], K3, K21 [22], K5, K8, K30/K69 [21], K11, K25, K35, K64, KN5 [23], K63 [20], KN1, KN3, KN4, and K56 [25] have been isolated and characterized. Pan et al. [26] identified nine PS-depolymerases in a broad host range bacteriophage infecting *K. pneumoniae* and correlated the activities of the gene products to all host strains that belonged to capsule types K1, K11, K21, K25, K30/K69, K35, K64, KN4, and KN5. However, depolymerases, specific for *K.pneumoniae* of K57 capsule type, have not been characterized yet.

Previously, we reported on the isolation and genome characterization of bacteriophage KpV767 that can produce polysaccharide depolymerase and infect hypermucoviscous *K.pneumoniae* of K57 capsular type [24]. In this study, we describe a new lytic phage KpV79 against K57 capsule type *K. pneumoniae* and characterize two K57 specific polysaccharide depolymerases from phages KpV79 and KpV767.

2. Results and Discussion

2.1. Bacteriophages KpV79 and KpV767 Are Specific for Klebsiella pneumoniae of Capsular Type K57

Bacteriophages KpV767 and KpV79 were originally isolated from sewage in Russia in November 2015 and February 2016, respectively. We evaluated the lytic spectrum of KpV767 and KpV79 on a collection of *K. pneumoniae* clinical isolates (N = 250) including 21 strains of capsular type K57 (Table 1). Both phages multiplied and lysed *K. pneumoniae* capsular type K57 to form plaques with a halo expanding during incubation, which we suspected is related to a putative phage-derived depolymerase (Figure 1A). The phages did not multiply with plaque formation on some K57-type strains. However, in these cases, a translucent spot resembling the plaque halo was observed in the spot test by using undiluted phage suspension (Table 2). It should be noted that both phages were not active against others (non-K57 type) strains of the test collection ($n = 229$), as indicated by the absence of plaques of lysis or translucent spots on a bacterial lawn. These results indicate that phages KpV767 and KpV79 exhibit specificity to K57 capsular polysaccharides.

Table 1. General features of bacteriophages KpV79 and KpV767.

Characteristics	KpV79	KpV767
Isolation source	sewage	sewage
Isolation date	Feb-2016	Nov-2015
Genome length, bp	47,760	40,395
Total genes	75	52
Predicted PS depolymerase gene	kpv79_42	kpv767_46
The number of strains lysed by phage		
K57 type ($n = 21$)	18 [b] + 3 [c]	10 [b] + 10 [c]
Not-K57 type ($n = 229$) [a]	0	0

[a] Strains of capsular type K1, K2, K3, K13, K20, K22, K23, K24, K27, K28, K31, K39, K47, K54, K60, K62, K64, LK107 ($n = 116$), as well as 113 untyped strains of non-K57 type were tested. [b]. The number of strains on which the bacteriophage multiply and form plaques with a translucent halo. [c]. The number of strains on which the bacteriophage does not multiply but forms a translucent spot without plaque formation.

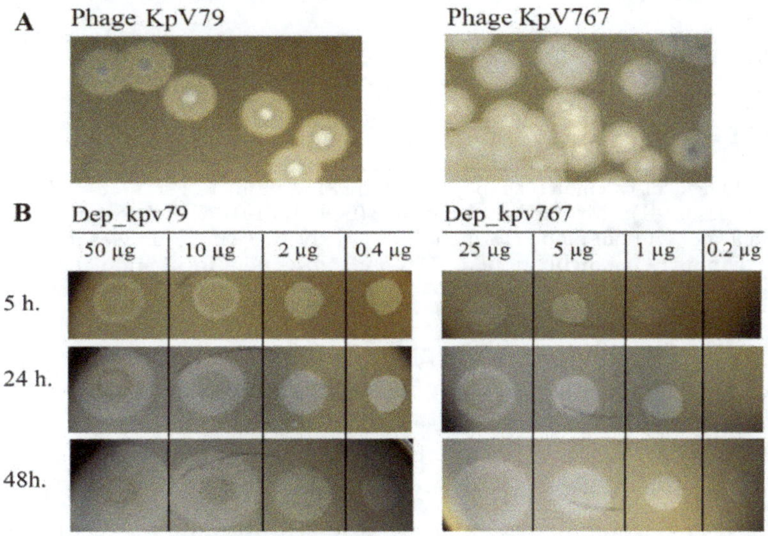

Figure 1. The activity of depolymerases Dep_kpv79 and Dep_kpv767 against *Klebsiella pneumoniae* strains. (**A**) Haloes surrounding phage KpV79 and KpV767 plaques, suggesting the presence of phage depolymerases. (**B**) Spot tests of purified depolymerases Dep_kpv79 and Dep_kpv767 on *K. pneumoniae* KPi8289 lawn. Aliquots (10 μL) of serial five-fold dilutions of Dep_79 and Dep_767 were spotted onto a plate containing the *K. pneumoniae* strain. The plates were observed for 5 h, 24 h, and 48 h of incubation at 37 °C.

Table 2. The activity of bacteriophages KpV79 and KpV767 and their PS-depolymerases Dep79 and Dep767 against the *K. pneumoniae* strains of capsular type K57.

#	Strain	HV Phenotype	The Efficiency of Plating (EOP) [a]		Spot-Test [b]	
			KpV79	KpV767	Dep79	Dep767
1	KPB1106-2	–	H	H	++	+++
2	KPB156	+	H	H	++	+
3	KPB335	–	L	T	++	+
4	KPB500	–	L	-	+/–	–
5	KPB542-15	–	M	H	++	++
6	KPB550	–	L	T	++	++
7	KPB567	+	L	T	++	++
8	KPB584	–	T	T	++	+
9	KPB612-1	+	L	T	++	++

Table 2. *Cont.*

#	Strain	HV Phenotype	The Efficiency of Plating (EOP) [a]		Spot-Test [b]	
			KpV79	KpV767	Dep79	Dep767
10	KPB690	+	L	T	+++	++
11	KPB697-1	–	T	T	++	++
12	KPB742	–	H	H	++	++
13	KPB757	+	L	T	+++	+++
14	KPB774-1	–	H	H	+++	+++
15	KPB811	+	T	T	++	++
16	KPBP1	–	M	H	+++	+++
17	KPi112	–	H	H	++	+
18	KPi4605	–	H	H	+++	+++
19	KPi4891	–	H	H	+++	+++
20	KPi8289	+	H	L	+++	+++
21	KPX4	–	H	T	++	+

[a] Efficiency of plating (EOP) was calculated as the ratio of the PFU on the test strain to PFU on the host strain. "H", high efficiency (EOP values 0.1–1); "M", medium efficiency (EOP 0.001-0.01); "L", low efficiency (EOP values 0.0001 or less); "T", translucent spot without plaque formation; "–", no lysis. [b] The result was recorded after 15 min (+++), an hour (++), and two hours (+) of incubation at 37 °C and subsequent incubation at 8 °C overnight (+/–). "–", no activity.

2.2. *Genomes of Bacteriophages KpV79 and KpV767, Representing Distinct Taxonomic Groups, Contain Genes Encoding Polysaccharide Depolymerases with a High Level of Similarity at the Amino Acid Level*

In a previous study [24], we presented the sequence of the bacteriophage KpV767 genome, consisting of 40,395 bp with a G + C content of 52.28%. The KpV79 phage genome has also been sequenced and analyzed. This phage has a linear double-stranded DNA genome of 47,760 bp. Its G + C content is 48.96%, and 75 open reading frames (ORF) are identified and annotated. In both phage genomes, putative functions were assigned to ORF products involved in phage DNA replication, transcription, packaging DNA into the capsid, host lysis, head, and tail morphogenesis (Figure 2).

In a number of bacteriophages, the proteins containing the depolymerase domain are usually located in tail spikes and tail fibers attached to the base plate, though they may also be located in other components of the virion, such as internal head or viral membrane [13]. Such proteins identified in genomes of the studied phages, as well as some hypothetical proteins with unknown functions, were analyzed for the presence of amino-acid and structural homology with carbohydrate-hydrolyzing enzymes in protein databases using PSI-BLAST and HHpred tools. To all appearances, the products of genes kpv79_42 and kpv767_46 of phages KpV79 and KpV767, respectively, are structural proteins with predicted PS depolymerase activity involved in the phage interaction with host bacterium. The proteins have significant homology within about 360 aa in the regions 309–671 aa of kpv767_46 protein and 177–539 aa of kpv79_42 protein, respectively (Figure 2).

A high level of similarity at the amino acid level in the central domain of proteins may indicate that these domains determine phage specificity and are responsible for the catalytic cleavage of polysaccharides. In contrast, differences in their N-terminal domains that are responsible for binding to phage particles reflect the fact that phages KpV79 and KpV767 are representatives of different families.

To confirm whether the predicted proteins have polysaccharide-degrading activity, we cloned genes kpv767_46 (phage KpV767) and kpv79_42 (phage KpV79) into the pET22b+ expression vector. His-tag fusion proteins Dep_kpv767 and Dep_kpv79 with a predicted molecular mass of 64 kDa and 77 kDa, respectively, were expressed and purified.

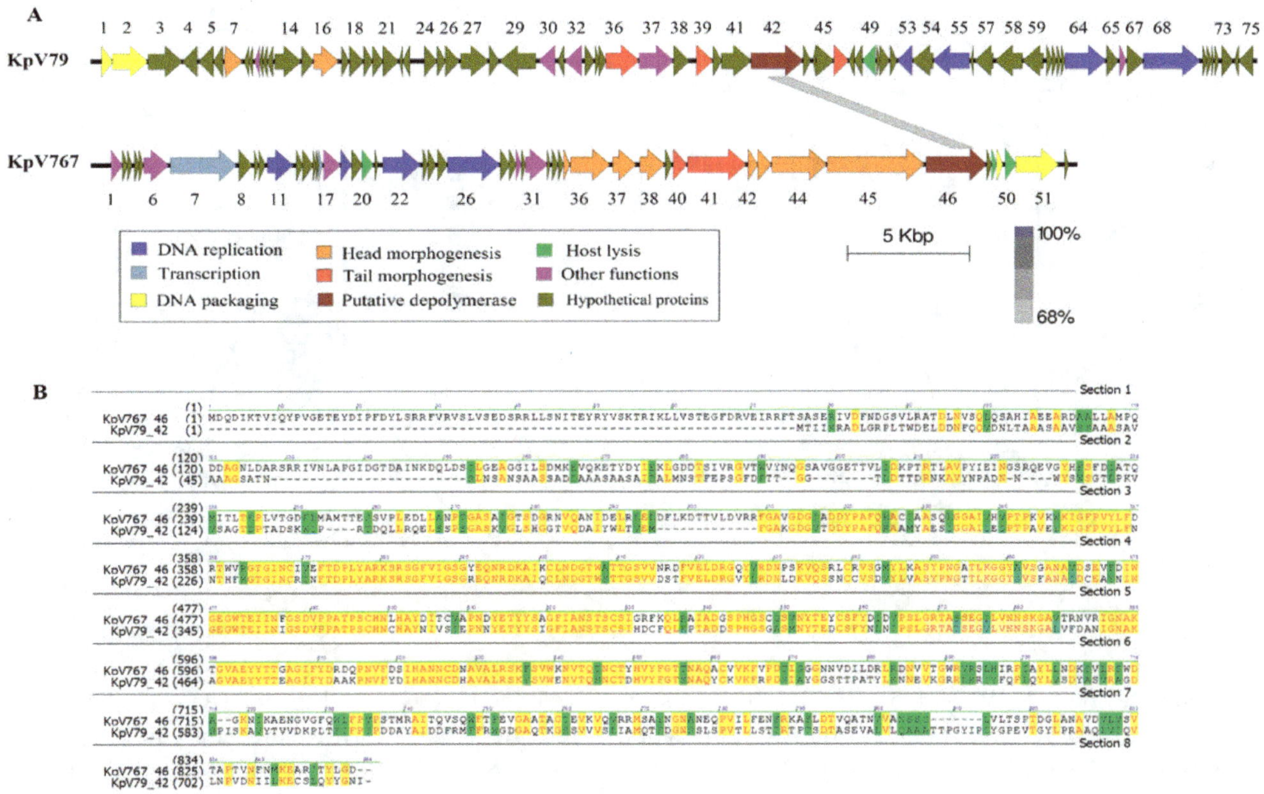

Figure 2. Genomic analysis of bacteriophages KpV79 and KpV767. (**A**) Open reading frame organization and comparison of phages genomes. (**B**) Comparison of amino acid sequences encoded by the PS-depolymerase genes kpvV79_42 and kpv767_46 from bacteriophages KpV79 and KpV767, respectively.

The activities of Dep_kpv767 and Dep_kpv79 were tested on 50 *K. pneumoniae* strains including 21 strains of K57 type. On spotting to a plate inoculated with *K. pneumoniae* K57 strains, the recombinant proteins generated a translucent spot resembling the plaque halo (Table 2; Figure 1B). At the same time, translucent spots were not formed on any bacterial lawn of non-K57 type *K. pneumoniae* strains. The presented results show that depolymerases Dep_kpv79 and Dep_kpv767, like parental phages KpV79 and KpV767, exhibit specificity to K57-capsulated *K. pneumoniae*.

2.3. Depolymerases Dep_767 and Dep_79 Inhibit the Adsorption of Corresponding Bacteriophages on a Capsule of Bacterial Cells

To study the interaction of bacterial viruses with surface receptors of a bacterial cell, quite simple methods are often used, such as inhibiting the adsorption of bacteriophages on the surface of bacteria by proteins of the phage receptor apparatus, including PS depolymerase, as well as phage inactivation by primary bacterial receptors such as capsule polysaccharides. We used a competition assay to assess the role of Dep_kpv79 and Dep_kpv767 in the KpV79 phage-cell interaction at the initial stages of the infection process. According to our preliminary results, at least 90% of the phage KpV79 are adsorbed to the *K. pneumoniae* cells in 5 min, and the phage latent period is about 30 min. The phage was added to the *K. pneumoniae* cells pre-treated with PS-depolymerases, and after 5 min of incubation at 37 °C, the titer of the phage that remained free in solution was measured. In one result, both K57 specific depolymerases Dep_kpv79 and Dep_kpv767 inhibited the phage-host binding. In the same conditions, K1-specific depolymerase Dep_kpv71 [24] did not affect phage adsorption (Figure 3A).

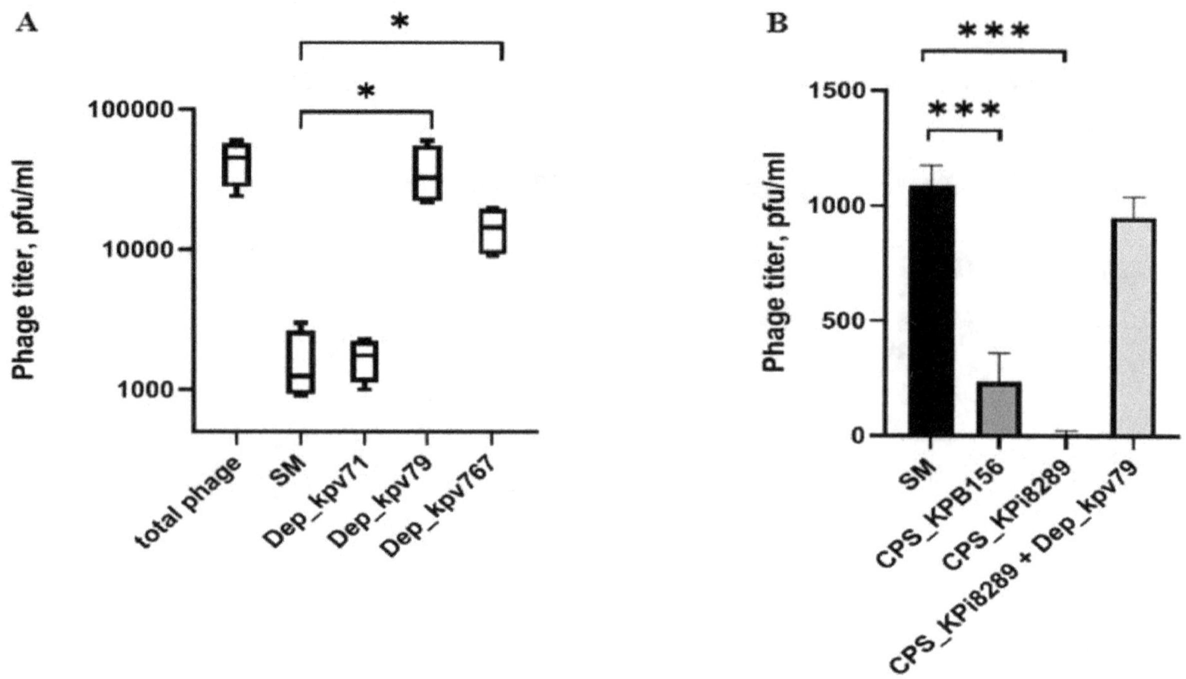

Figure 3. Phage KpV79 adsorption inhibition assay and phage inactivation by CPS. (**A**) From left to right, a titer of the total phage sample and titers of the free phage particles after the 5 min incubation with KpV79-sensitive *K. pneumoniae* KPi8289 cells in SM buffer and in the presence of Dep_kpv71, Dep_kpv79, and Dep_kpv767. (**B**) Purified CPS's from *K. pneumoniae* KPB156 and KPi8289, as well as CPS KPi8289 pre-treated with PS-depolymerase Dep_kpv79, were added to identical phage samples and, after 10 min of incubation, aliquots were taken and titrated. Pure SM buffer was used as a negative control. Average titers from three independent experiments are shown and compared using Student's *t*-test analysis. * $p < 0.05$, *** $p < 0.001$.

Primary bacterial receptors usually competitively inhibit phage adsorption on a bacterial cell. To evaluate the inhibitory effect of capsular polysaccharides on the bacteriophage KpV79, we conducted the phage inactivation by CPS. As a result, it turned out that the CPS of both strains can inactivate the bacteriophages. Furthermore, phage KpV79 inactivation did not occur if the CPS is pre-treated with Dep_kpv79 depolymerase (Figure 3B).

The data on specificity of the bacteriophages KpV79 and KpV767 to *K. pneumoniae* of the capsular type K57, as well as the inhibition of the phages by the K57-CPS, suggest that the capsule is the primary receptor for these phages. Both Dep_767 and Dep_79 inhibit the phage adsorption to *K. pneumoniae* cells, indicating that the phage and depolymerases compete for the same set of receptors on the host cell surface, that are capsular polysaccharides. These results are consistent with that Dep_kpv79 and Dep_kpv767 proteins were bioinformatics predicted as putative tail fibers/spikes that are responsible for host cell recognition. Based on adsorption experiments, we suppose that Dep_kpv79 and Dep_kpv767 proteins are responsible for the reversible binding and subsequent degradation of capsular polysaccharides of *K. pneumoniae* cells. This mechanism, followed by a stage of irreversible adsorption to the main receptors of the bacterial outer membrane, has been described for other capsule-specific phages [11].

2.4. Depolymerases Dep_kpv79 and Dep_kpv767 Specifically Cleave the β-Galactosidic Linkages in the K57 Capsular Polysaccharide of K. pneumoniae

Depending on the way of degradation of bacterial polysaccharides, phage-associated enzymes fall into the two main groups—glycoside hydrolases (EC 3.2.1.) and lyases (EC 4.2.2.)—which are divided into different subclasses [13]. Usually, the mechanism of the enzymatic action of phage enzymes is predicted according to amino-acid and structural homology with carbohydrate- degrading enzymes in protein databases. At the same time, only a few experimental data that relate to the enzymatic mechanisms of phage depolymerase action are available [17,27,28].

In this work, we elucidated the mechanism and characterized for the first time the product of cleavage of the *K. pneumoniae* K57 CPS by phage depolymerases Dep_kpv79 and Dep_kpv767.

The composition and structure of the K57 CPS of *K. pneumoniae* have been known [29]. CPS structure of the strain KPi8289 was confirmed by the assignment of the ^{1}H and ^{13}C NMR spectra of the CPS using two-dimensional ^{1}H,^{1}H COSY, ^{1}H,^{1}H TOCSY, and ^{1}H,^{13}C HSQC experiments (Table 3). The CPS is built up of branched tetrasaccharide repeats (K units) containing two residues of D-mannose (units B and D) and one residue each of D-galactose (unit A) and D-galacturonic acid (unit C) (Figure 4).

Table 3. ^{1}H and ^{13}C NMR chemical shifts (δ, ppm) of the K57 CPS from *K. pneumoniae* KPi8289 and oligosaccharide 1 (OS) derived by cleavage of the CPS with depolymerase Dep_kpv79.

Residue	H-1	H-2	H-3	H-4	H-5	H-6
	C-1	*C-2*	*C-3*	*C-4*	*C-5*	*C-6 (6a, 6b)*
CPS						
→3)–β–δ–Γαλπ–(1→	4.66	3.64	3.72	4.01	3.63	3.77, 3.83
A	*105.7*	*71*	*78*	*65.9*	*76.3*	*62.5*
→2)–α–δ–Μανπ–(1→	5.21	4.05	4.02	3.77	4.01	3.75, 3,84
B	*96.2*	*81.2*	*71.6*	*68.4*	*74.1*	*62.3*
→3,4)–±–δ–ΓαλπΑ–(1→	5.21	4.07	4.16	4.63	4.57	
C	*102.1*	69.5	*78.4*	*79*	*72.2*	*174.7*
α-d-Man*p*-(1→	4.99	3.87	3.82	3.69	4.02	3.80, 3,89
D	*101.5*	*71.4*	*71.6*	*68*	*74*	*62.2*
OS						
→3)–α–δ–Γαλπ	5.28	3.91	3.94	4.19	n.f.	3.73–3.88
Aα	*93.6*	*68.1*	*74.5*	*66.3*	n.f.	*62.5*
→3)–β–δ–Γαλπ	4.61	3.56	3.73	4.12	3.65	3.73–3.88
Aβ	*97.3*	*71.7*	*77.9*	*65.8*	*76.3*	*62.4*
→2)–α–δ–Μανπ–(1→	5.22	4.02	4.03	3.78	3.84	3.73–3.88
B	*96.1*	*81*	*71.7*	*68.4*	*74*	*62.2*
→4)–α–δ–ΓαλπΑ–(1→	5.18	3.87	4.02	4.4	4.39	
C	*102.1*	*70.1*	*70.3*	*80.5*	*72.6*	n.f.
α-d-Man*p*-(1→	4.91	3.92	3.85	3.67	3.98	3.73–3.88
D	102.4	*71.5*	*71.5*	*68*	*74.2*	*62.2*

^{13}C NMR chemical shifts are italicized. n.f., not found.

The CPS was cleaved with depolymerase Dep_kpv79 and the products were fractionated by gel permeation chromatography on Fractogel TSK HW-40S to give two oligosaccharides 1 and 2 (Figure 4). The same oligosaccharides were obtained by cleavage of the CPS_KPi8289 with Dep_kpv767 depolymerase (data not shown).

Analysis by negative ion mode high-resolution electrospray ionization mass spectrometry (HR ESI MS) showed that a lower oligosaccharide 1 corresponded to the K unit of the CPS and oligosaccharide 2 to a dimer of the K unit (the spectra showed $[M-H]^{-}$ ion peaks for Hex_3HexA_1 and Hex_6HexA_2 at *m/z* 679.1841 and 1341.3834; calculated values *m/z* 679.1938 and 1341.3844, respectively).

The ^{1}H and ^{13}C NMR spectra of oligosaccharide 1 were assigned as described above for the CPS (Table 3). As in the K unit of the CPS, four monosaccharide residues (units A-D) were identified. Whereas the ^{13}C NMR chemical shifts of B and D were essentially the same in the CPS and 1, those of units A and C were different. The monosaccharide residue at the reducing end of 1 that was identified as the Gal residue (unit A) occurred in two anomeric pyranose forms (δ_H 5.28 and 4.61, δ_C 93.6 and 97.3 for the α- and β-anomers; compare published data [30] δ_H 5.22 and 4.53, δ_C 93.18 and 97.37 for α-Gal*p* and β-Gal*p*, respectively). This residue is β-linked in the CPS. The GalA residue (unit C) is 3,4-disubstituted in the CPS and 4-substituted in 1, as followed from a marked displacement of the signal for C-3 of unit C from δ 78.4 in the CPS to δ 70.3 in 1.

```
       α-D-Manp  D
           1
           ↓
           4
→3)-α-D-GalpA-(1→2)-α-D-Manp-(1→3)-β-D-Galp-(1→   CPS
         C                  B            A

α-D-Manp  D
    1
    ↓
    4
α-D-GalpA-(1→2)-α-D-Manp-(1→3)-D-Gal   1
    C              B         A

α-D-Manp                                    α-D-Manp
    1                                           1
    ↓                                           ↓
    4                                           4
α-D-GalpA-(1→2)-α-D-Manp-(1→3)-β-D-Galp-(1→3)-α-D-GalpA-(1→2)-α-D-Manp-(1→3)-D-Gal   2
```

Figure 4. Structures of the K57 CPS of *K. pneumoniae* KPi8289 and oligosaccharides derived by cleavage of the CPS with depolymerase Dep_kpv79. CPS, tetrasaccharide repeat, containing two residues of D-mannose (units **B** and **D**) and one residue each of D-galactose (unit **A**) and D-galacturonic acid (unit **C**). 1 and 2, oligosaccharides obtained by cleavage of the CPS_KPi8289 with Dep_kpv79 (the same oligosaccharides were obtained by CPS splitting with Dep_kpv767).

Therefore, taking into account the CPS structure, oligosaccharide 1 has the structure shown in Figure 4. The same oligosaccharide has been obtained by mild acid hydrolysis of the K68 CPS of *K. pneumoniae*, which differs from the K57 CPS in the presence of a pyruvic acid acetal linked to the terminal Man residue [31].

The ^{1}H and ^{13}C NMR spectra of oligosaccharide 2 showed two series of signals, one corresponding to oligosaccharide 1 and the other to the K unit within the CPS. Particularly, there were signals for two units A, one β-linked as in the CPS and the other at the reducing end as in 1, and two units C, one 3,4-disubstituted and the other 4-substituted. Therefore, 2 represented a dimer of the K unit having the structure shown in Figure 4.

Thus, the data obtained indicated that depolymerases Dep_kpv79 and Dep_kpv767 are β-glycosidases that cleave specifically the β-D-Gal*p*-(1→3)-D-Gal*p*A (A→C) linkages in the CPS of *K. pneumoniae* KPi8289 by the hydrolytic mechanism.

2.5. Capsule Depolymerases Are Effective Therapeutics in Mouse Infection Models

Clinical manifestations of diseases caused by *K. pneumoniae* are extremely diverse [1,2]. Therefore, it is advisable to evaluate the therapeutic efficacy of bacteriophages on different laboratory models that differ in the route of infection and the infectious process development. We used two previously described mouse models of *K. pneumoniae* primary sepsis [32] and thigh soft tissue infection [33]. Two strains KPB156 and KPB550, differing in the degree of lysis by bacteriophages KpV79 and KpV767, were used to modeling *K. pneumoniae* infections in mice. As noted in Table 2, strain KPB550 was sensitive only to phage KpV79, whereas strain KPB156 was lysed by both phages with high efficiency of plating. Three groups of infected mice with 10 animals in each were used in each experiment: control (depolymerase untreated mouse) and two experimental (30 min after infection, the depolymerase Dep_kpv79 or Dep_kpv767 at a dose of 50 μg/mouse was injected intraperitoneally to mice of these groups).

As a result, nine of 10 (90%) and six of 10 (60%) mice infected with an intraperitoneal injection of *K. pneumoniae* KPB550 and KPB156 strains, respectively (primary sepsis model), and eight of 10 (80%)

mice infected with an intramuscular injection of *K. pneumoniae* KPB550 (soft tissue infection model) died in the control groups (without PS-depolymerase treatment). In these cases, a culture of the infecting strain was isolated from all mice, those that died and those that survived. In contrast, 80% to 100% of the animals survived in the experimental groups treated with Dep_kpv79 and Dep_kpv767 depolymerase (Figure 5). It should be noted that no pathology was detected in the organs and tissues of the animals that survived the challenge with *K. pneumoniae*. In total, a culture of *K. pneumoniae* was isolated from five of the 53 surviving animals that were treated with Dep_kpv79 or Dep_kpv767 depolymerase.

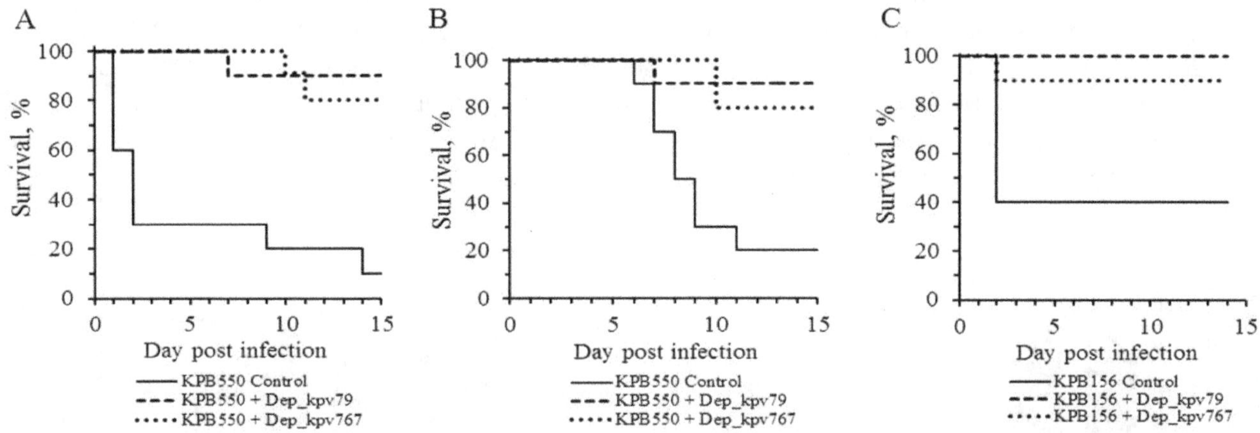

Figure 5. Protective effects of K57-specific depolymerases Dep_kpv79 and Dep_kpv767 of bacteriophages KpV79 and KpV767, respectively, in mice challenged with *K. pneumoniae* strains. Depolymerases (50 μg per mouse) were administered intraperitoneally 30 min after infection. (**A**) Primary sepsis caused by *K. pneumoniae* KPB550 (intraperitoneal injection). (**B**) Thigh soft tissue infection caused by *K. pneumoniae* KPB550 (intramuscular injection). (**C**) Primary sepsis caused by *K. pneumoniae* KPB156 (intraperitoneal injection).

The bacteria of strain KPB156 are less virulent compared to KPB550. Therefore, when mice were infected in the thigh muscles with a culture of strain KPB156 no death of animals was observed. However, seven of ten infected mice showed soft tissue edema at the injection site. After depolymerase treatment, edema was observed on day 3 in three of 10 mice, followed by its disappearance on day 7. In the control group, 30% of mice remained edematous for up to 10 days after infection.

Therefore, Dep_kpv79 and Dep_kpv767 depolymerases that cleave the K57 type CPS, have a therapeutic effect against *K. pneumoniae*–induced infections in mice.

Previously, a few works have been published on the in vivo therapeutic efficacy of CPS specific depolymerases against *K. pneumoniae*-induced infections [19,20,22]. Lin et al. demonstrated that treatment with the recombinant depolymerase provided increased survival in mice infected with *K.pneumoniae* of capsular type K1 [19]. Experiments by Majkowska-Skrobek et al. have shown that depolymerizing enzymes were effective against *K. pneumoniae* strains, representing K3, K21, and K63 capsular types, in a *G. mellonella* infecting model. In particular, the application of K3 and K21 specific depolymerases considerably increased the lifespan of *G. mellonella* larvae in a time- and *K. pneumoniae* strain–dependent manner [22]. The other depolymerase was effective against a native capsule of *K. pneumoniae* strains, representing the K63 type, and significantly inhibited Klebsiella-induced mortality of *G. mellonella* larvae in a time-dependent manner [20].

Our results on the therapeutic efficacy of two K57 specific depolymerases against *K. pneumoniae*-induced infection in mice provide one more confirmation that phage polysaccharide depolymerases represent a promising tool for antimicrobial therapy.

2.6. *Depolymerases Increase Sensitivity to Serum-Complement-Mediated Killing*

It is believed that the therapeutic efficacy of depolymerases is due to their antivirulence potential [20]. Since the polysaccharide capsule is one of the major virulence factors of *K. pneumoniae*, the CPS destruction makes bacteria less virulent and more susceptible to innate host defense including pathogen phagocytosis and complement-mediated killing [19,22,23,34].

In our work, the influence of the depolymerases on the *K. pneumoniae* serum resistance was studied. As shown in Figure 6, the depolymerases did not affect bacterial survival in the absence of serum. Also, serum alone did not affect the killing of *K. pneumoniae* strains KPB550 and KPi8289. However, treatment of *K. pneumoniae* cells with depolymerase Dep_kpv79 or Dep_kpv767 resulted in a decrease of the bacteria survival by four orders of magnitude after 3-h incubation. The pre-heated serum did not affect the viability of the *K. pneumoniae* cells. Therefore, increased bactericidal serum activity is associated with complement-mediated killing.

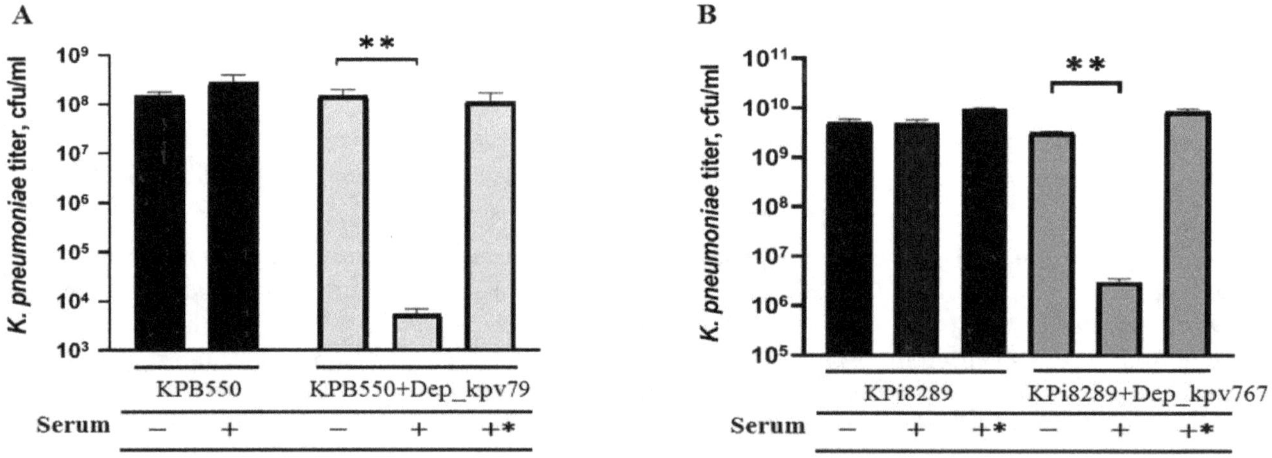

Figure 6. Serum sensitivity assay. Bacteria of serum-resistant *K. pneumoniae* strains KPB550 (**A**) and Kpi8289 (**B**) were pre-incubated with or without depolymerase Dep_kpv79 and Dep_kpv767, respectively (100 μg/mL, overnight at 4 °C). Mixtures were diluted 10 times, incubated for 3 h at 37 °C with 50% sheep serum (+), heat-inactivated serum (+*) or PBS (-), and then plated to determine colony-forming units. Average titers from three independent experiments are shown and compared by Student's t-test analysis. ** $p < 0.01$.

These results show that the depolymerases increase *K. pneumoniae* sensitivity to serum complement-mediated killing, consistent with other findings [19,22,34]. In addition, our data confirmed that the CPS is responsible for the complement resistance since other virulence factors of *K. pneumoniae* were unchanged after depolymerase treatment.

3. Materials and Methods

3.1. *Bacterial Strains and Bacteriophages*

Bacterial strains used in the experiments were obtained from the State Collection of Pathogenic Microorganisms and Cell Cultures (SCPM, Obolensk, Russia). All bacteria were routinely cultured in the Nutrient Medium No.1 (SRCAMB, Obolensk, Russia), or on LB broth agar (Difco Laboratories, Detroit, MI, USA) plates at 37 °C stationary or with shaking.

Bacteriophages were propagated using liquid cultures of *K. pneumoniae* sensitive strains as previously described [24]. Phage suspensions were titrated by a standard soft agar overlay method and stored at 4 °C until use.

3.2. Determination of the Host Range of Phage and Polysaccharide Depolymerase

Bacteriophages were screened against *K. pneumoniae* strains using the efficiency of the plating micro method [24]. PS depolymerase activity was assessed using the spot assay on *K. pneumoniae* strains of different capsular types and varying the enzyme concentrations. For *K. pneumoniae* lawn formation, 0.3 mL of bacterial suspension (~3×10^8 CFU/mL) was mixed with 4 mL of soft agar (LB broth supplemented with 0.6% agarose) at 45 °C and poured onto the surface of the Nutrient Medium No.1 plate. The plate was incubated at 37 °C overnight. These bacterial lawns as well as lawns inactivated by chloroform vapor were used for depolymerase testing.

3.3. Analysis of the Phage Genomes

Phage genomes were sequenced using the Ion Torrent PGM (Life Technologies, Carlsbad, CA, USA), according to the manufacturer's instructions. De novo assembly of sequence reads was performed using Newbler v.2.9. The correctness of assembly was checked using the SeqMan NGen software (DNAStar, Madison, WI, USA).

Protein-encoding genes and their putative functions were automatically predicted by RAST [35] and compared with known gene and protein sequences using the BLASTx and BLASTp algorithms [36] at the NCBI website. Linear genome comparisons of the phages and visualization of the coding regions were performed with Easyfig [37].

To identify putative polysaccharide depolymerase, a tail fiber, tail spike, and hypothetical proteins were analyzed by PSI-BLAST at the NCBI website and were screened through the HHpred web server [38].

The annotated genome sequences of bacteriophages KpV79 and KpV767 were deposited in the GenBank database and are available under accession numbers MF663761, and KX712070, respectively.

3.4. Cloning of CPS Depolymerase Genes and Preparation of Recombinant Proteins

Coding sequences for genes kpv79_42 (GenBank: ATI16495.1) and kpv767_46 (GenBank: AOZ65519.1) containing the putative depolymerase motifs were amplified from DNA of phages KpV79 and KpV767, respectively, by PCR using the following oligonucleotide primer pairs 79_XhoI: 5′-CTTACTCGAGGTTACCATAATACTGCAGAGAA-3′ and 79_NcoI: 5′-TATACCATGGACATTA TCAAACGCGCAGAC-3′ (kpv79_42 gene cloning); 767_XhoI; 5′-TTAGTCCTCGAGATACGTGATG CGAGCCTC-3′ and 767_NcoI: 5′-CTTCACCATGGCAATGACCACAGAGTCTAG-3′ (kpv767_46 gene cloning).

The amplified genes were cloned into pET22b expression vector (Novagen, Madison, WI, USA) via the NcoI and XhoI sites by using *Escherichia coli* strain DH5α. The kpv79_42 gene was cloned entirely; kpv767_46 gene was cloned without N-terminus fragment encoding the first 252 amino acid residues. Recombinant plasmids were verified by Sanger sequencing and then transformed into *E. coli* BL21 (DE3) for protein expression.

The production of recombinant proteins in LB broth (Difco Laboratories, Detroit, MI, USA) supplemented with 100 μg/mL ampicillin was induced by the addition of isopropyl-β-D-thiogalactopyranoside (IPTG) to the final concentration of 0.5 mM for 20 h at 22 °C. The recombinant proteins were separated from contaminating proteins by affinity chromatography using a Ni-iminodiacetic acid-Sepharose 5 mL column. The CPS-degrading activity of the recombinant proteins was assayed in the spot-test using *K. pneumoniae* strains of different capsular types [24].

3.5. Determination of the Structures of the Oligosaccharides Derived by the Treatment of K. pneumoniae CPS with Phage Depolymerases

Polysaccharides were isolated from *K. pneumoniae* KPi8289 cells by the modified water—phenol extraction method described previously [39].

Gel-permeation chromatography was carried out on a column (85 × 2.5 cm) of Fractogel TSK HW-40S (Merck, Germany). Elution was performed with 0.1% HOAc at a flow rate of 0.5 mL/min and monitored with a differential refractometer (Knauer, Germany).

^{1}H and ^{13}C NMR spectra were recorded on a Bruker Avance II 600 MHz spectrometer (Germany). Before measurement, samples were freeze-dried from 99.9% D_2O and then examined in 99.95% D_2O at 60 °C using sodium 3-(trimethylsilyl)propanoate-2,2,3,3-d$_4$ (δ_H 0, δ_C −1.6) as the internal reference for calibration. Two-dimensional NMR spectra were obtained using standard Bruker software, and the TopSpin 2.1 program (Bruker, Germany) was employed to acquire and to process the data. A MLEV-17 spin-lock time of 60 ms was used in the TOCSY experiment.

HR ESI MS was performed in the negative ion mode using a micrOTOF II instrument (Bruker Daltonics). Oligosaccharide samples (~50 ng/μL) were dissolved in a 1:1 (*v/v*) water/acetonitrile mixture and injected with a syringe at a flow rate of 3 μL/min. The capillary entrance voltage was set at 3200 V, and the interface temperature at 180 °C. Nitrogen was used as the drying gas. The mass range was from *m/z* 50 to 3500 Da. Internal calibration was done with ESI Calibrant Solution (Agilent).

3.6. Phage Adsorption Inhibition Test and Phage Inactivation by CPS

The effect of PS-depolymerases Dep_767 and Dep_79 on the bacteriophage adsorption to *K. pneumoniae* cells was evaluated as described previously [40] with some modifications. In brief, a bacterial cell suspension (~2 × 10^8 cfu/mL) in SM buffer (8 mM $MgCl_2$, 100 mM NaCl, 50 mMTris-HCl pH 7.5) with added depolymerase (400 μg/mL) and a phage sample at 10^5 pfu/mL were first brought to 37 °C. Then equal volumes of the phage and cell suspension were mixed. After 5 min incubation at 37 °C, bacterial cells with adsorbed phage were precipitated by centrifugation at 14,000 *g* for 1 min, and the titer of non-adsorbed phages was determined in the supernatant.

The phage inactivation by CPS was performed as follows. Phage samples (10^3 pfu/mL) in SM buffer with added CPS at a final concentration of 1 μg/mL were incubated at 37 °C for 10 min. After incubation, the phage titer was assessed by the double agar layer technique.

3.7. In Vivo K. pneumoniae Infection Models

Female white outbred mice (age 6–8 weeks) were used in experiments. Mice were housed under standard conditions of light, temperature, and water and food availability.

Two previously described mouse models of *K. pneumoniae* primary sepsis [32] and thigh soft tissue infection [33] were used to evaluate the therapeutic efficacy of PS-depolymerases. In the first case, *K. pneumoniae* strain KPB550 (LD_{50} = 6 × 10^6 cfu) or KPB156 (LD_{50} = 1 × 10^8 cfu) was used for intraperitoneal infection of mice (2 × 10^8 cfu in a volume of 0.5 mL). In the second model, mice were injected intramuscularly (in the left thigh) with a culture of *K. pneumoniae* strains and at the same dose in a volume of 0.1 mL. In each case, control and experimental groups of mice were used (10 animals each). PS-depolymerases were injected into the mice of the experimental group half an hour after infection (once, intraperitoneal, 50 μg per mouse). Mice of the control groups were not treated with depolymerases. All animals were sacrificed two weeks after infection, and blood and parenchymal organs were examined for the presence of a *K. pneumoniae* culture. Mice that died during the experiment were also autopsied and examined for bacterial dissemination.

Mouse experiments were performed following the Directive 2010/63/EU of the European Parliament and of the Council of 22 September 2010 on the protection of animals used for scientific purposes. The experiments were approved by the Animal Ethics Committee of State Research Center for Applied Microbiology and Biotechnology.

3.8. Serum Complement Activity

A log-phase bacterial culture (~10^8 cfu/mL) were preincubated with or without depolymerase (at the final concentration of 100 μg/mL) overnight at 4 °C. Mixtures were diluted 10 times, incubated for 3 h at 37 °C with 50% sheep serum, heat-inactivated serum (56 °C, 30 min) or phosphate-buffered saline (PBS), and then plated to determine colony-forming units. Each test was performed at least three independent experiments.

3.9. Statistical Analysis

Computation of data and statistical tests were performed using GraphPad Prism 8.0 software (GraphPad Software, Inc., La Jolla, CA, USA). Statistical significance was determined using a *t*-test. $p < 0.05$ was considered statistically significant.

4. Conclusions

In this work, we characterized two new PS-depolymerases Dep_kpv79 and Dep_kpv767 encoded by myovirus KpV79 and phage KpV767 of the Autographiviridae family both specific to *K. pneumoniae* K57. The depolymerases have a high level of similarity at the amino acid level in the central domain, which is responsible for the catalytic cleavage of polysaccharides. To the best of our knowledge, this is the first report of two identical depolymerases with the same CPS specificity derived from phages of a different family.

The phages recognize the K57 CPS as the primary receptor for adsorption, which is the first step in the phage infection process. These results were confirmed by a phage inactivation assay performed with the CPS isolated from *K. pneumoniae* strains of the K57 capsular type.

For the first time, we elucidated the mechanism and characterized the products of cleavage of the *K. pneumoniae* K57 CPS by depolymerases Dep_kpv79 and Dep_kpv767 encoded by phages with different tail structures. Both depolymerases were identified as specific glycosidases that cleave the CPS of *K. pneumoniae* strains by the hydrolytic mechanism.

Due to the ability to cleave the CPS, one of the main virulence factors of *K. pneumoniae*, depolymerases show a therapeutic effect in the treatment of *K. pneumoniae*–induced infections. Previously it has been reported that CPS-specific depolymerases successfully rescued mice and *G. mellonella* larvae infected by *K. pneumoniae* strains of capsular type K1 [19], K3, K21 [22], K63 [20], and K64 [34]. The Dep_kpv79 and Dep_kpv767 depolymerases were highly effective at combating sepsis and hip infection caused by *K. pneumoniae* in lethal mouse models. Both depolymerases exerted a therapeutic effect in mice infected with strains that differ in virulence and phage sensitivity. Indeed, 80–100% of animals were protected against death by a single dose (i.p., 50 μg/mouse) of the enzyme injected 0.5 h after infection by *K. pneumoniae* strains of K57 capsular type. The therapeutic effect of depolymerases is because they strip capsules and expose the underlying bacterium to immune attack such as complement-mediated killing.

Our study provides another confirmation that phage-borne depolymerases are able to target with high specificity such virulence factors as a bacterial capsule. Therefore, they represent a promising tool for typing and treatment of infections caused by bacterial pathogens.

Author Contributions: Conceptualization, N.V.V., and Y.A.K.; methodology, N.V.V., A.I.B., V.V.V., and A.S.S.; formal analysis, A.M.S., E.V.K. and V.M.K., investigation, A.M.S., A.I.B., V.M.K., E.V.K., A.S.S. and V.V.V.; writing—original draft preparation, review, and editing, N.V.V. and Y.A.K.; supervision and funding acquisition, N.V.V. All authors have read and agreed to the published version of the manuscript.

Acknowledgments: We are grateful to Olga Korobova, Tatyana Kombarova, and Irina Ufimtseva for assistance in mouse experiments. We also thank Angelina Kislichkina for help with DNA sequencing and Alexander Chizhov for measuring high-resolution mass spectra.

References

1. Podschun, R.; Ullmann, U. *Klebsiella* spp. as nosocomial pathogens: Epidemiology, taxonomy, typing methods, and pathogenicity factors. *Clin. Microbiol. Rev.* **1998**, *11*, 589–603. [CrossRef] [PubMed]
2. Paczosa, M.K.; Mecsas, J. *Klebsiella pneumoniae*: Going on the Offense with a Strong Defense. *Microbiol. Mol. Biol. Rev.* **2016**, *80*, 629–661. [CrossRef] [PubMed]
3. Wang, J.H.; Liu, Y.C.; Lee, S.S.; Yen, M.Y.; Chen, Y.S.; Wang, J.H.; Wann, S.R.; Lin, H.H. Primary liver abscess due to *Klebsiella pneumoniae* in Taiwan. *Clin. Infect. Dis.* **1998**, *26*, 1434–1438. [CrossRef]
4. Turton, J.F.; Englender, H.; Gabriel, S.N.; Turton, S.E.; Kaufmann, M.E.; Pitt, T.L. Genetically similar isolates of *Klebsiella pneumoniae* serotype K1 causing liver abscesses in three continents. *J. Med. Microbiol.* **2007**, *56*, 593–597. [CrossRef] [PubMed]
5. Cortes, G.; Borrell, N.; Ãstorza, B.; Gómez, C.; Sauleda, J.; Albertí, S. Molecular analysis of the contribution of the capsular polysaccharide and the lipopolysaccharide O side chain to the virulence of *Klebsiella pneumoniae* in a murine model of pneumonia. *Infect. Immun.* **2002**, *70*, 2583–2590. [CrossRef] [PubMed]
6. Domenico, P.; Salo, R.J.; Cross, A.S.; Cunha, B.A. Polysaccharide capsule-mediated resistance to opsonophagocytosis in *Klebsiella pneumoniae*. *Infect. Immun.* **1994**, *62*, 4495–4499. [CrossRef] [PubMed]
7. Ørskov, I.; Fife-Asbury, M.A. New Klebsiella capsular antigen K82 and the deletion of five of those previously assigned. *Int. J. Syst. Bacteriol.* **1977**, *27*, 386–387. [CrossRef]
8. Pan, Y.J.; Lin, T.L.; Chen, C.T.; Chen, Y.Y.; Hsieh, P.F.; Hsu, C.R.; Wu, M.C.; Wang, J.T. Genetic analysis of capsular polysaccharide synthesis gene clusters in 79 capsular types of *Klebsiella* spp. *Sci. Rep.* **2015**, *5*, 15573. [CrossRef]
9. Wyres, K.L.; Wick, R.R.; Gorrie, C.; Jenney, A.; Follador, R.; Thomson, N.R.; Holt, K.E. Identification of *Klebsiella* capsule synthesis loci from whole genome data. *Microb. Genom.* **2016**, *2*, e000102. [CrossRef]
10. Shon, A.S.; Bajwa, R.P.; Russo, T.A. Hypervirulent (*hypermucoviscous*) *Klebsiella pneumoniae*: A new and dangerous breed. *Virulence* **2013**, *4*, 107–118. [CrossRef]
11. Rakhuba, D.V.; Kolomiets, E.I.; Dey, E.S.; Novik, G.I. Bacteriophage receptors, mechanisms of phage adsorption and penetration into host cell. *Pol. J. Microbiol.* **2010**, *59*, 145–155. [CrossRef]
12. Latka, A.; Maciejewska, B.; Majkowska-Skrobek, G.; Briers, Y.; Drulis-Kawa, Z. Bacteriophage-encoded virion-associated enzymes to overcome the carbohydrate barriers during the infection process. *Appl. Microbiol. Biotechnol.* **2017**, *101*, 3103–3119. [CrossRef] [PubMed]
13. Pires, D.P.; Oliveira, H.; Melo, L.D.R.; Sillankorva, S.; Azeredo, J. Bacteriophage-encoded depolymerases: Their diversity and biotechnological applications. *Appl. Microbiol. Biotechnol.* **2016**, *100*, 2141–2151. [CrossRef] [PubMed]
14. Brzozowska, E.; Pyra, A.; Pawlik, K.; Janik, M.; Górska, S.; Urbańska, N.; Drulis-Kawa, Z.; Gamian, A. Hydrolytic activity determination of Tail Tubular Protein A of *Klebsiella pneumoniae* bacteriophages towards saccharide substrates. *Sci. Rep.* **2017**, *7*, 18048. [CrossRef] [PubMed]
15. Lin, H.; Paff, M.L.; Molineux, I.J.; Bull, J.J. Therapeutic application of phage capsule depolymerases against K1, K5, and K30 capsulated *E. coli* in mice. *Front. Microbiol.* **2017**, *8*, 2257. [CrossRef]
16. Chen, Y.; Sun, E.; Yang, L.; Song, J.; Wu, B. Therapeutic Application of Bacteriophage PHB02 and its Putative Depolymerase Against Pasteurella multocida Capsular Type A in Mice. *Front. Microbiol.* **2018**, *9*, 1678. [CrossRef]
17. Olszak, T.; Shneider, M.M.; Latka, A.; Maciejewska, B.; Browning, C.; Sycheva, L.V.; Cornelissen, A.; Danis-Wlodarczyk, K.; Senchenkova, S.N.; Shashkov, A.S.; et al. The O-specific polysaccharide lyase from the phage LKA1 tailsike reduces *Pseudomonas* virulence. *Sci. Rep.* **2017**, *7*, 16302. [CrossRef]
18. Hernandez-Morales, A.C.; Lessor, L.L.; Wood, T.L.; Migl, D.; Mijalis, E.M.; Cahill, J.; Russell, W.K.; Young, R.F.; Gill, J.J. Genomic and Biochemical Characterization of Acinetobacter Podophage Petty Reveals a Novel Lysis Mechanism and Tail-Associated Depolymerase Activity. *J. Virol.* **2018**, *92*, e01064-17. [CrossRef]
19. Lin, T.L.; Hsieh, P.F.; Huang, Y.T.; Lee, W.C.; Tsai, Y.T.; Su, P.A.; Pan, Y.J.; Hsu, C.R.; Wu, M.C.; Wang, J.T. Isolation of a bacteriophage and its depolymerase specific for K1 capsule of *Klebsiella pneumoniae*: Implication in typing and treatment. *J. Infect. Dis.* **2014**, *210*, 1734–1744. [CrossRef]
20. Majkowska-Skrobek, G.; Łątka, A.; Berisio, R.; Maciejewska, B.; Squeglia, F.; Romano, M.; Lavigne, R.; Struve, C.; Drulis-Kawa, Z. Capsule-Targeting Depolymerase, Derived from Klebsiella KP36 Phage, as a Tool for the Development of Anti-Virulent Strategy. *Viruses* **2016**, *8*, 324. [CrossRef]

21. Hsieh, P.F.; Lin, H.H.; Lin, T.L.; Chen, Y.Y.; Wang, J.T. Two T7-like Bacteriophages, K5-2 and K5-4, Each Encodes Two Capsule Depolymerases: Isolation and Functional Characterization. *Sci. Rep.* **2017**, *7*, 4624. [CrossRef]
22. Majkowska-Skrobek, G.; Latka, A.; Berisio, R.; Squeglia, F.; Maciejewska, B.; Briers, Y.; Drulis-Kawa, Z. Phage-Borne depolymerases decrease *Klebsiella pneumoniae* resistance to innate defense mechanisms. *Front. Microbiol.* **2018**, *9*, 2517. [CrossRef]
23. Pan, Y.J.; Lin, T.L.; Lin, Y.T.; Su, P.A.; Chen, C.T.; Hsieh, P.F.; Hsu, C.R.; Chen, C.C.; Hsieh, Y.C.; Wang, J.T. Identification of capsular types in carbapenem-resistant *Klebsiella pneumoniae* strains by wzc sequencing and implications for capsule depolymerase treatment. *Antimicrob. Agents Chemother.* **2015**, *59*, 1038–1047. [CrossRef]
24. Solovieva, E.V.; Myakinina, V.P.; Kislichkina, A.A.; Krasilnikova, V.M.; Verevkin, V.V.; Mochalov, V.V.; Lev, A.I.; Fursova, N.K.; Volozhantsev, N.V. Comparative genome analysis of novel Podoviruses lytic for hypermucoviscous *Klebsiella pneumoniae* of K1, K2, and K57 capsular types. *Virus Res.* **2018**, *243*, 10–18. [CrossRef]
25. Pan, Y.J.; Lin, T.L.; Chen, Y.Y.; Lai, P.H.; Tsai, Y.T.; Hsu, C.R.; Hsieh, P.F.; Lin, Y.T.; Wang, J.T. Identification of three podoviruses infecting Klebsiella encoding capsule depolymerases that digest specific capsular types. *Microb. Biotechnol.* **2019**, *12*, 472–486. [CrossRef]
26. Pan, Y.J.; Lin, T.L.; Chen, C.C.; Tsai, Y.T.; Cheng, Y.H.; Chen, Y.Y.; Hsieh, P.F.; Lin, Y.T.; Wang, J.T. Klebsiella Phage ΦK64-1 Encodes Multiple Depolymerases for Multiple Host Capsular Types. *J. Virol.* **2017**, *91*, e02457-16. [CrossRef]
27. Knirel, Y.A.; Shneider, M.M.; Popova, A.V.; Kasimova, A.A.; Senchenkova, S.N.; Shashkov, A.S.; Chizhov, A.O. Mechanisms of Acinetobacter baumannii Capsular Polysaccharide Cleavage by Phage Depolymerases. *Biochemistry* **2020**, *85*, 567–574. [CrossRef]
28. Lee, I.M.; Tu, I.F.; Yang, F.L.; Ko, T.P.; Liao, J.H.; Lin, N.T.; Wu, C.Y.; Ren, C.T.; Wang, A.H.; Chang, C.M.; et al. Structural basis for fragmenting the exopolysaccharide of *Acinetobacter baumannii* by bacteriophage ΦAB6 tailspike protein. *Sci. Rep.* **2017**, *17*, 42711. [CrossRef]
29. Kamerling, J.P.; Lindberg, B.; Lönngren, J.; Nimmich, W. Structural studies of the Klebsiella type 57 capsular polysaccharide. *Acta Chem. Scand.* **1975**, *29*, 593–598. [CrossRef] [PubMed]
30. Jansson, P.E.; Kenne, L.; Widmalm, G. Computer-assisted structural analysis of polysaccharides with an extended version of CASPER using 1H- and 13C-n.m.r. data. *Carbohydr. Res.* **1989**, *188*, 169–191. [CrossRef]
31. Dutton, G.G.S.; Parolis, H.; Parolis, L.A.S. The structural elucidation of the capsular polysaccharide of Klebsiella K68. *Carbohydr. Res.* **1986**, *152*, 249–259. [CrossRef]
32. Borzilov, A.I.; Myakinina, V.P.; Korobova, O.V.; Kombarova, T.I.; Krasilnikova, V.M.; Verevkin, V.V.; Volozhantsev, N.V. Evaluation of preventive and therapeutic efficacy of *Klebsiella pneumoniae* bacteriophage vB_KpnP_KpV289 on the model of acute sepsis in mice. *Bacteriology* **2017**, *2*, 73–77. [CrossRef]
33. Borzilov, A.I.; Volozhantsev, N.V.; Korobova, O.V.; Kombarova, T.I.; Myakinina, V.P.; Krasilnikova, V.M.; Verevkin, V.V.; Svetoch, E.A.; Dyatlov, I.A. The effectiveness of bacteriophage KpV289 in treatment of acute pneumonia and a hip infection caused by *Klebsiella pneumoniae* in mice. *Infekc. Bolezni (Infect. Dis.)* **2017**, *15*, 48–56. [CrossRef]
34. Liu, Y.; Leung, S.; Huang, Y.; Guo, Y.; Jiang, N.; Li, P.; Chen, J.; Wang, R.; Bai, C.; Mi, Z.; et al. Identification of Two Depolymerases From Phage IME205 and Their Antivirulent Functions on K47 Capsule of Klebsiella pneumoniae. *Front. Microbiol.* **2020**, *11*, 218. [CrossRef] [PubMed]
35. Aziz, R.K.; Bartels, D.; Best, A.A.; DeJongh, M.; Disz, T.; Edwards, R.A.; Formsma, K.; Gerdes, S.; Glass, E.M.; Kubal, M.; et al. The RAST server: Rapid annotations using subsystems technology. *BMC Genomics* **2008**, *9*, 75. [CrossRef]
36. Altschul, S.F.; Gish, W.; Miller, W.; Myers, E.W.; Lipman, D.J. Basic local alignment search tool. *J. Mol. Biol.* **1990**, *215*, 403–410. [CrossRef]
37. Sullivan, M.J.; Petty, N.K.; Beatson, S.A. Easyfig: A genome comparison visualizer. *Bioinformatics* **2011**, *27*, 1009–1010. [CrossRef]
38. Söding, J.; Biegert, A.; Lupas, A.N. The HHpred interactive server for protein homology detection and structure prediction. *Nucleic Acids Res.* **2005**, *33*, 244–248. [CrossRef]

39. Volozhantsev, N.V.; Shpirt, A.M.; Kislichkina, A.A.; Shashkov, A.S.; Verevkin, V.V.; Fursova, N.K.; Knirel, Y.A. Structure and gene cluster of the capsular polysaccharide of multidrug resistant carbapenemase OXA-48-producing *Klebsiella pneumoniae* strain KPB536 of the genetic line ST147. *Res. Microbiol.* **2020**, *171*, 74–79. [CrossRef]
40. Prokhorov, N.S.; Riccio, C.; Zdorovenko, E.L.; Shneider, M.M.; Browning, C.; Knirel, Y.A.; Leiman, P.G.; Letarov, A.V. Function of bacteriophage G7C esterase tailspike in host cell adsorption. *Mol. Microbiol.* **2017**, *105*, 385–398. [CrossRef]

3

Structure and Analysis of R1 and R2 Pyocin Receptor-Binding Fibers

Sergey A. Buth [1,†], Mikhail M. Shneider [1,2], Dean Scholl [3] and Petr G. Leiman [1,*,†]

1 Institute of Physics of Biologic Systems, École Polytechnique Fédérale de Lausanne (EPFL), BSP-415, 1015 Lausanne, Switzerland; sebuth@utmb.edu (S.A.B.); mm_shn@mail.ru (M.M.S.)
2 Shemyakin Ovchinnikov Institute of Bioorganic Chemistry, 16/10 Mikluho Maklaya Str., Moscow 117997, Russia
3 Pylum Biosciences, 385 Oyster Point Blvd., Suite 6A, South San Francisco, CA 94080, USA; dean@avidbiotics.com
* Correspondence: pgleiman@utmb.edu
† Current address: Department of Biochemistry and Molecular Biology, University of Texas Medical Branch, Basic Sciences Building 6.600D, 301 University Blvd., Galveston, TX 77555-0647, USA.

Abstract: The R-type pyocins are high-molecular weight bacteriocins produced by some strains of *Pseudomonas aeruginosa* to specifically kill other strains of the same species. Structurally, the R-type pyocins are similar to "simple" contractile tails, such as those of phage P2 and Mu. The pyocin recognizes and binds to its target with the help of fibers that emanate from the baseplate structure at one end of the particle. Subsequently, the pyocin contracts its sheath and drives the rigid tube through the host cell envelope. This causes depolarization of the cytoplasmic membrane and cell death. The host cell surface-binding fiber is ~340 Å-long and is attached to the baseplate with its N-terminal domain. Here, we report the crystal structures of C-terminal fragments of the R1 and R2 pyocin fibers that comprise the distal, receptor-binding part of the protein. Both proteins are ~240 Å-long homotrimers in which slender rod-like domains are interspersed with more globular domains—two tandem knob domains in the N-terminal part of the fragment and a lectin-like domain at its C-terminus. The putative substrate binding sites are separated by about 100 Å, suggesting that binding of the fiber to the cell surface causes the fiber to adopt a certain orientation relative to the baseplate and this then triggers sheath contraction.

Keywords: R-type pyocin; bacteriocin; contractile injection systems; *Pseudomonas aeruginosa*; X-ray crystallography; receptor-binding protein

---

1. Introduction

The chromosomes of many *Pseudomonas* species carry a cluster of genes for one or both of the two types of high-molecular-weight pyocins, the R type and the F type [1]. For example, the "laboratory" strain *P. aeruginosa* PAO1 contains a cluster of both pyocins (the R upstream of the F, genes *pa0610-pa0648*) between the *trpE* and *trpG* genes of its tryptophan operon [1]. The cluster is controlled by a common 5′-end regulatory element comprising the activator PrtN (PA0610) and its repressor PrtR (PA0611) [2]. Production of pyocins is triggered by UV irradiation or mitomycin C treatment that cause activation of RecA, which cleaves the repressor, PrtR, allowing the positive regulator, PrtN, to initiate transcription.

Morphologically and functionally, the R-type pyocins resemble the contractile tails of *Myoviridae* bacteriophages [3–6]. These systems recognize the target cell with the help of fibers or, more generally, receptor-binding proteins that emanate from the baseplate of the particle. Attachment of the fibers to the target cell surface causes structural changes in the baseplate that, in turn, trigger contraction of the external sheath, which drives the internal rigid tube through the host cell envelope. As the pyocins have no capsid, the cell's cytoplasm becomes open to the external milieu, which causes uncontrollable leakage of ions, destroys the membrane potential, and results in cell death [7]. Mass-spectrometry and bioinformatics show that the mature particle contains 12 proteins that are orthologous to those comprising the conserved core part of the phage T4 tail (Table S1) [8,9].

The killing mechanism of the pyocins is not specific to *Pseudomonas*, and their spectrum is fully determined by the fibers [10–12]. The N-terminal part of the fiber attaches it to the baseplate whereas the rest of its structure participates in target cell recognition. The latter can be replaced with a receptor-binding protein from an *Escherichia coli*, *Salmonella enterica* or, *Yersinia pestis* bacteriophage, resulting in a chimerical pyocin particle with a killing spectrum that is the same or wider than that of the donor phage [10–12]. The folding of phage and pyocin fibers and, in some cases, their attachment to the particle are controlled by chaperones, which are often encoded by a gene immediately downstream from the fiber gene [13]. In the case of the pyocin with chimerical fibers, both the donor fiber chaperone gene (if present) and the cognate pyocin fiber chaperone gene are required for particle assembly [10].

Five R-type pyocins, called R1 to R5, each with a unique killing spectrum, have been described [14,15]. Their spectra can be represented by a "spectrum tree" with two branches in which R5 is at the root, R1 is one branch, and R2, R4, and R3 form another branch, in that order [16]. The phylogenetic tree of their fiber sequences is roughly similar and contains two branches—R1- and R2-type (Figure S1). The amino acid sequences of the fiber proteins of all the five subtypes are nearly identical from the N-terminus through about one half of the protein. The second half contains stretches of 100% sequence identity and completely dissimilar regions [10], and averages to have slightly greater than 50% identity. Interestingly, the chaperones of these fibers display a significantly greater sequence diversity [10]. It was shown that the L-Rha residue and two distinct D-Glc residues of the outer core of the *P. aeruginosa* lipopolysaccharide (LPS) are part of the receptor sites for R1-, R2-, and R5-pyocins, respectively [17].

The process of host cell recognition and attachment by a bacteriophage or pyocin remains poorly understood. The initial and reversible interaction of receptor-binding proteins with the host cell surface is somehow converted into an irreversible attachment of the particle to the host [18]. It is clear, however, that the overall conformation of receptor-binding proteins changes little upon ligand binding even in proteins that display an enzymatic activity towards cell surface polysaccharides [19–21]. Instead, changes in the supramolecular conformation, such as reorientation or other types of global movement of receptor-binding proteins, relative to the rest of the particle appear to initiate the cascade of structural transformations that commit the particle to irreversible host cell binding [22,23].

Here, we present the crystal structures of the R1 and R2 pyocin fiber fragments comprising about two thirds of the fiber and lacking the particle-binding N-terminal domain. These structures represent some of the most complete atomic models of fibrous proteins ever studied in tailed phages or pyocins [8,24–28]. We found that both R1 and R2 pyocin fiber fragments form a ~240 Å-long homotrimer that contains a rod-like and a shaft-like domain, two tandem knob domains, and the C-terminal lectin-like domain. The most diverse regions of the fiber amino acid sequence, which likely determine the killing spectrum, map onto the second knob and C-terminal domains, which are about 100 Å away from each other. Such a distribution of binding sites makes it possible to orient the fiber relative to the target cell surface and fix it in an orientation that will lead to subsequent conformational changes in the baseplate.

2. Materials and Methods

2.1. Engineering and Choice of the Construct for Expression

The R2 pyocin fiber and chaperone genes (*pa0620* and *pa0621* of *P. aeruginosa* PAO1, respectively) were cloned into two separate and compatible expression vectors, pTSL and pATE, respectively.

pTSL (Am^R) is described elsewhere (NCBI database accession number KU314761) [8]. pATE (Cm^R) is a pACYCDuet-1 derivative with a single instance of the multiple cloning site and a Tobacco Etch Virus (TEV) protease cleavage site downstream from a 5′-end His-tag (GenBank database accession number MH593819). The PA0620 fiber was expressed as a downstream fusion to the following construct: an N-terminal His-tag, a SlyD domain (FKBP-type peptidyl-prolyl cis-trans isomerase folding chaperone that was introduced to enhance protein solubility), and a linker containing a TEV protease cleavage site [8]. The PA0621 chaperone was tagless.

The full-length protein was soluble but showed a tendency to aggregation and therefore was unsuitable for structural studies. Treatment of this protein with trypsin resulted in two stable fragments, the compositions of which were analyzed by mass spectrometry. These fragments had intact C-terminus and comprised the residues, 173–691 and 124–691. We called these fragments PA0620d1 and PA0620d2, respectively. Both fragments were cloned into the same plasmid system, expressed, and purified. The PA0620d2 mutant behaved similarly to the full-length protein. The PA0620d1 mutant was soluble and could be crystallized although the crystals diffracted poorly and this direction was eventually abandoned.

It was shown that a pyocin particle with a chimeric fiber that is composed of residues 1–164 of the cognate R2 pyocin fiber and residues 291–669 of the phage P2 fiber is active against *E. coli* C (phage P2 host) [10]. This P2 fiber fragment corresponds to residues 323–691 of the R2 pyocin fiber in the sequence alignment. We cloned this R2 pyocin fiber fragment (which we called PA0620d3) into the pTSL expression system and found that this fragment was soluble when co-expressed with the PA0621 chaperone, could be purified to homogeneity, and gave diffraction quality crystals.

We then cloned a PA0620d3-like fragment of the R1 pyocin fiber PLES_06171 (residues 323–701) and its chaperone PLES_RS03160 from *P. aeruginosa* LESB58 into the same dual plasmid expression system. The R1 fiber fragment was a more difficult protein to work with than PA0620d3, but it was also eventually crystallized and gave reasonably well diffracting crystals.

2.2. *Protein Expression and Purification*

Protein expression was performed in the B834 (DE3) strain of *E. coli* (a methionine auxotroph). The transformed cells were grown at 37 °C in the 2xTY medium (supplemented with ampicillin at a concentration of 200 μg/mL) with aeration by orbital shaking at 200 rpm until the optical density at 600 nm (OD_{600}) reached a value of 0.6. The medium was cooled on ice to a temperature of 18–20 °C and the protein expression was induced by an addition of IPTG to a final concentration of 1 mM. After further incubation at 18 °C overnight (approximately 16 h), the cells were harvested by centrifugation at 5180× *g* for 15 min at 4 °C. The cell pellet was resuspended in 1/50th of the original culture volume of a 20 mM Tris-HCl pH 8.0, 300 mM NaCl, 5 mM imidazole buffer. The cells were lysed by sonication with the sample kept on ice. The cell debris was removed by centrifugation at 35,000× *g* for 15 min at 4 °C. The supernatant was loaded onto a Ni^{2+}-precharged 5 mL GE HisTrap FF Crude column (GE Healthcare Life Sciences, Chicago, IL, United States) that was equilibrated with a 20 mM Tris-HCl pH 8.0, 300 mM NaCl buffer. The non-specifically bound material was removed by washing the column with a buffer containing 50 mM Tris-HCl pH 8.0, 250 mM NaCl, and 50 mM imidazole. The affinity-bound material was eluted with a buffer containing 20 mM Tris-HCl pH 8.0, 250 mM NaCl, and 200 mM imidazole. Fractions containing the target protein were pooled together and set up at 25 °C for an overnight digestion with the TEV protease at a protease/protein ratio of 1/100 (*w*/*w*). This reaction mixture was simultaneously dialyzed against a 10 mM Tris-HCl pH 8.0, 3 mM DTT, 1.5 mM EDTA buffer, resulting in cleavage of the His-SlyD expression tag. The digested protein was filtered and purified by ion-exchange chromatography using a GE MonoQ 10/100 GL column and 0 to 650 mM NaCl gradient in a 20 mM Tris-HCl pH 8.0 buffer. Relevant fractions were combined and concentrated using Sartorius (Sartorius AG, Göttingen, Germany) ultrafiltration devices with a molecular weight cutoff of 50,000 to a volume of ~5 mL. This sample was loaded onto a GE HiLoad 16/60 Superdex 200 size-exclusion column equilibrated with 10 mM Tris-HCl pH 8.0, 150 mM NaCl. Fractions containing the pure protein were combined and concentrated with the help of a similar Sartorius ultrafiltration unit. Proteins were stored in the same buffer at +4 °C until they were used for crystallization. All purification buffers and the final protein solution contained NaN_3 at a concentration of 0.02% (*w*/*v*).

To produce a Se-methionine (SeMet) derivative of R2 PA0620d3, the cells were first grown in the 2xTY medium until OD_{600} of 0.3, then pelleted by centrifugation at 3315× *g* for 10 min at 20 °C, and transferred to the SelenoMet Medium (Molecular Dimensions, Newmarket, Suffolk, UK) prepared according to the manufacturer's instructions and supplemented with ampicillin at a concentration of 200 µg/mL. All the subsequent steps including the expression at low temperature and protein purification were the same as for the native protein.

2.3. Crystallization, Data Collection, and Structure Determination

For crystallization, purified R1 and R2 pyocin fiber fragments were concentrated to 9 and 11 mg/mL, respectively. The initial crystallization screening was carried out by the sitting drop method in 96 well 2-lens MRC plates (SWISSCI AG, Neuheim, Switzerland) using JBScreens (Jena Bioscience, Jena, Germany) crystallization screens. Optimization of crystallization conditions was performed in Jena Bioscience SuperClear pregreased 24 well plates by hanging drop vapor diffusion. Crystallization drops of the 24 well plate setup contained 1.5 µL of the protein solution in 10 mM Tris-HCl pH 8.0, 150 mM NaCl mixed with an equal volume of the well solution. Best R1 and R2 fiber crystals were obtained at +18 °C with the protein concentration at 7.4 and 7 mg/mL equilibrated against 750 µL of the well solution containing 4% PEG 4000, 100 mM PIPES pH 6.1, 140 mM $Na_2(C_3H_2O_4)$ (sodium malonate) and 5% PEG 6000, 100 mM Tris-HCl pH 8.5, 180 mM KCl, respectively. For data collection, the crystals were incubated for 20–45 s in a cryo solution that contained all the well solution components and, additionally, 30% (*v*/*v*) of glycerol for the R1 and 25% (*v*/*v*) of ethylene glycol for the R2 fiber, and flash frozen in a vaporized nitrogen stream at 100 K.

Crystallographic data collection was carried out at the X06SA PXI beam line of the Swiss Light Source (SLS) at the Paul Scherrer Institute (SLS, Villigen, Switzerland). Although best crystals of R2 PA0620d3 diffracted to better than 1.7 Å resolution, the resolution of the collected diffraction data had to be limited to 1.9 Å even when using a PILATUS 6M detector (424 × 435 mm^2, 2463 × 2527 pixels) because of the excessive spot overlap due to a large unit cell and a relatively high mosaicity (Table 1). The structure was solved by the single-wavelength anomalous diffraction (SAD) technique using a SeMet derivative [29]. The SAD data were collected at a wavelength of the maximum Se adsorption (K-line) that was established with a fluorescent scan. For both proteins (R1 and R2) the native data were collected at a wavelength of 1 Å (Table 1).

The diffraction data was indexed, integrated, and scaled with the program XDS [30,31]. The Se sites were located with the SHELX_CDE program suite [32] using the HKL2MAP interface [33]. The program SOLVE [34] was used for the refinement of sites found by SHELX and for phasing. The SOLVE phases were improved by non-crystallographic symmetry (NCS) averaging and solvent flattening with the program RESOLVE [34]. The program ARP/wARP [35] was used for automated model building. Further refinement of the atomic model was performed with the programs Coot [36], Refmac5 [37], and PHENIX [38] with TLS [39]. Details of data reduction and refinement are given in Table 1.

The structure of the R1 pyocin fiber was solved by the molecular replacement method [40] with the PHASER [41] program as part of the CCP4 program package [42] using the PA0620d3 structure as a search model. Similar to the R2 fiber structure, Coot and PHENIX with TLS were used for the refinement of the atomic model of the R1 fiber.

Table 1. X-ray data collection, reduction, and refinement statistics for R2 and R1 pyocin fiber structures.

Crystal	R1 Fiber Native	R2 Fiber Native	R2 Fiber SeMet
Data collection			
Wavelength (Å)	1.0000	1.0000	0.9797
Number of frames	720	780	1440
Frame width (°)	0.25	0.25	0.25
Space group	$P2_12_12_1$	$P2_12_12_1$	$P2_12_12_1$
Cell dimensions			
a, b, c (Å)	144.08, 154.46, 198.83	56.12, 126.35, 431.25	56.20, 126.82, 433.43
Resolution (Å)	50.0–2.3 (2.46–2.32)	50.0–1.9 (2.01–1.90)	50.0–2.4 (2.54–2.40)
R_{meas} (%)	10.5 (61.4)	13.0 (55.0)	7.1 (18.0)
I/σ_I	8.8 (2.1)	8.8 (2.6)	20.8 (8.2)
$CC_{1/2}$ (%)	99.5 (81.0)	99.4 (81.3)	99.8(98.0)
Completeness (%)	99.2 (96.2)	99.6 (98.5)	99.3 (96.1)
Redundancy	3.4 (3.2)	3.8 (3.8)	6.7 (6.2)
Anomalous signal *	0.77 (0.65)	0.81 (0.75)	1.42 (0.78)
Refinement			NA **
Resolution	47.69–2.32	49.96–1.90	
No. reflections used in refinement	192,247	246,066	
No. atoms (non-H)	18,886	20,843	
Protein	16,848	16,720	
Ligand/ion	10	98	
Water	2028	4025	
R_{work}/R_{free}	0.169/0.206	0.160/0.208	
Average B factor (Å^2)	60.7	24.8	
Protein	61.3	22.2	
Ligand/Ion	91.1	29.5	
Water	56.1	35.7	
R.m.s. deviations			
Bond lengths (Å)	0.003	0.009	
Bond angles (°)	0.600	0.886	
Ramachandran plot statistics			
Favored (%)	97.49	97.24	
Allowed (%)	2.51	2.76	
Outliers (%)	0.00	0.00	
Protein Data Bank accession code	6CL5	6CL6	

Statistics for the highest resolution shell are shown in the parentheses. * As defined by the program XDS [30]. ** Neither refinement nor Protein Data Bank (PDB) deposition of the SeMet derivative structure of the R2 fiber was pursued.

2.4. Molecular Graphics, Analysis of Surface Properties, and Bioinformatics

All molecular graphics figures were prepared with the program UCSF Chimera [43]. The electrostatic potential was calculated with the program APBS [44,45]. The sequence diversity analysis was performed with a non-redundant set of protein sequences that were identified with the help of the Basic Local Alignment Search Tool (BLAST) [46]. The molecular surfaces were calculated with the program MSMS [47] as implemented in UCSF Chimera. Bioinformatic analysis was performed with the web servers BLAST [46], HHpred [48,49], and Phylogeny.fr [50].

2.5. Pyocin Fiber Competition Assay

10^6 CFU/mL of *P. aeruginosa* strain 13s was incubated for 10 min at 37 °C with varying amounts of the R2 fiber fragment PA0620d3 that carried the His-SlyD expression tag. 10^8 KU/mL of the R2 pyocin that was purified from the *P. aeruginosa* PAO1 strain [10] was then added and incubated for an additional 20 min (KU, a killing unit, is defined as the amount of activity required to kill a single cell; in this case, it corresponds to a single pyocin particle). The cells were then diluted and spotted onto an LB agar plate. See [10] for KU calculation, strains specification, and pyocins purification protocol.

3. Results

3.1. R1 and R2 Fibers Are Structurally Similar

The R1 and R2 fiber fragments have a similar domain organization and overall structure (Figure 1). Each protein is a ~240 Å-long fiber that is formed by three intertwined polypeptide chains comprising five domains. Four copies of the "helix-plus-turn" motif create the N-terminal "Rod" domain (residues 328–356). It is followed by two tandem "Knob" domains (Knob1, residues 357–439, and Knob2, residues 445–528) that have similar folds. The two Knobs are followed by a "Shaft" domain (residues 529–598) that is ~93 Å long and contains a buried iron ion approximately in the middle. The remaining C-terminal domain (residues 599–691) is a β-sandwich with a lectin-like fold. About 45% of the total surface area of each monomer is buried in the trimeric interface (32,719 $Å^2$ for R1 and 31,680 $Å^2$ for R2 fiber). The estimated dissociation energy for R1 and R2 fiber trimers is 196.1 kcal/mol and 192.7 kcal/mol, respectively. This is consistent with the fiber being an SDS-resistant trimer at room temperature as is the case for many other interdigitated fibrous proteins [51,52].

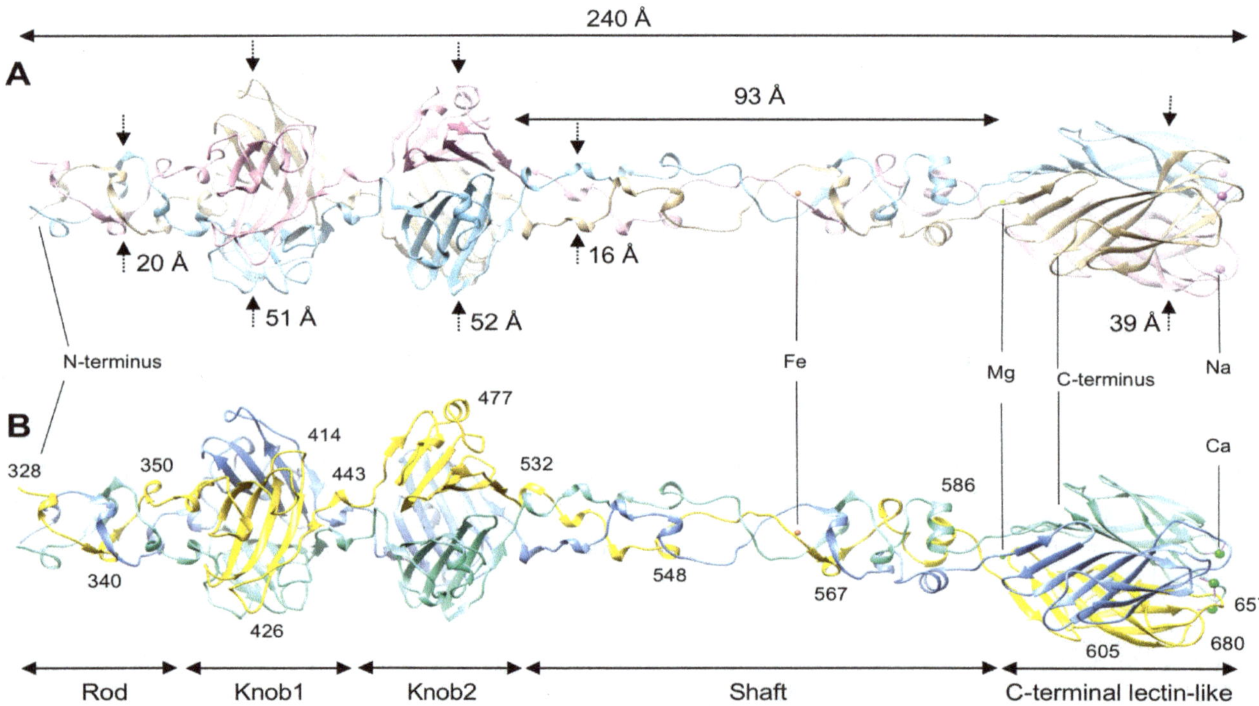

Figure 1. The structure and domain organization of the R1 (panel A) and R2 (panel B) fibers. Domain dimensions, position of the metal ion sites, and residues at strategic positions are indicated. Polypeptide chains are showed in the ribbon representation. Monomers are colored in plum, tan, and sky blue for the R1 molecule (**A**), and gold, cornflower blue, and aquamarine for the R2 molecule (**B**).

3.2. The Rod and Shaft Domains Are Built Using a Helix-Plus-Turn Motif

The folds of the Rod and Shaft domains are characterized by a low secondary structure content and complex topology in which the three polypeptide chains extensively interdigitate (Figure 2). Both contain a repeating unit, which we termed the "helix-plus-turn" motif, that consists of a very short α-helix (3–4 residues) followed by a sharp clockwise turn (Figure 2A,B). The motifs are connected by linkers of variable lengths (two to seven residues). The α-helical part of the helix-plus-turn motif is identified as the "niche4r" motif by PDBeMotif [53].

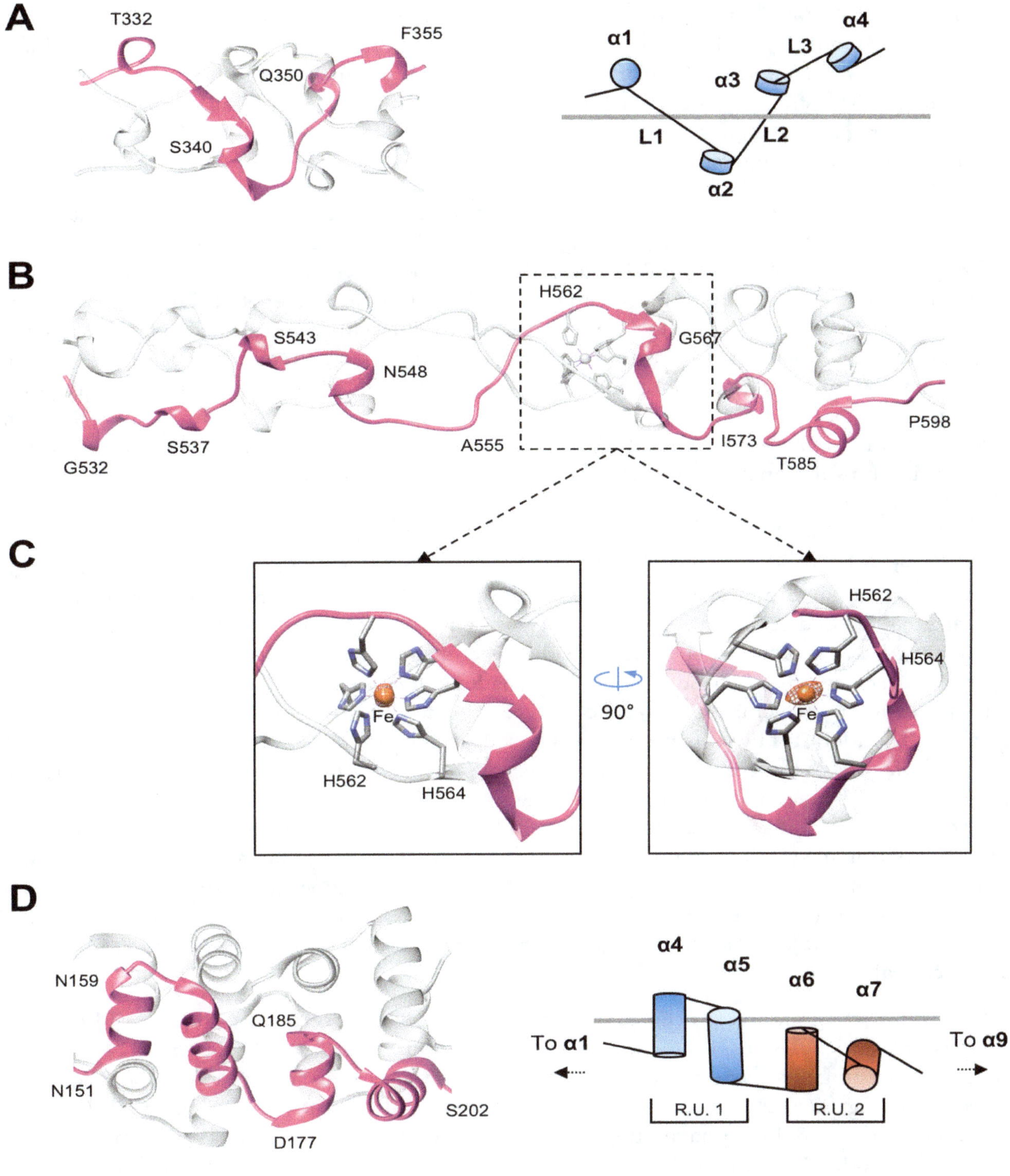

Figure 2. Detailed analysis of the Rod and Shaft domains. The fold and schematic topology overview of the Rod (**A**) and Shaft (**B**) domains. Linkers of variable length connecting the α-helices (α1–α4) are labeled L1, L2, and L3. (**C**) Side view and N-to-C end-on view of the iron-binding motif that is located approximately in the middle of the Shaft domain. The sidechains of histidines constituting the binding site are showed in a stick representation. The centrally positioned iron ion is shown as a rust-colored sphere. The Bijvoet difference Fourier synthesis map is shown as a red mesh and contoured at 4.0 standard deviations above the mean. His-Nε2-Fe coordination bonds are showed with dashed lines. One monomer is colored in magenta and the other two are colored in transparent grey. The R2 fiber model was used for illustration. (**D**) The fold and schematic topology overview of the stem domain fragment of phage phi29 head fiber. R.U. stands for Repeating Unit.

The helix-plus-linker motif is topologically similar to the helix-turn-helix motif of the stem domain of the phage phi29 head fiber Figure 2D [54]. The phi29 stem domain is highly regular: Each turn raises the subsequent α-helix by approximately 10 Å. Additionally, the double-helix repeat units of the super helix region (helices 4–9) related by a 45° turn and a translation of 19.5 Å along the helical axis (Figure 2D). The structure of the pyocin fiber is not as regular because of the varying length of the linkers connecting the helix-plus-turn motifs (Figure 2A,B). Nevertheless, the Shaft and Rod domains have a nearly constant diameter throughout.

Repeats are a common theme in the organization of many fibrous proteins [8,55]. In many cases they can be detected at the amino acid sequence level, such as the heptad repeats in α-helical coiled coils [56] or valine-glycine repeats in β-helices [57,58]. In other cases, where the repeat is short, such as the one described here (Figure 2A,B), it can only be identified at the level of the protein structure. A repeating structure allows for adjustment of the length of the fiber and shows that most fibers are evolving by reusing the same structural element by means of domain or motif duplication events. The function of many fibrous proteins, such as pyocin and phage fibers, involves the binding of cell surface moieties that represent an extremely diverse set of ligands. A structure built with repeats makes it possible to fine tune this binding both spatially and temporarily to the ligand at hand and coordinate it with subsequent conformational changes in the pyocin/phage particle.

3.3. R1 and R2 Fibers Contain Buried and Solvent Exposed Metal Ions

The interior of the Shaft domain in R1 and R2 fibers contains two metal ions—an iron roughly in the middle and a hydrated magnesium in its very C-terminal part where the Shaft transitions into the C-terminal lectin domain. The presence of the iron ion was first detected with X-ray fluorescent spectroscopy. Its location in the structure was then established with the help of the Bijvoet difference Fourier synthesis (Figure 2C). In addition to the buried metal ions, the R1 and R2 fibers contain solvent exposed sodium and calcium ions, respectively, bound to the tip of the C-terminal lectin domain (Figure 3). The identification of these ions is based on the analysis of the coordination geometry, temperature factors of these ions, and those of the surrounding residues, as well as on the correlation with the electron density map. The corresponding statistics are given in Table 2 and Table S2.

The parameters of the magnesium, sodium, and calcium binding sites suggests that these ions are bound to the protein structure in their most common oxidation state (Table 2). The CheckMyMetal web server [59,60] indicates that the valence and therefore the oxidation state of iron ions incorporated into pyocin fibers is (II). As Fe(III) is more stable in solution, it is likely that iron is reduced to Fe(II) after it binds to the protein. The crystallographic refinement statistics for the Fe(II) and Fe(III) ions are nearly identical.

The iron-binding site has an octahedral geometry and is composed of six histidine residues—three symmetry-related copies of His561 and His563 in R1 and His562 and His564 in R2 (Figure 2C). Similar iron-binding sites created by the same HxH motif are found in the structure of the T4 long tail fiber protein gp37 (PDB code 2XGF) [26] and in the central spike proteins of phage P2 (gpV, PDB codes 3QR7 and 3QR8) [52], phage phi92 (gp138, PDB codes 3PQI and 3PQH) [52], and R2 pyocin (PA0616, PDB codes 4S36 and 4S37). Phage T4 short tail fiber protein gp12 also contains an HxH motif, but its crystal structure contains a zinc ion in place of the more common iron [24,25]. However, the purification process of gp12 involved a heat denaturation step and reconstitution in the presence of a zinc salt, so it is unclear whether the zinc ion is the natural ligand of that site in the wild type gp12 protein.

The water shell of the buried hydrated magnesium ion has a nearly perfect octahedral geometry (Figure 3A–C,E, Table 2). The ion is coordinated by three symmetry-related copies of the main chain oxygen of Val596 and side chain of Asp612 in the R1 fiber (Val597 and Asp613 in R2) (Figure 3C,E). The structure of this site is somewhat similar to that of the phage T4 cell-puncturing gp5 protein, where a hydrated magnesium ion is buried in a hydrophobic cavity where it is coordinated by three glutamate residues [61].

The function of the buried iron and magnesium ions is most likely related to protein folding. They might form "reference points" for three nascent protein chains that are about to fold into a trimer [52,62], although there is no experimental data to support this hypothesis at this point. The metal-binding sites could also give the fiber the required balance of stiffness and flexibility because crystal packing forces can elastically bend, but not break, the iron-containing Shaft domain (see Figure S2).

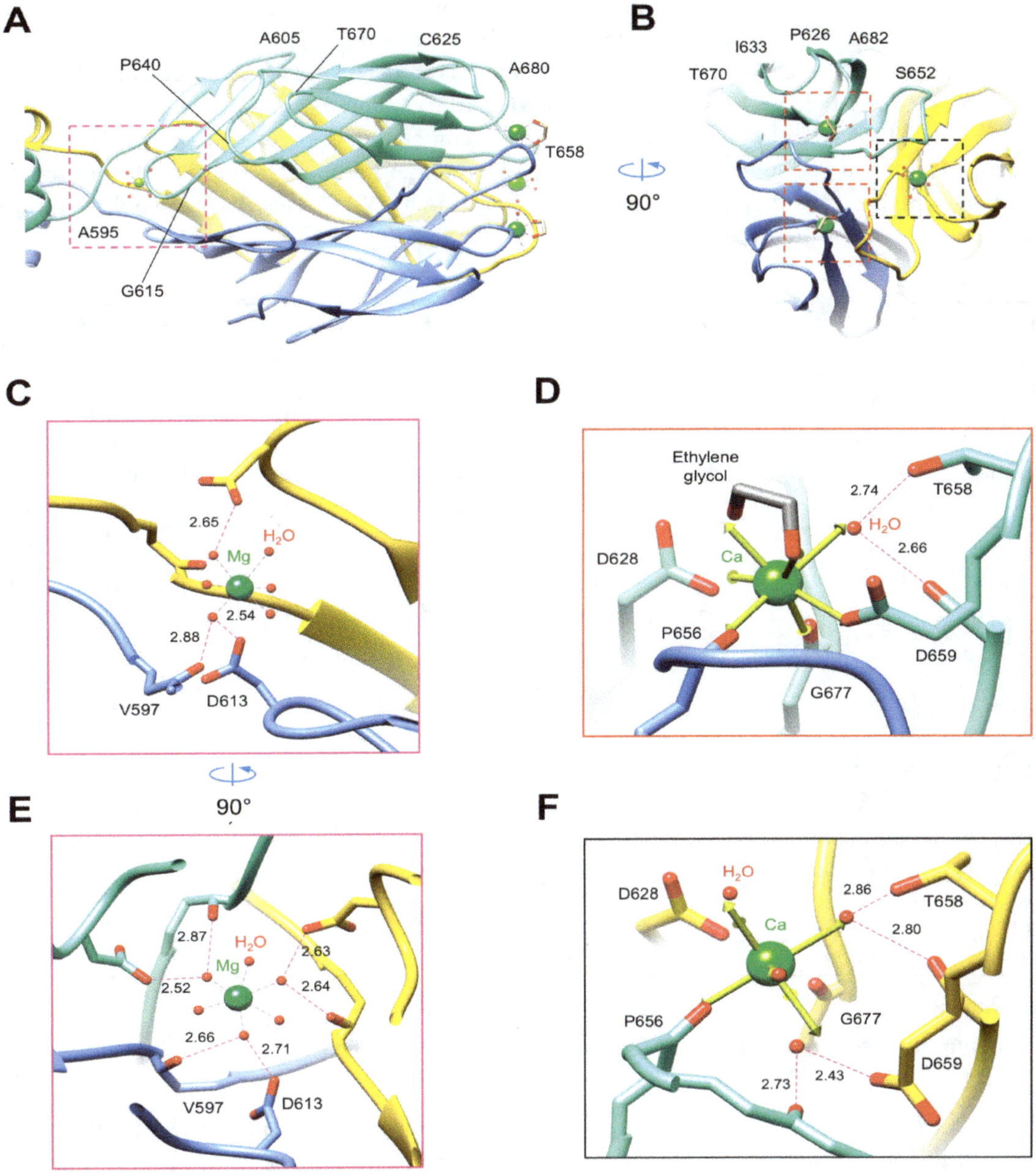

Figure 3. Structure of the R2 fiber C-terminal domain with bound metal ions and ligands. Side view (**A**) and C-to-N end-on view (**B**) of the R2 pyocin receptor binding domain. Cut-away side view (**C**) and C-to-N end-on view (**E**) of the magnesium site. (**D**,**F**) Structure of two different calcium ion binding sites. The three protein chains comprising the trimer are colored in gold, cornflower blue, and aquamarine. The dashed rectangles in the top panels mark the area of detail enlarged in the bottom panels. The magenta rectangle outlines the location of the magnesium ion. The red and black rectangles mark two different calcium-binding sites. The arrows in panel (**C**) represent the ideal coordination geometry for a calcium ion. The hydrogen and coordination bonds are showed as dashed lines. The distances of H-bonds between the outer and inner coordination shells are indicated in Å.

Table 2. Validation and coordination geometry analysis of metal-binding sites in R1 and R2 pyocin fiber structures with the CheckMyMetal web server [59,63]. ABC and DEF indicate the three chains of the two independent trimers comprising the asymmetric unit. The **B-factor** column gives the isotropic atomic displacement factor for the metal ion and its valence-weighted environmental average in parentheses. **Ligands** are the elemental composition of the coordination sphere. **Valence** is the overall bond valence, the average of individual bond valence values for all metal-ligand bonds. For an ideal site, the valence parameter should be the same as the charge of the metal ion. **nVECSUM** is a summation of the ligand vectors, weighted by bond valence values, and normalized by the overall valence. In a complete coordination sphere, the ligand vectors should sum to zero. The value increases as the coordination sphere becomes less complete. **Geometry** is a pattern of three-dimensional arrangement of ligands around the metal ion, as defined by the coordination chemistry: OH—Octahedral; TB—Trigonal Bipyramidal; PB—Pentagonal Bipyramidal. **gRMSD** is the overall root mean square deviation of the geometry angles (L-M-L angles) compared to the ideal geometry, in degrees. **Vacancy** is the percentage of unoccupied sites in the coordination sphere for a given geometry. The values of parameters are scored as "acceptable", "outlier", or "borderline" according to [59]. The outliers are bold-highlighted, and "borderline" values are underlined.

Metal Ion	B-Factor (Å2)	Ligands	Valence	nVECSUM	Geometry	gRMSD (°)	Vacancy
R1 fiber							
Fe^{2+} 1 (ABC)	81.0 (102.7)	N_6	1.8	0.05	OH	5.7	0
Fe^{2+} 2 (DEF)	68.7 (79.7)	N_6	1.9	0.10	OH	4.2	0
Mg^{2+} 3 (ABC)	153.5 (87.5)	O_6	1.6	0.15	OH	5.8	0
Mg^{2+} 4 (DEF)	89.8 (63.9)	O_6	1.5	0.10	OH	11.5	0
Na^{+} 5 (ABC)	65.6 (78.0)	O_5	0.9	0.18	OH	13.0	16%
Na^{+} 6 (ABC)	75.8 (83.8)	O_5	1.0	**0.25**	TB	13.2	0
Na^{+} 7 (ABC)	77.9 (94.7)	O_6	0.9	0.18	OH	**23.0**	0
Na^{+} 8 (DEF)	84.6 (66.2)	O_5	0.9	0.17	OH	14.0	16%
Na^{+} 9 (DEF)	138.4 (98.4)	O_5	1.2	0.08	OH	16.1	16%
Na^{+} 10 (DEF)	75.5 (91.1)	O_5	0.9	0.21	OH	17.3	16%
R2 fiber							
Fe^{2+} 1 (ABC)	41.6 (33.3)	N_6	1.5	0.04	OH	4.4	0
Fe^{2+} 2 (DEF)	34.7 (35.2)	N_6	1.6	0.11	OH	4.2	0
Mg^{2+} 3 (ABC)	23.5 (22.8)	O_6	2.0	0.07	OH	9.0	0
Mg^{2+} 4 (DEF)	26.1 (21.2)	O_6	1.8	0.07	OH	9.2	0
Ca^{2+} 5 (ABC)	16.7 (20.4)	O_7	2.1	0.11	PB	11.4	0
Ca^{2+} 6 (ABC)	13.9 (16.8)	O_7	2.2	0.11	PB	9.7	0
Ca^{2+} 7 (ABC)	15.1 (20.0)	O_7	2.4	0.09	PB	9.9	0
Ca^{2+} 8 (DEF)	19.7 (22.9)	O_7	1.9	0.16	PB	11.5	0
Ca^{2+} 9 (DEF)	18.0 (19.3)	O_7	1.9	0.11	PB	10.4	0
Ca^{2+} 10 (DEF)	19.5 (21.6)	O_7	2.0	0.11	PB	10.2	0

3.4. A Small Compound is Buried in the Hydrophobic Interior of the Knob2 and Shaft Domains

The Knob2 and Shaft domains of R1 and R2 fibers display 53% and 74% sequence identity and have a nearly identical main chain traces, which can be superimposed with a root mean square deviation (RMSD) of 0.73 and 0.82Å, respectively. However, their interchain hydrophobic cores display an interesting structural difference. A flat and nearly perfectly triangular density is located on the axis of the Knob2 of the R1 fiber, but no such feature is present in the R2 (Figure S3). A similar triangular density is buried in the N-terminal part of the Shaft of the R2 fiber (residues Phe534-Tyr539), but there is no such "molecule" in the Shaft domain of the R1 fiber (Figure S4). Remarkably, in both cases, the threefold axis of the "molecule" is not parallel to the axis of the trimer and the "molecule" thus does not interact with its threefold environment in a threefold symmetric fashion. The deviation of the "molecule's" axis from the threefold axis of the fiber is more prominent in the Knob2 domain of the R1 fiber (Figure S3C). All residues of the Knob2 domain cavity in the R2 structure are conserved except Ile491 of R1 is replaced with His491, which gives the cavity a slightly different configuration. In the R2

Shaft domain cavity, the situation is more mysterious as the residues forming the cavity in both fibers are identical (Figure S4G), but the cavity of the R1 fiber is empty.

We attempted to determine the identity of the "molecule" corresponding to the triangular electron density by crystallographic refinement of different compounds possessing a threefold symmetry (nitrate and carbonate ions) or pseudo threefold symmetry (acetate ion and acetone), as well as three water molecules placed in the vertices of the electron density feature. Unfortunately, none of the ligands were a clear favorite because all gave acceptable refinement statistics, agreed with the electron density, and did not distort the surrounding protein residues (Figures S3 and S4). The height of the resulting map peaks, map correlation coefficients, and temperature factors for each refined ligand are given in Table S3. In the structures of R1 and R2 pyocin fibers deposited to the Protein Data Bank, these electron density features are interpreted as water molecules.

3.5. Knob-Like Domains Are Found in Saccharide-Binding Tail Fibers and Tailspikes

The tandem Knob domains, Knob1 and Knob2, have a similar fold that is, essentially, an antiparallel β-sheet with five or six strands (Knob1 and Knob2, respectively) that are connected by loops of variable lengths (Figure 4). One such loop in the Knob2 domain is 16 residues long (S471–R486 in the R2 fiber) and contains a six residue-long α-helix. The loops curve toward the β-sheet and create a partially closed structure. The loops form the outer surface of the molecule whereas the β-sheets are buried in the trimeric interface (Figure 4B). As a consequence, both sides of the β-sheet display hydrophobic side chains. One set of hydrophobic residues points towards the axis of the fiber and mediates interactions between the three chains comprising the trimer. The other, together with the side chains donated by the loops, creates the actual hydrophobic core of the Knob domain (intra-chain interactions). Because of their unusual multi-hydrophobic core structure, Knob-like domains are likely to play an important role in folding of this and other trimeric proteins. The Knob1 and Knob2 domains are connected by a helix-plus-turn motif that is similar to those of the Rod and Shaft domains (Figure 2).

The two Knob domains of the R2 fiber show 20.0% sequence identity on superposition with 201 aligned C_α atoms (out of 254) and a root mean square deviation (RMSD) of 2.13 Å (Figure 4B). These domains have clearly evolved from a single ancestor as a result of yet another gene duplication event. Some residues have been retained, but repurposed, in this process. F445 plays a very important role in the structure of Knob2. Its side chain is fully buried and positioned so that it constitutes an integral part of both the inter- and intra-chain hydrophobic cores of the Knob2 domain. The main chain of F445 is exposed to the solution, "sealing" both cores. The equivalent residue in the Knob1 domain is F335. Its side chain is partially exposed to solution, and it is a component of the last helix-plus-turn element (V354–R357) preceding the Knob1 domain.

A search for protein structures similar to Knob1 and Knob2 domains (performed with DALI [64]) identified similar trimeric domains in other viral proteins, such as the putative receptor-binding proteins gp45 (PDB code 5EFV, Z = 9.8, RMSD = 2.2 Å) [65] and gp144 (PDB code 5M9F, Z = 8.2, RMSD = 2.1 Å) of Staphylococcus phages Phi11 and K, respectively, and T4 proximal long tail fiber protein gp34 (PDB code 5NXF, Z = 7.6, RMSD = 2.5 Å) [27]. In particular, the C-terminal "tower" of the putative receptor-binding protein of *Staphylococcus* phage Phi11 (PDBID 5EFV) features two such domains in tandem, identical to the R-type pyocin fiber organization (Figure 5).

Endosialidase tailspike proteins of phages K1F (PDB code 3JU4) [66,67], and Phi92 (PDB code 4HIZ) [51] as well as the KflA tailspike of K5A (PDB code 2X3H) [68] and gp42 tailspike of *A. baumannii* phage AS12 (PDB code 6EU4, N. I. M. Taylor, M. M. Shneider, P. G. Leiman, unpublished data) all also possess a Knob-like domain. It is called a β-prism in the K1F endosialidase and is located downstream from the sialidase domain (Figure 5B). The Knob-like β-prism domain is further extended by a triple-stranded β-helix. The module comprising the β-prism and the triple-stranded β-helix of endosialidases is, in turn, structurally similar to the C-terminal, membrane-puncturing module of central spike proteins of contractile tail bacteriophages [52,69,70]. In all these proteins, the transition between the Knob-like β-prism and triple-stranded β-helix is very smooth—a ladder-like trace of the polypeptide chain of the Knob-like β-prism domain is continued by the triple-stranded β-helix without interruption. Thus, β-prisms, Knob-like fiber domains and the triple-stranded β-helices are likely to have a common ancestor.

The presence of Knob-like domains in host cell recognition and binding proteins of pyocins and phages is clearly dictated by their conserved function. A Knob-like domain of phage K1F endosialidase was shown to bind a fragment of sialic acid [66,67] (Figure 5C,F), which is the host cell surface molecule that is recognized by K1F during infection [71]. Knob-like domains are likely to be involved in host cell recognition and binding in other pyocin/phage systems.

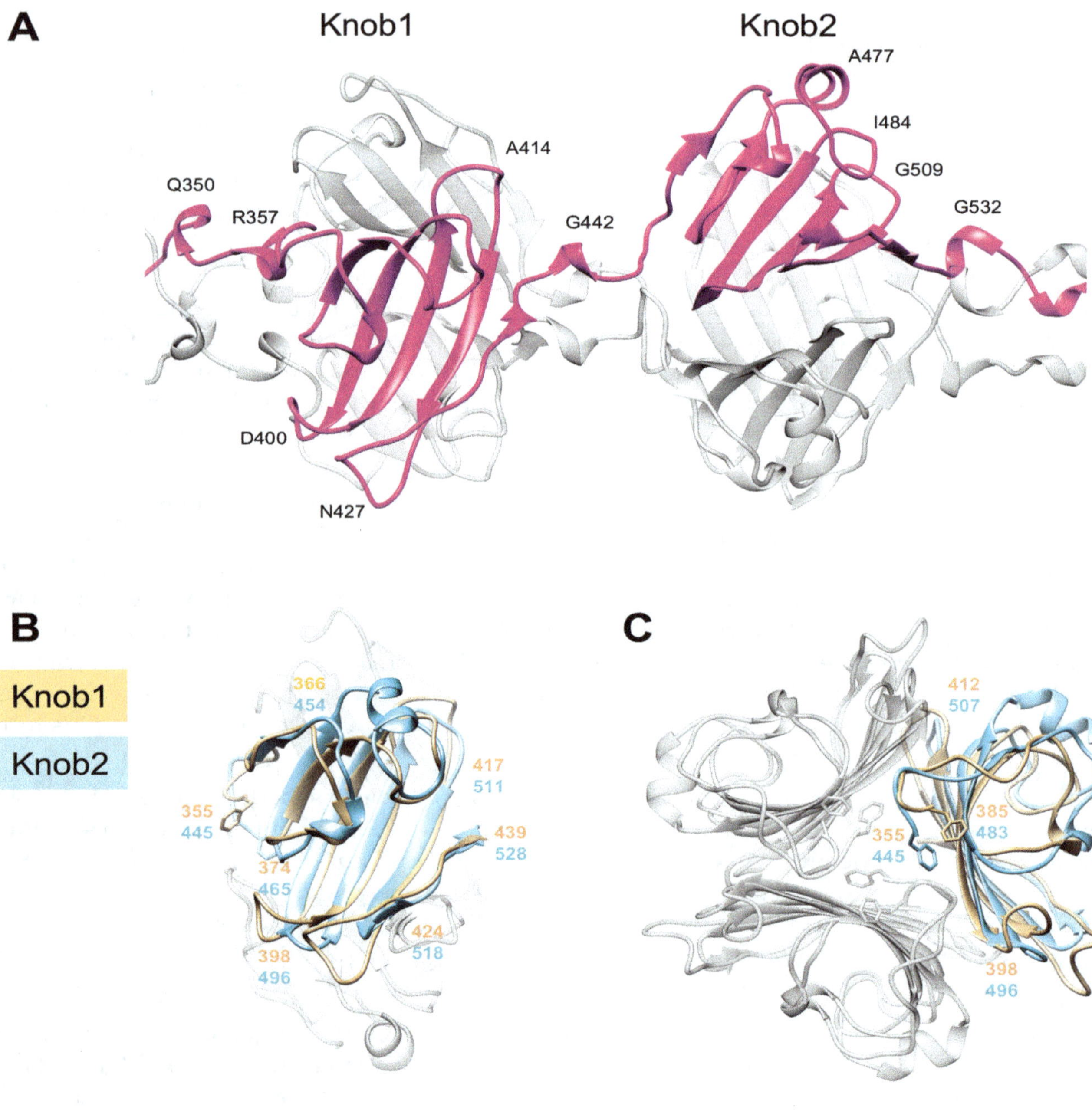

Figure 4. Structure of the two tandem Knob domains Knob1 and Knob2. (**A**) The trace of one of the three chains comprising the trimer. A side view (**B**) and an N-to-C end-on view (**C**) of the superimposed Knob1 and Knob2 domains. Residues are numbered with the color code of the corresponding domain. The R2 fiber model was used in all panels.

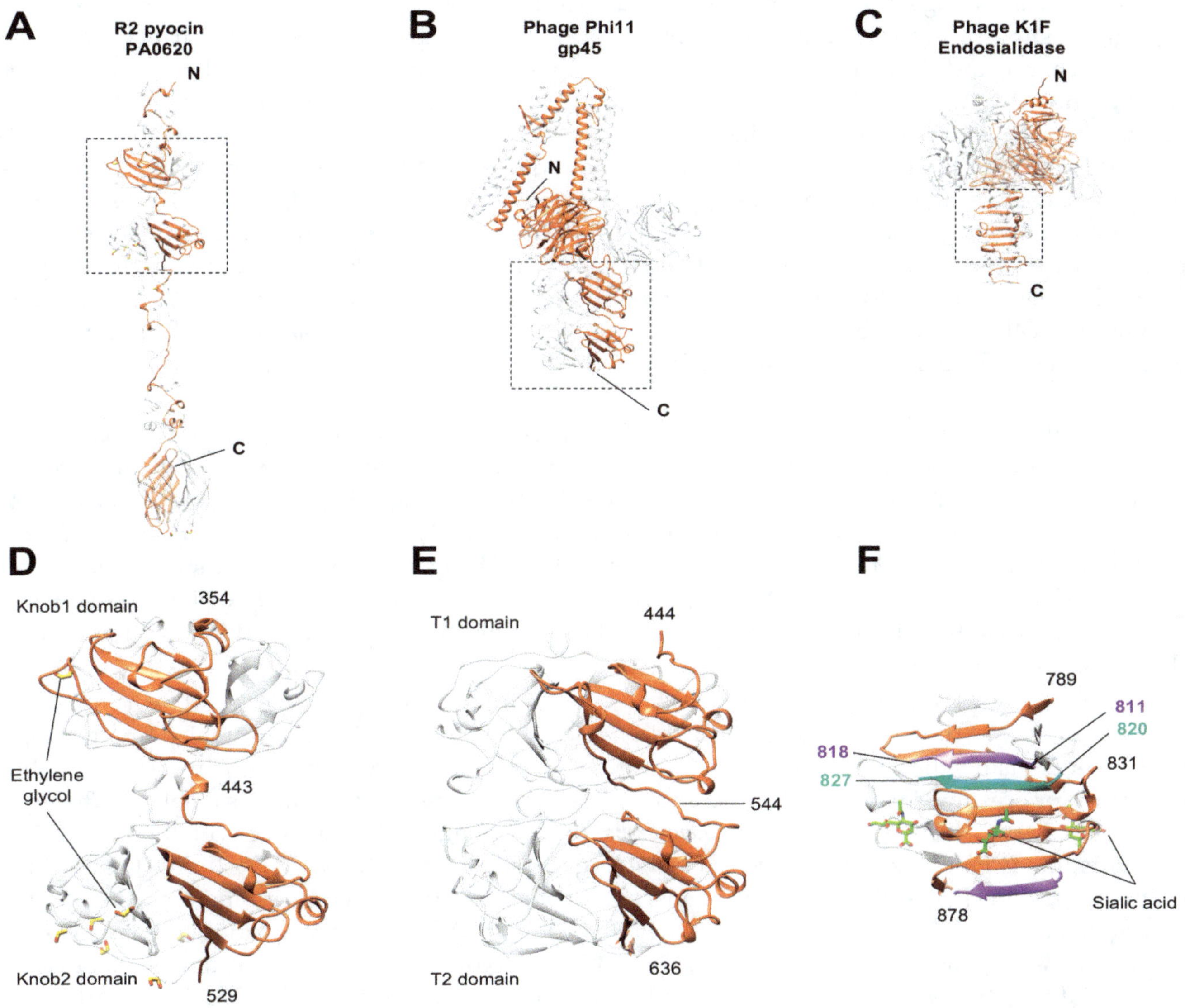

Figure 5. Comparison of the Knob domains with their orthologs from other receptor-binding proteins. Location of the Knob-like domains in the pyocin fiber (**A**), in the putative receptor-binding protein gp45 of phage phi11 [65], and (**B**) in the endosialidase tailspike of phage K1F (where it is called β-prism) [66,67] (**C**). The dashed-line rectangles outline the area of detail shown enlarged in panels (**D–F**). Each face of the Knob-like β-prism of the K1F endosialidase (panel **F**) is composed of three polypeptide chains (three different colors). The β-prism forms a smooth extension of the triple-stranded β-helix domain.

3.6. The C-Terminal β-Sandwich Domain Has a Lectin-Like Fold with a Negatively Charged Surface Groove

Residues 596–691 and 595–701 comprise the C-terminal β-sandwich domain of the R2 and R1 fibers, respectively. Six of the eight β-strands of the β-sandwich form a jellyroll fold-like structure giving rise to an asymmetric sandwich with five and three strands per "side" (Figure 3A). The folds of the R1 and R2 fiber domains are similar and the two structures can be superimposed with an RMSD of 1.35 Å between 249 equivalent Cα atoms in the alignment (out of 318 atoms in R1 and 285 in R2) (Figure S5). The sequence identity of this superposition is 40%.

The greatest difference in the two structures is at the tip of the fiber because the loops connecting the β-strands have different lengths and conformations (Figure S5). The molecular surfaces of the fibers feature prominent surface grooves in this region, which might be involved in binding to the lipopolysaccharide (LPS) (Figure 6). In the R1 fiber, the grooves start at the threefold axis and extend radially outwards (Figure 6A) whereas in the R2 fiber the grooves are connected at the threefold axis, creating a "supercavity" at the tip of the fiber (Figure 6A). In both fibers, the cavity displays a strong negative charge and contains a metal-binding site—the R1 fiber contains three sodium ions, and the R2 fiber three calcium ions (Figures 3 and 6, Table 2 and Table S2). The calcium cavity of R2 is located on the interface of two polypeptide chains and is much deeper than the sodium cavity of R1, which is formed by residues belonging to the same polypeptide chain. On superposition of the two fibers, the distance between the calcium and sodium ions is about 8 Å (Figure S5). The calcium ions of the R2 fiber are significantly more solvent exposed than the sodium ions of R1. The coordination polyhedra of the calcium ions are mostly complete, with some water molecules replaced by ethylene glycol molecules that diffused into crystal during cryo protection (Table 2). On the contrary, the coordination polyhedra of the sodium ions are mostly incomplete, which can be due to the poor quality of the electron density map in those regions (Table 2).

The role of the sodium and calcium ions in the structure and function of pyocin fibers is unclear at the moment. The sodium ion could be incorporated into the structure of R1 fiber during protein purification or crystallization. The calcium ion in the R2 structure takes its origins in the cell cytoplasm as the protein was never exposed to calcium containing salts during purification or crystallization. On the one hand, these ions could be important for the folding or stability of the structure. On the other hand, these ions occupy a strategic position in the fiber and might mediate binding to the cell surface LPS. Notably, a chelating agent that is used during the purification procedure does not remove the calcium ion from the R2 fiber.

A search for protein domains with folds resembling that of the C-terminal domain of the pyocin fiber using the Dali server [64] identified several structures that are trimeric and either confirmed or presumed to be able to bind and/or degrade surface polysaccharides (Table 3). The most similar structures were those of *Acinetobacter baumannii* phage AP22 tail fiber (PDB code 4MTM) and tailspike (PDB code 4Y9V), *Helix Pomatia* agglutinin (PDB code 2CGZ) [72] and Discoidin-II (PDB code 2VM9) [73] lectin-binding proteins, and another two phage receptor-binding proteins—the C-terminal domain of the phage Sf6 tailspike (PDB code 2VBK) [74] and the C-terminal domain of the *Lactococcus lactis* phage bIL170 fiber (PDB code 2FSD) [75]. None of these domains, however, display a prominent negatively charged cavity at its tip that is indicative of metal ion binding. Nevertheless, the structural similarity of the C-terminal domain of the pyocin fiber to other sugar-binding domains and phage fibers shows that it is likely involved in binding to the saccharide portion of the LPS molecule.

Table 3. Structural orthologs of the C-terminal domain of the R1 and R2 pyocin fibers. In all columns, the first and second number corresponds to the R1 and R2 structures, respectively. **RMSD** is the root mean square deviation of all aligned C_α atoms. **Lali** stands for the number of residues used in the superposition; **# res** is the number of residues in the compared structure; **% id** is the percent of sequence identity in the compared structures.

PDB Code	Z-Score	RMSD (Å)	Lali	# res	% id	Hit Description
4MTM	11.2/12.6	2.3/2.0	90/91	137	26/30	Tail fiber of bacteriophage AP22
3WMP	8.0/7.6	2.7/2.2	82/73	94	13/15	SLL-2, galactose-binding lectin
2VME	7.8/7.8	4.1/3.7	88/80	256	17/13	Discoidin-2; oligosaccharide-binding lectin
2CGZ	7.8/6.5	2.6/2.1	81/71	101	20/15	*Helix pomatia* agglutinin; oligosaccharide-binding lectin
4Y9V	6.9/6.5	3.3/3.1	87/80	603	13/13	Tailspike of bacteriophage AP22
2VBK	5.9/6.1	3.3/3.2	79/75	511	13/8	Tailspike of bacteriophage SF6
2FSD	4.7/5.5	3.2/3.0	82/83	110	10/11	Tail fiber of bacteriophage bIL170

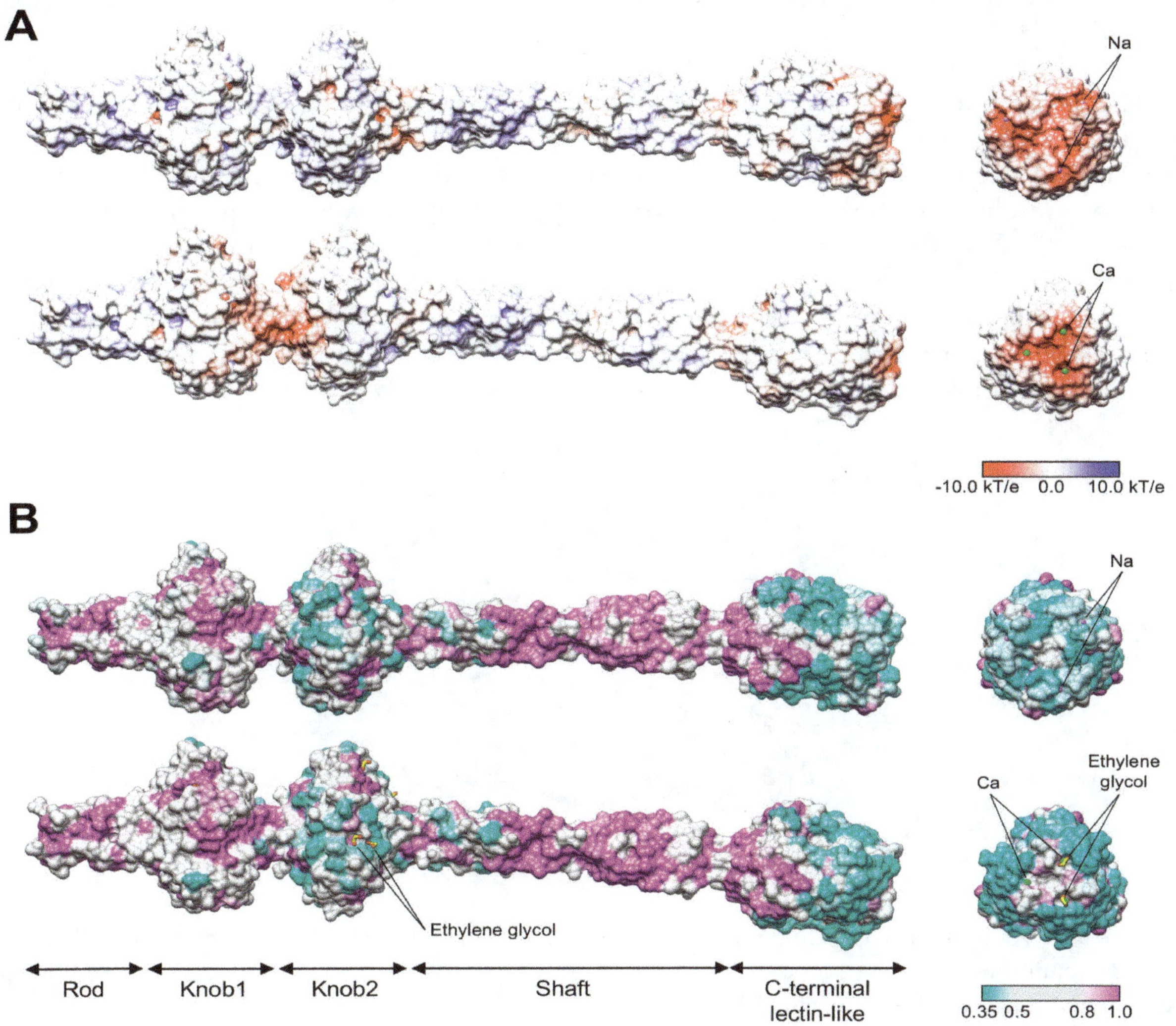

Figure 6. Surface properties of pyocin fibers. (**A**) Surface-mapped electrostatic potential of the R1 (top panel) and R2 (lower panel) fibers. (**B**) Sequence diversity of the surface residues of the R1 (top panel) and R2 (lower panel) fibers. The color code is given as a color bar. Panels on the left: side view. Panels on the right: C-to-N end-on view.

Based on pairwise similarity, the fiber sequences of the five previously characterized subtypes of R-pyocins can be divided into two groups—R1-like (R1 and R5) and R2-like (R2, R3, and R4) (Figure S1). As the overall folds of R1 and R2 fibers are very similar, the differences in substrate specificity must be determined by the surface properties of the proteins. We compiled a list of diverse sequences of R1-like and R2-like fibers available in the GenBank, aligned them, and mapped the degree of sequence diversity and conservation onto the structure of R1 and R2 pyocins fibers (Figure 6B). The most diverse regions are in the Knob2 domain and in the C-terminal domain. A similar trend is displayed when the two known R1-like (R1 and R5) and three R2-like fibers (R2, R3 and R4) are compared to each other, although in this case the surface of the Knob2 domain showed greater diversity than that of the C-terminal domain. Furthermore, the C-terminal domains of R2-like fibers had only one conserved amino acid substitution at the very apex of the structure: His655 of R2 and R4 was replaced with Gln655 in R3. Thus, the Knob2 domain appears to play a greater role than the C-terminal lectin-like domain in determining the spectrum of R-type pyocins. In the structure of the R2 pyocin fiber, ethylene glycol molecules that mimic various parts of the LPS bind to the Knob2 and C-terminal domains. In the latter case, the interaction of the ethylene glycol molecule with the protein involves calcium ions, further supporting the role of calcium ions in binding to the LPS (Figure 3D, Figure S5).

3.7. The Binding Constant of the Pyocin to the Pseudomonas Cell Surface Is Three Orders of Magnitude Greater than That of the Free Fiber

Earlier experiments showed that the killing spectrum of the pyocin is determined by the fibers [10–12]. We nevertheless decided to investigate the nature of fiber-cell surface interaction with the help of a competition experiment where binding of the pyocin particle to the cell is blocked by free fiber present in the same reaction mixture (Figure 7). We assayed the number of R2-sensitive *P. aeruginosa* 13s cells that survived coincubation with the R2 pyocin in the presence of varying concentrations of the PA0620d3 R2 fiber fragment used for structure determination (which carried the SlyD expression tag). We found that an increasing concentration of the fiber protects the cells from the pyocin. This experiment allows us to compare the interaction between the pyocin particle and its free fiber with the cell surface.

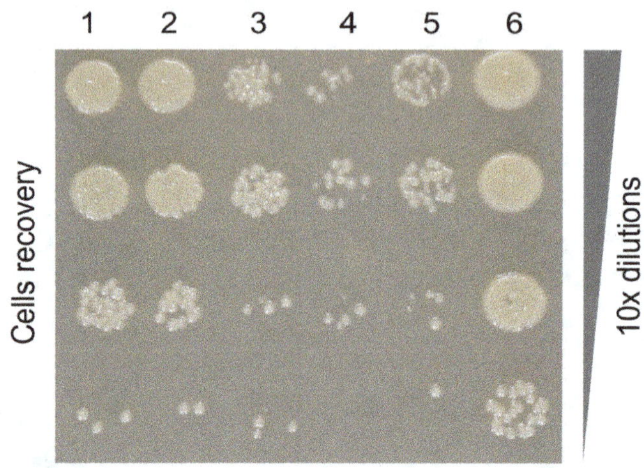

Figure 7. Pyocin fiber—pyocin competition assay. Recovery of viable *P. aeruginosa* 13s cells after pre-incubation with different concentrations of Ni-NTA purified fragment of R2 pyocin fiber followed by incubation with R2 pyocins. The concentration of the fiber in different columns varied as follows: 1—50 μg/mL; 2—5.0 μg/mL; 3—0.5 μg/mL; 4—0.05 μg/mL; 5—none, pyocin treatment only; 6—*P. aeruginosa* 13s cells alone. The rows are a 10-fold dilution series of the plated cells.

In the experimental conditions shown in Figure 7, the inhibition effect of the R2 fiber becomes apparent at a concentration of 0.5 μg/mL, which is equivalent to 2.7 nM (assuming the SlyD-fiber construct is trimeric with a molecular weight of 183 kDa) or ~10^{12} fibers per ml. This is six orders of magnitude greater than the amount of cells used in this assay (10^6 CPU/mL) and four orders of magnitude greater than the amount of pyocin particles (10^8 killing units per mL). In other words, there are 10^6 fibers and 10^2 pyocin particles (with six fibers each) per each bacterial cell in this system. Considering that the recombinant fibers and the fibers on the pyocin particle are identical and they compete for binding to the same substrate, the three orders of magnitude difference (10^6 vs. 6×10^2) in the molar amounts represents the difference in the equilibrium binding constant. The latter is likely due to a higher avidity of the pyocin to the substrate compared to that of the fiber [76]. The fiber is a trimer with several substrate binding sites, and thus is also likely to possess avidity towards the substrate. However, each pyocin particle carries six fibers that emanate from the baseplate and the binding of one fiber to the substrate promotes the binding of the others because of their spatial arrangement. The affinity of the pyocin to its substrate is therefore not a sum of the affinities of its six fibers, but is instead a cooperative non-linear function of thereof.

Notably, in this system, the fraction of the fiber that can be immobilized on the cell surface at any given time is very small. The area of the bacterial cell surface is ~4 μm^2 or 4×10^8 Å^2. Depending on the orientation, the fiber can occupy an area of 3000 to 15,000 Å^2 of the cell surface, which means that 10^4–10^5 fibers will cover the cell surface as a continuous layer. As this constitutes only 1 to 10% of the total amount of the fiber per cell present in the system, it is possible that the fibers do cover the cell surface as a continuous layer. The pyocin particle thus needs to outcompete several surface-bound fibers for successful attachment to the cell surface. The pyocin's avidity is likely to play an important role in this process.

4. Discussion

Similar to other cell-surface binding proteins of phages, pyocin fibers are unlikely to change their structure upon cell surface binding [20,51,74,77]. At the same time, this binding initiates a cascade of structural changes that commits the phage or pyocin particle to irreversible binding to the cell (e.g., in the case of R-type pyocin—to sheath contraction). This creates an obvious paradox: The cell surface recognition signal must be transmitted to the particle along the length of fiber (about 340 Å) while the structure of the fiber changes little upon cell surface binding. Structural analysis and bioinformatics suggest Knob2 and C-terminal domain of the pyocin fibers, which are separated by a distance of about 100 Å, are likely involved in cell surface binding. This finding makes it possible to explain how the signal of cell surface binding is transmitted to the rest of the particle (Figure 8).

Because of the two sets of spatially separated ligand binding sites on the Knob2 and C-terminal domains, the fiber likely adapts a certain orientation upon interaction with the cell surface. This fiber remains bound to the cell surface while the pyocin particle is buffeted around by the surrounding solvent because of Brownian motion and/or cell swimming. Eventually, the second, third, etc. fiber binds to the surface receptor in the same or similar orientation as the first fiber. In this configuration, the pyocin particle becomes oriented perpendicular to the cell surface. At the same time, because the fibers act as rigid bodies or levers, they "unravel" the baseplate, causing it to initiate sheath contraction. We propose that this generic mechanism is employed by all contractile tail-like phages and pyocins that carry only one set of fibers (e.g., phage P2, Mu, etc.).

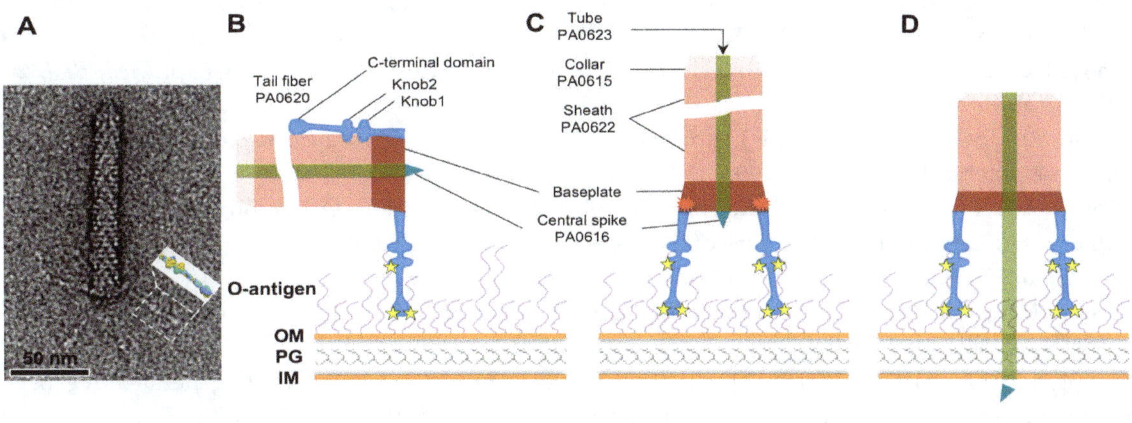

Figure 8. Function of the fiber in triggering sheath contraction. (**A**) A particle of R2 pyocin negatively-stained with uranyl acetate and imaged in a transmission electron microscope. The inset shows the molecular surface of the fiber fragment structure determined by X-ray crystallography. (**B**) Two spatially separated substrate-binding sites cause the fiber to bind at a certain angle relative to the cell surface. Here, the fiber is shown to be perpendicular to the cell surface, but the actual angle is determined by the structure of the substrate. (**C**) Binding of the second and subsequent fibers to the host cell surface requires their orientation relative to the cell surface to be identical or very similar to that of the first fiber. This requirement also orients the pyocin particle to be perpendicular to the cell surface. The configuration of the baseplate with all the fibers pointed towards the cell surface is unstable and causes it to initiate sheath contraction. (**D**) The sheath contracts and drives the tube through the cell envelop. The central spike protein dissociates to initiate leakage of ions from the cytoplasm. The interaction of the fibers with the bacterial surface polysaccharides is indicated with yellow stars. The fiber-to-baseplate "triggering signal" is shown with a red multipointed star. Bacterial O-antigen, outer membrane (OM), peptidoglycan layer (PG) and inner membrane (IM) are labeled.

Supplementary Materials
Figure S1. Phylogenetic tree of fiber sequences of the five R-type pyocins. Figure S2. Crystal packing of R1 and R2 pyocin fibers. Figure S3. Location and identification of the ligand buried within the hydrophobic core of the Knob2 domain of R1 pyocin fiber. Figure S4. Location and identification of the ligand buried within the hydrophobic core of the Shaft domain of R2 pyocin fiber. Figure S5. Comparison of the structure of the C-terminal lectin-like domains of R1 and R2 fibers. Table S1. Composition of the R2 pyocin particle as determined by mass-spectrometry. Table S2. Refinement statistics and final electron density map correlation for metal ions contained in the R1 and R2 fibers. Table S3. Identification of buried compounds in the Knob2 domain of the R1 fiber and the Shaft domains of the R2 fiber.

Author Contributions: Conceptualization, P.G.L. and D.S.; Methodology, S.A.B., M.M.S., D.S. and P.G.L.; Validation, S.A.B. and P.G.L.; Investigation, S.A.B., M.M.S., D.S. and P.G.L.; Writing—Original Draft Preparation, S.A.B.; Writing—Review & Editing, P.G.L., D.S., S.A.B., M.M.S.; Visualization, S.A.B. and P.G.L.; Supervision, P.G.L.; Funding Acquisition, P.G.L.

Acknowledgments: We acknowledge the Paul Scherrer Institut, Villigen, Switzerland for provision of synchrotron radiation beamtime at the beamline X06SA PXI of the SLS and would like to thank Meitian Wang and Vincent Olieric for assistance.

References

1. Nakayama, K.; Takashima, K.; Ishihara, H.; Shinomiya, T.; Kageyama, M.; Kanaya, S.; Ohnishi, M.; Murata, T.; Mori, H.; Hayashi, T. The R-type pyocin of Pseudomonas aeruginosa is related to P2 phage, and the F-type is related to lambda phage. *Mol. Microbiol.* **2000**, *38*, 213–231. [CrossRef] [PubMed]
2. Matsui, H.; Sano, Y.; Ishihara, H.; Shinomiya, T. Regulation of pyocin genes in Pseudomonas aeruginosa by positive (prtN) and negative (prtR) regulatory genes. *J. Bacteriol.* **1993**, *175*, 1257–1263. [CrossRef] [PubMed]
3. Ikeda, K.; Kageyama, M.; Egami, F. Studies of a Pyocin. II. Mode of Production of the Pyocin. *J. Biochem.* **1964**, *55*, 54–58. [CrossRef] [PubMed]
4. Kageyama, M. Studies of a Pyocin. I. Physical and Chemical Properties. *J. Biochem.* **1964**, *55*, 49–53. [CrossRef] [PubMed]
5. Kageyama, M.; Ikeda, K.; Egami, F. Studies of a Pyocin. III. Biological Properties of the Pyocin. *J. Biochem.* **1964**, *55*, 59–64. [CrossRef] [PubMed]
6. Leiman, P.G.; Shneider, M.M. Contractile tail machines of bacteriophages. *Adv. Exp. Med. Biol.* **2012**, *726*, 93–114. [PubMed]
7. Uratani, Y.; Hoshino, T. Pyocin R1 inhibits active transport in *Pseudomonas aeruginosa* and depolarizes membrane potential. *J. Bacteriol.* **1984**, *157*, 632–636. [PubMed]
8. Taylor, N.M.; Prokhorov, N.S.; Guerrero-Ferreira, R.C.; Shneider, M.M.; Browning, C.; Goldie, K.N.; Stahlberg, H.; Leiman, P.G. Structure of the T4 baseplate and its function in triggering sheath contraction. *Nature* **2016**, *533*, 346–352. [CrossRef] [PubMed]
9. Taylor, N.M.I.; van Raaij, M.J.; Leiman, P.G. Contractile injection systems of bacteriophages and related systems. *Mol. Microbiol.* **2018**, *108*, 6–15. [CrossRef] [PubMed]
10. Williams, S.R.; Gebhart, D.; Martin, D.W.; Scholl, D. Retargeting R-type pyocins to generate novel bactericidal protein complexes. *Appl. Environ. Microbiol.* **2008**, *74*, 3868–3876. [CrossRef] [PubMed]
11. Scholl, D.; Cooley, M.; Williams, S.R.; Gebhart, D.; Martin, D.; Bates, A.; Mandrell, R. An engineered R-type pyocin is a highly specific and sensitive bactericidal agent for the food-borne pathogen *Escherichia coli* O157:H7. *Antimicrob. Agents Chemother.* **2009**, *53*, 3074–3080. [CrossRef] [PubMed]
12. Gebhart, D.; Williams, S.R.; Scholl, D. Bacteriophage SP6 encodes a second tailspike protein that recognizes Salmonella enterica serogroups C2 and C3. *Virology* **2017**, *507*, 263–266. [CrossRef] [PubMed]
13. Montag, D.; Henning, U. An open reading frame in the Escherichia coli bacteriophage lambda genome encodes a protein that functions in assembly of the long tail fibers of bacteriophage T4. *J. Bacteriol.* **1987**, *169*, 5884–5886. [CrossRef] [PubMed]
14. Kageyama, M. Bacteriocins and bacteriophages in *Pseudomonas aeruginosa*. In *Microbial Drug Resistance*; Mitsuhashi, S., Hashimoto, H., Eds.; University of Tokyo Press: Tokyo, Japan, 1975; pp. 291–305.
15. Ito, S.; Kageyama, M.; Egami, F. Isolation and Characterization of Pyocins from Several Strains of *Pseudomonas-Aeruginosa*. *J. Gen. Appl. Microbiol.* **1970**, *16*, 205–214. [CrossRef]
16. Scholl, D.; Martin, D.W., Jr. Antibacterial efficacy of R-type pyocins towards *Pseudomonas aeruginosa* in a murine peritonitis model. *Antimicrob. Agents Chemother.* **2008**, *52*, 1647–1652. [CrossRef] [PubMed]

17. Kohler, T.; Donner, V.; van Delden, C. Lipopolysaccharide as shield and receptor for R-pyocin-mediated killing in *Pseudomonas aeruginosa*. *J. Bacteriol.* **2010**, *192*, 1921–1928. [CrossRef] [PubMed]
18. Walter, M.; Fiedler, C.; Grassl, R.; Biebl, M.; Rachel, R.; Hermo-Parrado, X.L.; Llamas-Saiz, A.L.; Seckler, R.; Miller, S.; van Raaij, M.J. Structure of the receptor-binding protein of bacteriophage det7: A podoviral tail spike in a myovirus. *J. Virol.* **2008**, *82*, 2265–2273. [CrossRef] [PubMed]
19. Steinbacher, S.; Seckler, R.; Miller, S.; Steipe, B.; Huber, R.; Reinemer, P. Crystal structure of P22 tailspike protein: Interdigitated subunits in a thermostable trimer. *Science* **1994**, *265*, 383–386. [CrossRef] [PubMed]
20. Steinbacher, S.; Baxa, U.; Miller, S.; Weintraub, A.; Seckler, R.; Huber, R. Crystal structure of phage P22 tailspike protein complexed with *Salmonella* sp. O-antigen receptors. *Proc. Natl. Acad. Sci. USA* **1996**, *93*, 10584–10588. [CrossRef] [PubMed]
21. Prokhorov, N.S.; Riccio, C.; Zdorovenko, E.L.; Shneider, M.M.; Browning, C.; Knirel, Y.A.; Leiman, P.G.; Letarov, A.V. Function of bacteriophage G7C esterase tailspike in host cell adsorption. *Mol. Microbiol.* **2017**, *105*, 385–398. [CrossRef] [PubMed]
22. Leiman, P.G.; Chipman, P.R.; Kostyuchenko, V.A.; Mesyanzhinov, V.V.; Rossmann, M.G. Three-dimensional rearrangement of proteins in the tail of bacteriophage T4 on infection of its host. *Cell* **2004**, *118*, 419–429. [CrossRef] [PubMed]
23. Hu, B.; Margolin, W.; Molineux, I.J.; Liu, J. Structural remodeling of bacteriophage T4 and host membranes during infection initiation. *Proc. Natl. Acad. Sci. USA* **2015**, *112*, E4919–E4928. [CrossRef] [PubMed]
24. Van Raaij, M.J.; Schoehn, G.; Burda, M.R.; Miller, S. Crystal structure of a heat and protease-stable part of the bacteriophage T4 short tail fibre. *J. Mol. Biol.* **2001**, *314*, 1137–1146. [CrossRef] [PubMed]
25. Thomassen, E.; Gielen, G.; Schutz, M.; Schoehn, G.; Abrahams, J.P.; Miller, S.; van Raaij, M.J. The structure of the receptor-binding domain of the bacteriophage T4 short tail fibre reveals a knitted trimeric metal-binding fold. *J. Mol. Biol.* **2003**, *331*, 361–373. [CrossRef]
26. Bartual, S.G.; Otero, J.M.; Garcia-Doval, C.; Llamas-Saiz, A.L.; Kahn, R.; Fox, G.C.; van Raaij, M.J. Structure of the bacteriophage T4 long tail fiber receptor-binding tip. *Proc. Natl. Acad. Sci. USA* **2010**, *107*, 20287–20292. [CrossRef] [PubMed]
27. Granell, M.; Namura, M.; Alvira, S.; Kanamaru, S.; van Raaij, M.J. Crystal Structure of the Carboxy-Terminal Region of the Bacteriophage T4 Proximal Long Tail Fiber Protein Gp34. *Viruses* **2017**, *9*, 168. [CrossRef] [PubMed]
28. Garcia-Doval, C.; van Raaij, M.J. Structure of the receptor-binding carboxy-terminal domain of bacteriophage T7 tail fibers. *Proc. Natl. Acad. Sci. USA* **2012**, *109*, 9390–9395. [CrossRef] [PubMed]
29. Hendrickson, W.A.; Ogata, C.M. Phase determination from multiwavelength anomalous diffraction measurements. *Macromol. Crystallogr. Part A* **1997**, *276*, 494–523.
30. Kabsch, W. XDS. *Acta Crystallogr. Sect. D Biol. Crystallogr.* **2010**, *66 Pt 2*, 125–132. [CrossRef]
31. Kabsch, W. Integration, scaling, space-group assignment and post-refinement. *Acta Crystallogr. Sect. D Biol. Crystallogr.* **2010**, *66 Pt 2*, 133–144. [CrossRef]
32. Sheldrick, G.M. Experimental phasing with SHELXC/D/E: Combining chain tracing with density modification. *Acta Crystallogr. Sect. D Biol. Crystallogr.* **2010**, *66 Pt 4*, 479–485. [CrossRef]
33. Pape, T.; Schneider, T.R. HKL2MAP: A graphical user interface for macromolecular phasing with SHELX programs. *J. Appl. Crystallogr.* **2004**, *37*, 843–844. [CrossRef]
34. Terwilliger, T.C. Automated structure solution, density modification and model building. *Acta Crystallogr. Sect. D Biol. Crystallogr.* **2002**, *58*, 1937–1940. [CrossRef]
35. Perrakis, A.; Harkiolaki, M.; Wilson, K.S.; Lamzin, V.S. ARP/wARP and molecular replacement. *Acta Crystallogr. Sect. D Biol. Crystallogr.* **2001**, *57*, 1445–1450. [CrossRef]
36. Emsley, P.; Cowtan, K. Coot: Model-building tools for molecular graphics. *Acta Crystallogr. Sect. D Biol. Crystallogr.* **2004**, *60*, 2126–2132. [CrossRef] [PubMed]
37. Murshudov, G.N.; Skubak, P.; Lebedev, A.A.; Pannu, N.S.; Steiner, R.A.; Nicholls, R.A.; Winn, M.D.; Long, F.; Vagin, A.A. REFMAC5 for the refinement of macromolecular crystal structures. *Acta Crystallogr. D. Biol. Crystallogr.* **2011**, *67 Pt 4*, 355–367. [CrossRef]
38. Adams, P.D.; Afonine, P.V.; Bunkoczi, G.; Chen, V.B.; Davis, I.W.; Echols, N.; Headd, J.J.; Hung, L.W.; Kapral, G.J.; Grosse-Kunstleve, R.W.; et al. PHENIX: A comprehensive Python-based system for macromolecular structure solution. *Acta Crystallogr. D Biol. Crystallogr.* **2010**, *66 Pt 2*, 213–221. [CrossRef]

39. Painter, J.; Merritt, E.A. Optimal description of a protein structure in terms of multiple groups undergoing TLS motion. *Acta Crystallogr. D Biol. Crystallogr.* **2006**, *62 Pt 4*, 439–450. [CrossRef]
40. Rossmann, M.G. *The Molecular Replacement Method: A Collection of Papers on the Use of Non-Crystallographic Symmetry*; Gordon and Breach: New York, NY, USA, 1972.
41. Mccoy, A.J.; Grosse-Kunstleve, R.W.; Adams, P.D.; Winn, M.D.; Storoni, L.C.; Read, R.J. Phaser crystallographic software. *J. Appl. Crystallogr.* **2007**, *40*, 658–674. [CrossRef] [PubMed]
42. Winn, M.D.; Ballard, C.C.; Cowtan, K.D.; Dodson, E.J.; Emsley, P.; Evans, P.R.; Keegan, R.M.; Krissinel, E.B.; Leslie, A.G.; McCoy, A.; et al. Overview of the CCP4 suite and current developments. *Acta Crystallogr. D Biol. Crystallogr.* **2011**, *67 Pt 4*, 235–242. [CrossRef]
43. Pettersen, E.F.; Goddard, T.D.; Huang, C.C.; Couch, G.S.; Greenblatt, D.M.; Meng, E.C.; Ferrin, T.E. UCSF Chimera—A visualization system for exploratory research and analysis. *J. Comput. Chem.* **2004**, *25*, 1605–1612. [CrossRef] [PubMed]
44. Dolinsky, T.J.; Czodrowski, P.; Li, H.; Nielsen, J.E.; Jensen, J.H.; Klebe, G.; Baker, N.A. PDB2PQR: Expanding and upgrading automated preparation of biomolecular structures for molecular simulations. *Nucleic Acids Res.* **2007**, *35*, W522–W555. [CrossRef] [PubMed]
45. Unni, S.; Huang, Y.; Hanson, R.M.; Tobias, M.; Krishnan, S.; Li, W.W.; Nielsen, J.E.; Baker, N.A. Web servers and services for electrostatics calculations with APBS and PDB2PQR. *J. Comput. Chem.* **2011**, *32*, 1488–1491. [CrossRef] [PubMed]
46. Altschul, S.F.; Gish, W.; Miller, W.; Myers, E.W.; Lipman, D.J. Basic local alignment search tool. *J. Mol. Biol.* **1990**, *215*, 403–410. [CrossRef]
47. Sanner, M.F.; Olson, A.J.; Spehner, J.C. Reduced surface: An efficient way to compute molecular surfaces. *Biopolymers* **1996**, *38*, 305–320. [CrossRef]
48. Soding, J.; Biegert, A.; Lupas, A.N. The HHpred interactive server for protein homology detection and structure prediction. *Nucleic Acids Res.* **2005**, *33*, W244–W248. [CrossRef] [PubMed]
49. Alva, V.; Nam, S.Z.; Soding, J.; Lupas, A.N. The MPI bioinformatics Toolkit as an integrative platform for advanced protein sequence and structure analysis. *Nucleic Acids Res.* **2016**, *44*, W410–W415. [CrossRef] [PubMed]
50. Dereeper, A.; Guignon, V.; Blanc, G.; Audic, S.; Buffet, S.; Chevenet, F.; Dufayard, J.F.; Guindon, S.; Lefort, V.; Lescot, M.; et al. Phylogeny.fr: Robust phylogenetic analysis for the non-specialist. *Nucleic Acids Res.* **2008**, *36*, W465–W469. [CrossRef] [PubMed]
51. Schwarzer, D.; Browning, C.; Stummeyer, K.; Oberbeck, A.; Muhlenhoff, M.; Gerardy-Schahn, R.; Leiman, P.G. Structure and biochemical characterization of bacteriophage phi92 endosialidase. *Virology* **2015**, *477*, 133–143. [CrossRef] [PubMed]
52. Browning, C.; Shneider, M.M.; Bowman, V.D.; Schwarzer, D.; Leiman, P.G. Phage pierces the host cell membrane with the iron-loaded spike. *Structure* **2012**, *20*, 326–339. [CrossRef] [PubMed]
53. Golovin, A.; Henrick, K. MSDmotif: Exploring protein sites and motifs. *BMC Bioinform.* **2008**, *9*, 312. [CrossRef] [PubMed]
54. Xiang, Y.; Rossmann, M.G. Structure of bacteriophage phi29 head fibers has a supercoiled triple repeating helix-turn-helix motif. *Proc. Natl. Acad. Sci. USA* **2011**, *108*, 4806–4810. [CrossRef] [PubMed]
55. Cerritelli, M.E.; Wall, J.S.; Simon, M.N.; Conway, J.F.; Steven, A.C. Stoichiometry and domainal organization of the long tail-fiber of bacteriophage T4: A hinged viral adhesin. *J. Mol. Biol.* **1996**, *260*, 767–780. [CrossRef] [PubMed]
56. Tao, Y.; Strelkov, S.V.; Mesyanzhinov, V.V.; Rossmann, M.G. Structure of bacteriophage T4 fibritin: A segmented coiled coil and the role of the C-terminal domain. *Structure* **1997**, *5*, 789–798. [CrossRef]
57. Kanamaru, S.; Gassner, N.C.; Ye, N.; Takeda, S.; Arisaka, F. The C-terminal fragment of the precursor tail lysozyme of bacteriophage T4 stays as a structural component of the baseplate after cleavage. *J. Bacteriol.* **1999**, *181*, 2739–2744. [PubMed]
58. Shneider, M.M.; Buth, S.A.; Ho, B.T.; Basler, M.; Mekalanos, J.J.; Leiman, P.G. PAAR-repeat proteins sharpen and diversify the type VI secretion system spike. *Nature* **2013**, *500*, 350–353. [CrossRef] [PubMed]
59. Zheng, H.; Chordia, M.D.; Cooper, D.R.; Chruszcz, M.; Muller, P.; Sheldrick, G.M.; Minor, W. Validation of metal-binding sites in macromolecular structures with the CheckMyMetal web server. *Nat. Protoc.* **2014**, *9*, 156–170. [CrossRef] [PubMed]

60. Zheng, H.; Cooper, D.R.; Porebski, P.J.; Shabalin, I.G.; Handing, K.B.; Minor, W. CheckMyMetal: A macromolecular metal-binding validation tool. *Acta Crystallogr. D Struct. Biol.* **2017**, *73 Pt 3*, 223–233. [CrossRef]
61. Buth, S.A.; Menin, L.; Shneider, M.M.; Engel, J.; Boudko, S.P.; Leiman, P.G. Structure and Biophysical Properties of a Triple-Stranded Beta-Helix Comprising the Central Spike of Bacteriophage T4. *Viruses* **2015**, *7*, 4676–4706. [CrossRef] [PubMed]
62. Van Raaij, M.J.; Mitraki, A.; Lavigne, G.; Cusack, S. A triple β-spiral in the adenovirus fibre shaft reveals a new structural motif for a fibrous protein. *Nature* **1999**, *401*, 935–938. [CrossRef] [PubMed]
63. Zheng, W.; Wang, F.; Taylor, N.M.I.; Guerrero-Ferreira, R.C.; Leiman, P.G.; Egelman, E.H. Refined Cryo-EM Structure of the T4 Tail Tube: Exploring the Lowest Dose Limit. *Structure* **2017**, *25*, 1436–1441 e2. [CrossRef] [PubMed]
64. Holm, L.; Laakso, L.M. Dali server update. *Nucleic Acids Res.* **2016**, *44*, W351–W355. [CrossRef] [PubMed]
65. Koc, C.; Xia, G.; Kuhner, P.; Spinelli, S.; Roussel, A.; Cambillau, C.; Stehle, T. Structure of the host-recognition device of Staphylococcus aureus phage varphi11. *Sci. Rep.* **2016**, *6*, 27581. [CrossRef] [PubMed]
66. Stummeyer, K.; Dickmanns, A.; Muhlenhoff, M.; Gerardy-Schahn, R.; Ficner, R. Crystal structure of the polysialic acid-degrading endosialidase of bacteriophage K1F. *Nat. Struct. Mol. Biol.* **2005**, *12*, 90–96. [CrossRef] [PubMed]
67. Schulz, E.C.; Neumann, P.; Gerardy-Schahn, R.; Sheldrick, G.M.; Ficner, R. Structure analysis of endosialidase NF at 0.98 A resolution. *Acta Crystallogr. D Biol. Crystallogr.* **2010**, *66 Pt 2*, 176–180. [CrossRef]
68. Thompson, J.E.; Pourhossein, M.; Waterhouse, A.; Hudson, T.; Goldrick, M.; Derrick, J.P.; Roberts, I.S. The K5 lyase KflA combines a viral tail spike structure with a bacterial polysaccharide lyase mechanism. *J. Biol. Chem.* **2010**, *285*, 23963–23969. [CrossRef] [PubMed]
69. Harada, K.; Yamashita, E.; Nakagawa, A.; Miyafusa, T.; Tsumoto, K.; Ueno, T.; Toyama, Y.; Takeda, S. Crystal structure of the C-terminal domain of Mu phage central spike and functions of bound calcium ion. *Biochim. Biophys. Acta* **2013**, *1834*, 284–291. [CrossRef] [PubMed]
70. Kanamaru, S.; Leiman, P.G.; Kostyuchenko, V.A.; Chipman, P.R.; Mesyanzhinov, V.V.; Arisaka, F.; Rossmann, M.G. Structure of the cell-puncturing device of bacteriophage T4. *Nature* **2002**, *415*, 553–557. [CrossRef] [PubMed]
71. Scholl, D.; Merril, C. The genome of bacteriophage K1F, a T7-like phage that has acquired the ability to replicate on K1 strains of *Escherichia coli*. *J. Bacteriol.* **2005**, *187*, 8499–8503. [CrossRef] [PubMed]
72. Lescar, J.; Sanchez, J.F.; Audfray, A.; Coll, J.L.; Breton, C.; Mitchell, E.P.; Imberty, A. Structural basis for recognition of breast and colon cancer epitopes Tn antigen and Forssman disaccharide by Helix pomatia lectin. *Glycobiology* **2007**, *17*, 1077–1083. [CrossRef] [PubMed]
73. Aragao, K.S.; Satre, M.; Imberty, A.; Varrot, A. Structure determination of Discoidin II from Dictyostelium discoideum and carbohydrate binding properties of the lectin domain. *Proteins* **2008**, *73*, 43–52. [CrossRef] [PubMed]
74. Muller, J.J.; Barbirz, S.; Heinle, K.; Freiberg, A.; Seckler, R.; Heinemann, U. An intersubunit active site between supercoiled parallel β helices in the trimeric tailspike endorhamnosidase of Shigella flexneri Phage Sf6. *Structure* **2008**, *16*, 766–775. [CrossRef] [PubMed]
75. Ricagno, S.; Campanacci, V.; Blangy, S.; Spinelli, S.; Tremblay, D.; Moineau, S.; Tegoni, M.; Cambillau, C. Crystal structure of the receptor-binding protein head domain from *Lactococcus lactis* phage bIL170. *J. Virol.* **2006**, *80*, 9331–9335. [CrossRef] [PubMed]
76. Kitov, P.I.; Bundle, D.R. On the nature of the multivalency effect: A thermodynamic model. *J. Am. Chem. Soc.* **2003**, *125*, 16271–16284. [CrossRef] [PubMed]
77. Barbirz, S.; Muller, J.J.; Uetrecht, C.; Clark, A.J.; Heinemann, U.; Seckler, R. Crystal structure of *Escherichia coli* phage HK620 tailspike: Podoviral tailspike endoglycosidase modules are evolutionarily related. *Mol. Microbiol.* **2008**, *69*, 303–316. [CrossRef] [PubMed]

4

Clostridium perfringens Virulent Bacteriophage CPS2 and its Thermostable Endolysin LysCPS2

Eunsu Ha [1,2,†], Bokyung Son [1,2,†] and Sangryeol Ryu [1,2,*]

1 Department of Food and Animal Biotechnology, Seoul National University, Seoul 08826, Korea; esha0521@gmail.com (E.H.); sonbk0722@gmail.com (B.S.)
2 Department of Agricultural Biotechnology, and Research Institute of Agriculture and Life Sciences, Seoul National University, Seoul 08826, Korea
* Correspondence: sangryu@snu.ac.kr
† These authors contributed equally to this work.

Abstract: *Clostridium perfringens* is one of the most common causes of food-borne illness. The increasing prevalence of multidrug-resistant bacteria requires the development of alternatives to typical antimicrobial treatments. Here, we isolated and characterized a *C. perfringens*-specific virulent bacteriophage CPS2 from chicken feces. The CPS2 phage contains a 17,961 bp double-stranded DNA genome with 25 putative ORFs, and belongs to the *Picovirinae,* subfamily of *Podoviridae*. Bioinformatic analysis of the CPS2 genome revealed a putative endolysin, LysCPS2, which is homologous to the endolysin of *Clostridium* phage phiZP2 and phiCP7R. The enzyme showed strong lytic activity against *C. perfringens* with optimum conditions at pH 7.5–10, 25–65 °C, and over a broad range of NaCl concentrations. Interestingly, LysCPS2 was found to be highly thermostable, with up to 30% of its lytic activity remaining after 10 min of incubation at 95 °C. The cell wall binding domain in the C-terminal region of LysCPS2 showed a binding spectrum specific to *C. perfringens* strains. This is the first report to characterize highly thermostable endolysin isolated from virulent *C. perfringens* bacteriophage. The enzyme can be used as an alternative biocontrol and detection agent against *C. perfringens*.

Keywords: bacteriophage; endolysin; *Clostridium perfringens*

---

1. Introduction

Clostridium perfringens is a Gram-positive anaerobic bacterium that has the ability to form spores [1]. *C. perfringens* is responsible for a wide range of diseases: including gas gangrene (clostridial myonecrosis), necrotic enteritis, and non-foodborne gastrointestinal infections [2]. Approximately 5% of all *C. perfringens* strains produce *C. perfringens* enterotoxin (CPE), which causes diarrhea and abdominal cramping symptoms. Most CPE-positive stains are classified as type A, and food poisoning by *cpe*-producing *C. perfringens* type A is the second most common foodborne illness in developed countries [3]. In addition, *C. perfringens* has become a significant problem in the poultry industry because it is a causative agent of necrotic enteritis, characterized by outbreaks with high mortality and small intestinal mucosal necrosis [4]. Increased mortality, economic loss, and contamination of poultry products for human consumption are important concerns regarding *C. perfringens* in the poultry industry.

Furthermore, the increasing incidence of antibiotic resistance of bacterial pathogens and the lack of novel antibiotics have become serious worldwide problems. Bacteriophages (phages) and gene products such as endolysins have been attracting considerable attention as alternatives to antibiotics. Endolysins are phage-encoded peptidoglycan hydrolases that break down the bacterial peptidoglycan at the end of their reproduction cycles to release the viral progeny [5]. The purified endolysin protein has potent hydrolytic activity against Gram-positive bacteria when applied exogenously. In addition, endolysins have significant advantages over classic antibiotics, such as narrow host specificity, high sensitivity, and low probability for development of resistant bacteria [6].

To date, there have been few reports on *C. perfringens* phage endolysins. Most studies have attempted to isolate endolysin from the prophage because of the relative difficulties in isolating the phage from anaerobic *C. perfringens*. *N*-acetylmuramidases of *C. perfringens* phage phiSM101 and *N*-acetylmuramoyl-L-alanine amidase of φ3626 phage were isolated from lysogenic *C. perfringens* strains and characterized [7,8]. CP25L endolysin was examined for the activity to kill *C. perfringens*, and endolysin delivery by *Lactobacillus johnsonii* was reported [9]. Two endolysins from the clostridial phages ΦCP39O and ΦCP26F have been reported to have lytic activity against *C. perfringens* stains [10]. However, detailed information about these endolysins have not been reported.

Here we isolated novel virulent *C. perfringens* bacteriophage CPS2 from chicken feces, and its predicted endolysin LysCPS2 was identified in the genome of bacteriophage CPS2. The gene was cloned and expressed in *Escherichia coli*, and the purified endolysin was biochemically characterized for its potential as an antimicrobial and a detection agent.

2. Materials and Methods

2.1. Bacterial Strains, and Growth Conditions

C. perfringens ATCC 13124 was used as a host strain for isolation and propagation of the bacteriophage CPS2. *E. coli* BL21 was grown in Luria-Bertani (LB) broth (Difco, Detroit, MI, USA) at 37 °C and used as the host for expression of the recombinant LysCPS2. Bacterial strains that were used for antimicrobial spectrum determination are listed in Table 1, along with the results. All of the bacterial strains were routinely grown at 37 °C in Brain Heart Infusion (BHI) broth medium (Difco) under anaerobic conditions.

2.2. Isolation and Propagation of Bacteriophage CPS2

To isolate a bacteriophage, we applied the same method as described in the previous study [11]. Isolated phages were amplified by serial propagation and concentrated by polyethylene glycol precipitation and subsequent CsCl density gradient ultracentrifugation (78,500× *g* at 4 °C for 2 h). The concentrated phages were dialyzed using 2 L of standard dialysis buffer (10 mM NaCl, 10 mM $MgSO_4$ and 1 M Tris-HCl; pH 8.0) for 2 h. The phage stock obtained was stored in glass vials at 4 °C.

2.3. Transmission Electron Microscopy (TEM) Analysis

Purified CPS2 (1×10^9 PFU/mL) was placed on carbon-coated copper grids and negatively stained with 2% aqueous uranyl acetate (pH 4.0) for 20 s. The morphology of CPS2 was analyzed by TEM (LEO 912AB transmission electron microscope; Carl Zeiss, Wezlar, Germany). Images were scanned at the National Academy of Agricultural Science (Jeonju, South Korea).

2.4. DNA Purification and Whole Genome Sequencing of Bacteriophage CPS2

To extract genomic DNA from CPS2, host DNA was removed by treatment of the virions with DNaseI and RNaseA (1 μg/mL each) at room temperature for 30 min. The virions were then lysed by reacting with proteinase K mixture (50 μg/mL proteinase K, 20 mM ethylenediaminetetraacetic acid (EDTA), 0.5% sodium dodecyl sulfate (SDS)) at 56 °C for 1 h. After lysis, the DNA was purified by the phenol-chloroform extraction [12] and concentrated by ethanol precipitation [13]. The purified genomic DNA of CPS2 was sequenced using the GS-FLX Titanium sequencer (Roche Holding AG, Basel, Switzerland). Total 23,050 sequencing reads obtained were assembled using the GS De Novo Assembler version 2.9 (Roche Holding AG, Basel, Switzerland) with default parameters. Additional DNA sequencing of phage was performed to identify end regions of genomic DNA sequence using primers (CPS2endF, 5′-CAC CCT GGA GCA TTT ACA C-3′; CPS2endR, 5′-TCC ATA ACA GAC AAT

AAA AAT TTT AAA T-3′) in Macrogen inc. (Seoul, South Korea). The position of open reading frames (ORFs) was predicted by bioinformatics tools, including Glimmer 3.02 [14] and Rapid Annotation using Subsystem Technology (RAST) software [15]. The function of each ORF was predicted using NCBI BLASTP and InterProScan [14] databases. Based on the information, each ORF's name was annotated manually. The gene encoding endolysin was identified, and its domain structure was investigated using InterProScan databases. The complete genome sequence of CPS2 phage was deposited in GenBank under accession number MH248069.

2.5. Cloning, Expression, and Purification of LysCPS2

The endolysin gene (*lysCPS2*) was amplified from the genomic DNA of the bacteriophage CPS2 by polymerase chain reaction (PCR) using primers lysCPS2F (5′-GCG GGA TCC ATG AAA ATA ATA CAA TCA AAT ATC CAT TTT-3′) and lysCPS2R (5′-CGC AAG CTT TTA GTC TTT TTT AAT ATA TTT TGC GGA-3′). The PCR product was cloned into pET28a (Novagen, Madison, WI, USA), which has an *N*-terminal hexahistidine (His)-tag sequence. Plasmid with a correct insert was transformed into competent *E. coli* BL21 (DE3). Expression of the recombinant LysCPS2 was induced with 0.5 mM isopropyl-b-D-thiogalactopyranoside, with adjusted OD600 to 0.6–0.8, followed by incubation for an additional 15 h at 30 °C. Bacterial cells were suspended in lysis buffer (50 mM Tris-HCl, 100 mM sodium chloride, pH 7.5) and disrupted by sonication (Branson Ultrasonics, Shanghai, China). After centrifugation at 15,000× *g* for 40 min, the supernatant was collected, mixed with 500 μL of nickel-nitrilotriacetic acid (Ni-NTA) agarose (Qiagen, Hilden, Germany) and incubated at 4 °C for 1 h with gentle shaking. The flow-through was discarded and the resin was serially washed with 10 mL of 10 mM imidazole and 5 mL of 20 mM imidazole. An elution buffer (50 mM Tris-HCl, 100 mM sodium chloride, and 250 mM imidazole; pH 7.5) was used to elute the protein. The purified protein was stored at −4 °C until use, after the buffer was changed to the storage buffer (50 mM Tris-HCl, 200 mM sodium chloride, pH 7.5, 30% glycerol) using PD Miditrap G-25 (GE Healthcare, Little Chalfont, UK).

2.6. Lytic Activity Assay

The lytic activity of the endolysin against bacterial cells was assayed by monitoring the decrease in OD_{600}. All tested bacteria were cultivated to the exponential phase. Cells were harvested and resuspended with reaction buffer (50 mM Tris-HCl, 200 mM NaCl, pH 7.5) to OD_{600} to 0.8–1.0. The endolysin (50 μL) was added to the cell suspension (950 μL), followed by incubation at room temperature, unless indicated otherwise. OD_{600} values were monitored over time. The lytic activity was calculated after 40 min as follows: {ΔOD_{600} test (endolysin added)-ΔOD_{600} control (buffer only)}/initial OD_{600}. Antimicrobial spectrum was tested by plate lysis assay as previously described [16]. In brief, 10 μL of diluted endolysin in reaction buffer (32 μM; final concentration) was spotted onto a freshly prepared bacterial lawn on BHI agar plates. Spotted plates were air-dried in a laminar flow hood for 15 min and incubated overnight at 37 °C. To evaluate the effect of pH on LysCPS2 enzymatic activity, the endolysin (162 nM) was added to *C. perfringens* cells suspended with a universal pH buffer [17]. The universal buffer consists of 50 mM KCl, 10 mM KH_2PO_4, 10 mM Na-citrate, and 10 mM H_3BO_4, and was adjusted to different pH values—between 4 and 10—using 5 M NaOH or 5 M HCl. Different temperatures (25–95 °C) were applied to test the effect of temperature on LysCPS2 (162 nM) enzymatic activity. To evaluate the stability of the endolysin, the lysis assays were performed against *C. perfringens* ATCC 13124 at room temperature and pH 7.5 after the enzyme was incubated for 10 min at different temperatures. The influence of NaCl on lytic activity of LysCPS2 (162 nM) was tested with the addition of concentrations of 0, 100, 200, 300, 400, and 500 mM NaCl. The effects of metal ions on lysis activity were determined as previously reported [6]. To chelate metal ions attached to the endolysin, thereby inhibiting its catalytic function, EDTA (100 mM; final concentration) was added to the endolysin (4.86 μM) and incubated at 37 °C for 1 h. The EDTA was removed by exchanging the endolysin into the reaction buffer using a PD trap G-25 column. The EDTA-treated enzyme was added to cell suspensions with metal ions ($MgCl_2$, $CaCl_2$, $MnCl_2$, $ZnCl_2$, $CuCl_2$; 1.0 mM final concentration), and the lysis activity was assayed in the reaction buffer.

2.7. CBD Binding and Fluorescence Microscopy

The gene encoding the putative CBD (between bases 457–681 of *lysCPS2*) was amplified from the genomic DNA of CPS2 by PCR using primers lysCPS2_CBDF (5′-GCG GGA TCC GGA AAC TTA GAT TTA AAC AAA TTA AGA ACA GAT GTA AAC-3′) and lysCPS2_CBDR (5′-CGC AAG CTT TTA GTC TTT TTT AAT ATA TTT TGC GGA-3′). The resulting PCR product was cloned using the BamHI and SalI restriction sites into pET28a-mCherry (*mCherry* gene inserted between NdeI and BamHI sites into pET28a) [18]. mCherry-tagged LysCPS2_CBD protein was produced in *E. coli* BL21 and purified by affinity chromatography, as described earlier [19]. Purified protein at a final concentration of 100 μM was added to *C. perfringens* cells, and the mixture was then incubated for 10 min at 25 °C. Subsequently, cells were collected by centrifugation and washed twice with Dulbecco's phosphate-buffered saline (PBS). For fluorescence microscopy, images were captured on a DE/Axio Imager A1 microscope (Carl Zeiss) with a charge-coupled-device camera using AxioVision release 4.7 (Carl Zeiss, Oberkochen, Germany).

3. Results and Discussion

3.1. Morphology of Phage CPS2

The *C. perfringens* phage CPS2 was isolated from chicken feces using *C. perfringens* isolate as a host stain. Morphological observation of phage CPS2 revealed that it belongs to the family *Podoviridae* due to the presence of short and non-contractile tails (Figure 1A). The diameter of the icosahedral head was approximately 40 nm and the length of non-contractile tails (from baseplate of virion to tail fiber) was 15 nm.

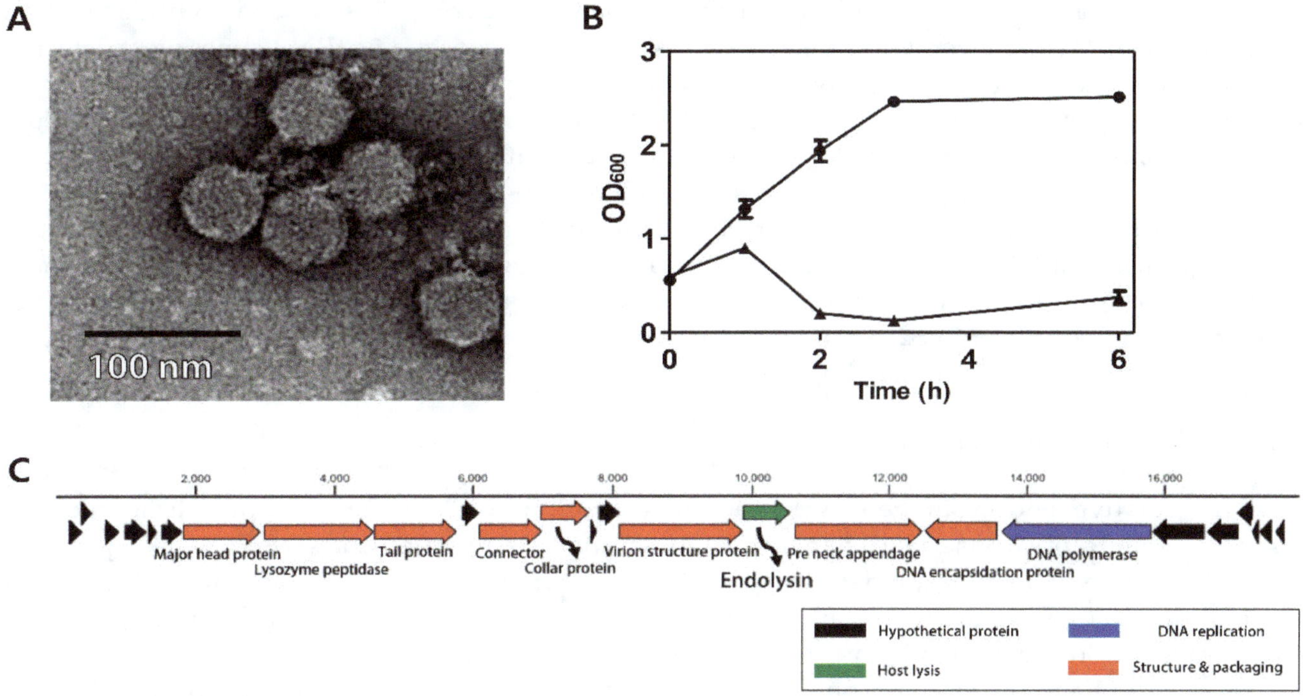

Figure 1. Characterization of *C. perfringens* virulent phage CPS2. (**A**) TEM image of CPS2; (**B**) bacterial challenge assay of CPS2 against *C. perfringens* ATCC 13124. The closed circle indicates non-phage-treated *C. perfringens* ATCC 13124, and the closed triangle indicates phage-treated *C. perfringens* ATCC 13124; (**C**) genome map of CPS2. Red: structure and packaging, blue: DNA replication, green: host lysis, black: hypothetical protein. Twenty-five putative ORFs were predicted in the CPS2 genome.

3.2. Antibacterial Properties of Phage CPS2

The bacterial challenge assay was performed to evaluate the growth inhibition ability of the CPS2 phage. When phage CPS2 was added to exponentially growing *C. perfringens* at a multiplicity of infection (MOI) of 1.0, complete inhibition of host cells was observed 1 h after phage infection, and growth inhibition was maintained for up to 6 h (Figure 1B). A host range test showed that CPS2 can inhibit the growth of only *C. perfringens* ATCC 13124, and *C. perfringens* human stool isolate 3 out of the 10 tested *C. perfringens* strains. CPS2 phage did not show lytic activity against other clostridial bacteria or other Gram-positive bacteria, indicating very narrow host specificity (Table 1).

Table 1. Antimicrobial spectra of the CPS2 and LysCPS2, and binding spectrum of LysCPS2_CBD.

Bacterial Strain	CPS2 Plaque Formation	Lysis Zone Formation by LysCPS2	Binding Activity of LysCPS2_CBD	Reference or Source
***Clostridium* strains**				
C. perfringens H3	Lysis from without	+	+	[20]
C. perfringens ATCC 3624	Lysis from without	+	+	ATCC [a]
C. perfringens ATCC 13124	+	+	+	ATCC
C. perfringens FORC25	−	+	+	This study
C. perfringens human stool isolate 1	−	+	+	This study
C. perfringens human stool isolate 2	−	+	+	This study
C. perfringens human stool isolate 3	+	+	+	This study
C. perfringens human stool isolate 4	Lysis from without	+	+	This study
C. histolyticum ATCC 19401	−	−	−	ATCC
C. indolis ATCC 25771	−	−	−	ATCC
Other Gram-positive bacteria				
Bacillus cereus ATCC 10987	−	−	−	ATCC
Bacillus subilis ATCC 23857	−	−	−	ATCC
Listeria monocytogenes EGD-e	−	−	−	[21]
Staphylococcus aureus RN4220	−	−	−	[16]

[a] ATCC, American Type Culture Collection; +, positive activity; −, negative activity.

3.3. Genomic Analysis of CPS2

The complete genome of *C. perfringens* bacteriophage CPS2 comprises 17,961-bp with an overall G + C content of 33.30%. Twenty-five putative ORFs were identified, and no tRNA genes were detected. The functional ORFs were clustered into 4 functional groups of DNA replication, host lysis, structure plus packaging, and additional function (Figure 1C). The sequence analysis showed the linear structure of CPS2 double-stranded DNA, as well as the inverted terminal repeats (ITR) of 366 nucleotide pairs on the phage genomic termini. Furthermore, the CPS2 genome encoded a putative Type-B DNA polymerase that is utilized in a protein-primed replication mechanism, implying the presence of terminal proteins for DNA replication [22]. These findings suggest that CPS2 belong to the *Podoviridae* sub-family *Picovirinae* [23]. Importantly, the toxin production- and bacterial virulence-associated genes were not identified in the CPS2 genome. Genes associated with lysogenization were not detected, suggesting that CPS2 is a virulent phage.

3.4. Identification and Expression of the LysCPS2 Endolysin

The putative endolysin gene was identified from the CPS2 genome and designated as LysCPS2. Pfam and Conserved Domain Database analysis revealed that LysCPS2 is a putative *N*-acetylmuramoyl-L-alanine amidase that consists of N-terminal amidase_2 domain (PF01520) as an enzymatically active domain (EAD) and a C-terminal SH3_3 (PF08239) as a cell wall binding domain (CBD). Amino acid sequence alignment revealed that LysCPS2 was highly homologous to endolysins of *Clostridium* phage phiZP2 and phiCP7R at the amino acid sequence level (Figure 2B) [22] The N-terminal region from LysCPS2 showed 51% sequence identity with previously reported *Clostridium* phage vB_CpeS-CP51 endolysin. Although BLAST analysis revealed many proteins homologous to LysCPS2, most of them were autolysins. The LysCPS2 was cloned and expressed in *E. coli* with an N-terminal His-tag. Sodium dodecyl sulfate polyacrylamide gel electrophoresis (SDS-PAGE) showed a single band of the purified endolysin (Figure 2C). Turbidity reduction assay revealed that

32 nM of the purified LysCPS2 showed saturated lysis activity against *C. perfringens* ATCC 13124 cells (Figure 2D). These results indicate that LysCPS2 is highly active compared to endolysin from *C. difficile* phage ΦCD27, in which a similar amount of endolysin had barely exhibited lytic activity against *C. difficile* [24].

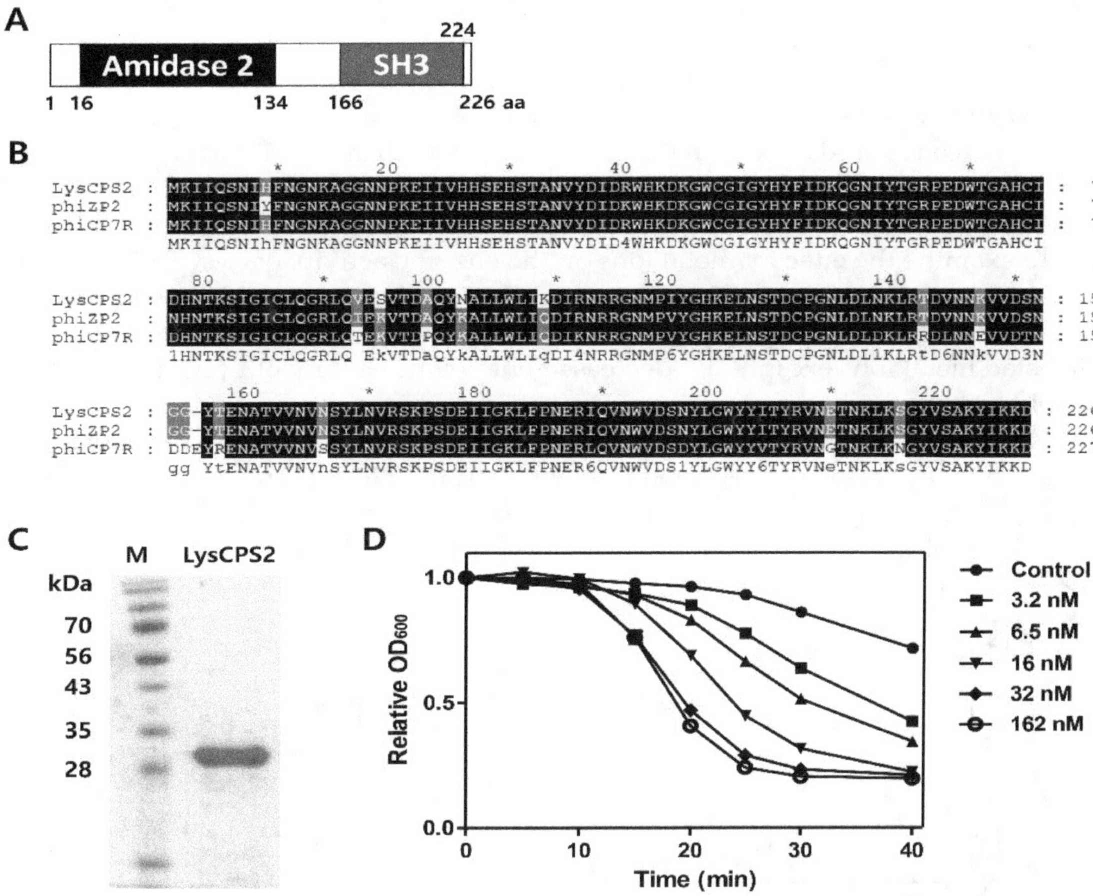

Figure 2. Modular structure and lytic activities of the LysCPS2 endolysin from CPS2. (**A**) Schematic representation of LysCPS2. The conserved amidase2 domain is shown; (**B**) Sequence alignment of various Clostridial phage endolysins; CPS2 phage endolysin, phiZP2 phage endolysin, phiCP7R phage endolysin; (**C**) SDS-PAGE analysis of purified LysCPS2. M: standard molecular weight marker, LysCPS2: purified LysCPS2 fraction; (**D**) Lysis of *C. perfringens* ATCC 13124 cells treated externally with various concentrations of recombinant LysCPS2. Optical density was measured periodically after LysCPS2 treatment.

3.5. Antimicrobial Spectrum of LysCPS2

Antimicrobial activity against *Clostridium* and other Gram-positive bacterial strains was examined (Table 1). All of the tested *C. perfringens* strains were susceptible to LysCPS2, indicating a much broader antimicrobial spectrum of LysCPS2, considering the narrow host range of the CPS2 phage. However, other *Clostridium* species such as *C. histolyticum* and *C. indolis* were not lysed by treatment of LysCPS2, meaning that its enzymatic activity requires species-specific moieties of the cell wall component. This enzyme did not show lytic activity against other Gram-positive bacteria, such as *Bacillus cereus, Bacillus subtilis, Listeria monocytogenes,* and *Staphylococcus aureus. B. cereus, B. subtilis,* and *L. monocytogenes* have an A1γ type peptidoglycan directly cross-linked with *meso*-diaminopimelic acid (m-DAP), and *S. aureus* has an A3α type cross-linked by penta-glycine bridges in its peptidoglycan. *C. perfringens* has a different type of peptidoglycan structure containing L,L-DAP and glycine instead of m-DAP in the A1γ type peptidoglycan [25].

3.6. pH, Temperature, NaCl, and Metal Effects on LysCPS2 Activity

To determine optimum conditions for the endolysin activity, the biochemical properties of LysCPS2 were evaluated. Analysis of lytic activity at different pH levels showed that LysCPS2 had the highest lytic activity at pH 7.5–10 (Figure 3A). More than 60% residual lytic activity was observed in the temperature range from 25 °C to 65 °C (Figure 3B). The effect of ionic strength on the lytic activity of LysCPS2 was assessed at various concentrations of NaCl, ranging from 0 to 500 mM (Figure 3C). LysCPS2 retained its lytic activity even at 500 mM NaCl, suggesting that LysCPS2 is highly stable in a broad range of NaCl concentrations in contrast to the *C. perfringens* phage phiSM101 endolysin that showed reduced activity in the presence of more than 200 mM of NaCl [7]. Even though phiSM101 endolysin showed 36% amino acid sequence identity with LysCPS2, it was the only *C. perfringens* endolysin that could be compared with the influence of NaCl on the lytic activity of LysCPS2. To examine the effect of metal ions on the enzyme activity of LysCPS2, metal ions were initially removed from the endolysin with 100 mM EDTA. Addition of EDTA showed little effect on the LysCPS2 activity, indicating that LysCPS2 activity is independent of the presence of divalent cations. All tested metal ions except Ca^{2+} decreased enzymatic activity of LysCPS2, and the enzyme was completely inactivated in the presence of Zn^{2+} (Figure 3D). Although many divalent ions have been reported to enhance the enzymatic activity of endolysin, the inhibition effect by certain metal ions on endolysin activity has also been reported in several studies [26,27]. The peptidoglycan hydrolase of *Burkholderia pseudomallei* phage ST79 showed reduced lytic activity in the presence of Zn^{2+}, Mg^{2+}, or Mn^{2+} [28]. The inhibition effect of Zn^{2+} is also found in the T5 phage endolysin [29].

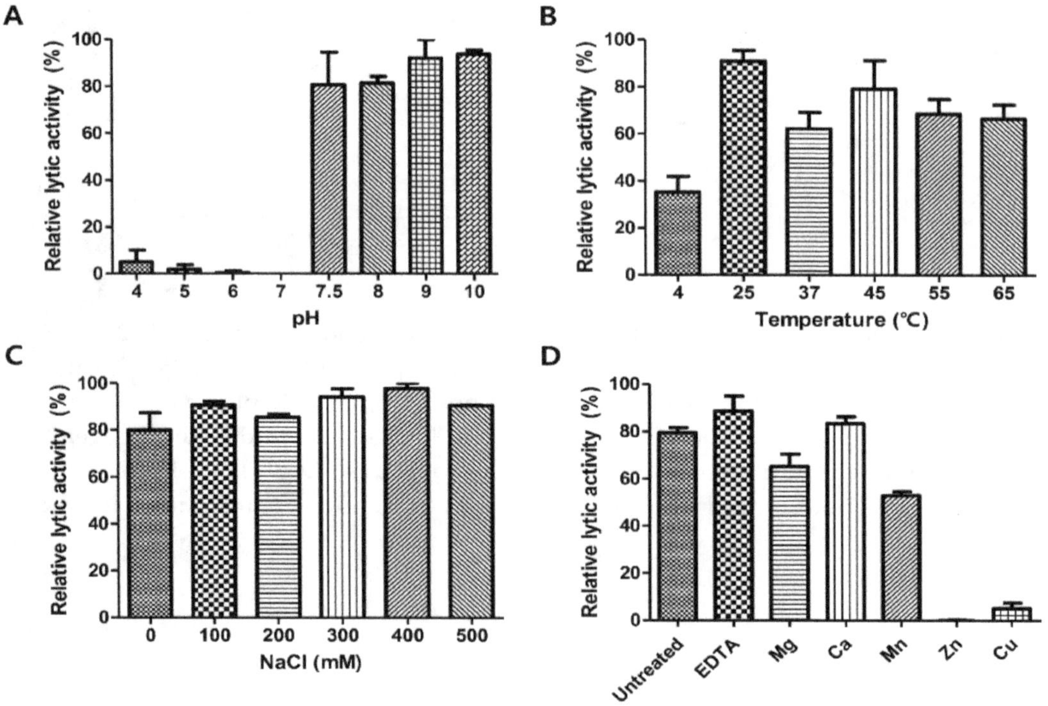

Figure 3. Effects of pH, temperature, NaCl and metal ions on the lytic activity of LysCPS2. Effects of (**A**) pH, (**B**) temperatures, (**C**) NaCl, and (**D**) metal ions on the lytic activity of LysCPS2 against *C. perfringens* ATCC 13124 cells. Each column represents the mean of triplicate experiments, and error bars indicate the standard deviation.

3.7. Determination of Thermal Stability

Generally, endolysins are not heat stable, and most lose their hydrolase activity above 50–60 °C [30]. However, few endolysins such as listerial phage endolysins HPL511, HPL118, and HPLP35 have been reported to be thermostable, showing residual activity after 10 min incubation at 90 °C [31].

Stenotrophomonas maltophilia phage endolysin P28 retained 55% of its activity after treatment at 70 °C for 30 min [32], and a thermostable chimeric endolysin, PlyGVE2CpCWB, targeting *C. perfringens* has exhibited 57% residual activity at 55 °C [33]. LysCPS2 showed remarkable heat stability. The lytic activity after pre-incubation of the enzyme at temperatures between 4 °C and 65 °C was not reduced compared with the heat-untreated control. LysCPS2 retained more than 30% of its activity after heating at 95 °C for 10 min and 60% after heating at 75 °C for 10 min (Figure 4). This is the first report to characterize highly thermostable endolysin isolated from virulent *C. perfringens* bacteriophage. This enzyme will provide powerful tools for many applications in molecular biology, biotechnology, and medicine [34].

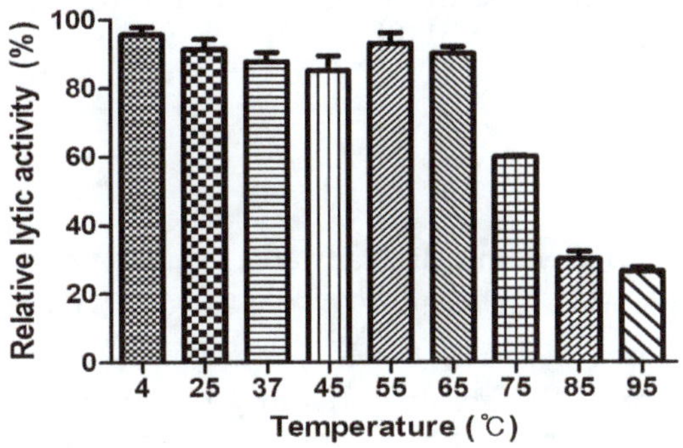

Figure 4. LysCPS2 thermal stability. LysCPS2 was incubated at different temperatures for 10 min, then cooled on ice before enzyme assay at 25 °C. Relative lytic activities are calculated using the activity of enzyme stored at 4 °C, which showed the maximal activity.

3.8. Binding Activity of LysCPS2_CBD

The specific bacterial binding activity of the putative LysCPS2_CBD was identified with an mCherry (red fluorescent protein) fusion protein (Figure 5). As shown in Table 1, all tested *C. perfringens* cells were decorated by mCherry_LysCPS2_CBD, whereas other *Clostridium* species and other Gram-positive bacteria could not be labeled by the fusion protein (Table 1). These results demonstrated that LysCPS2_CBD specifically binds to *C. perfringens* cells, which is consistent with the antimicrobial spectrum of the LysCPS2. The specificity and affinity of LysCPS2_CBD could be useful for developing an efficient bio-probe against *C. perfringens*.

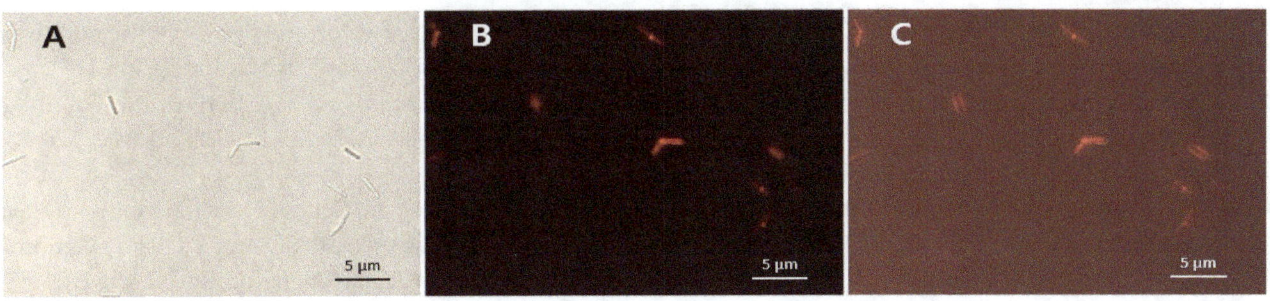

Figure 5. mCherry_LysCPS2_CBD binding activity. LysCPS2_CBD binds to *C. perfringens* human isolate cells. Bright field (**A**), fluorescence (**B**) and merged (**C**) images are shown.

4. Conclusions

In this study, we isolated a *C. perfringens* virulent phage CPS2 and identified a gene encoding LysCPS2 endolysin from CPS2 genome. LysCPS2 is a highly thermostable endolysin showing 30% of its lytic activity against *C. perfringens* even after 10 min of incubation at 95 °C. Optimum conditions for LysCPS2 were at pH 7.5–10, 25–65 °C, and over a broad range of NaCl concentrations, suggesting that LysCPS2 has high potential as an effective antibacterial agent to control *C. perfringens*. The cell wall binding domain in the C-terminal region of LysCPS2 showed a binding spectrum specific to *C. perfringens* strains. Taken together, LysCPS2 would be useful for the development of a powerful biocontrol and detection agent against *C. perfringens.*

Author Contributions: E.H. and B.S. carried out all the experiments. E.S., B.S. and S.R. conceived of the study, participated in its design and wrote the manuscript. All authors read and approved the final manuscript.

Acknowledgments: This work was supported by Basic Science Research Programs (NRF-2017R1A2A1A17069378) through the National Research Foundation of Korea (NRF) funded by the Ministry of Science, ICT and Future Planning and Korea Institute of Planning and Evaluation for Technology in Food, Agriculture, Forestry (IPET) through Agriculture, Food and Rural Affairs Research Center Support Program, funded by Ministry of Agriculture, Food and Rural Affairs (MAFRA) (710012-03-1-SB110).

References

1. Canard, B.; Saint-Joanis, B.; Cole, S. Genomic diversity and organization of virulence genes in the pathogenic anaerobe *Clostridium perfringens. Mol. Microbiol.* **1992**, *6*, 1421–1429. [CrossRef] [PubMed]
2. Myers, G.S.; Rasko, D.A.; Cheung, J.K.; Ravel, J.; Seshadri, R.; DeBoy, R.T.; Ren, Q.; Varga, J.; Awad, M.M.; Brinkac, L.M. Skewed genomic variability in strains of the toxigenic bacterial pathogen, *Clostridium perfringens. Genome Res.* **2006**, *16*, 1031–1040. [CrossRef] [PubMed]
3. Freedman, J.C.; Shrestha, A.; McClane, B.A. *Clostridium perfringens* enterotoxin: Action, genetics, and translational applications. *Toxins* **2016**, *8*, 73. [CrossRef] [PubMed]
4. Timbermont, L.; Haesebrouck, F.; Ducatelle, R.; van Immerseel, F. Necrotic enteritis in broilers: An updated review on the pathogenesis. *Avian Pathol.* **2011**, *40*, 341–347. [CrossRef] [PubMed]
5. Schmelcher, M.; Donovan, D.M.; Loessner, M.J. Bacteriophage endolysins as novel antimicrobials. *Future Microbiol.* **2012**, *7*, 1147–1171. [CrossRef] [PubMed]
6. Borysowski, J.; Weber-Dąbrowska, B.; Górski, A. Bacteriophage endolysins as a novel class of antibacterial agents. *Exp. Biol. Med.* **2006**, *231*, 366–377. [CrossRef]
7. Nariya, H.; Miyata, S.; Tamai, E.; Sekiya, H.; Maki, J.; Okabe, A. Identification and characterization of a putative endolysin encoded by episomal phage phiSM101 of *Clostridium perfringens. Appl. Microbiol. Biotechnol.* **2011**, *90*, 1973–1979. [CrossRef] [PubMed]
8. Zimmer, M.; Vukov, N.; Scherer, S.; Loessner, M.J. The murein hydrolase of the bacteriophage φ3626 dual lysis system is active against all tested *Clostridium perfringens* strains. *Appl. Environ. Microbiol.* **2002**, *68*, 5311–5317. [CrossRef] [PubMed]
9. Gervasi, T.; Horn, N.; Wegmann, U.; Dugo, G.; Narbad, A.; Mayer, M.J. Expression and delivery of an endolysin to combat *Clostridium perfringens. Appl. Microbiol. Biotechnol.* **2014**, *98*, 2495–2505. [CrossRef] [PubMed]
10. Seal, B.S. Characterization of bacteriophages virulent for *Clostridium perfringens* and identification of phage lytic enzymes as alternatives to antibiotics for potential control of the bacterium1. *Poult. Sci.* **2013**, *92*, 526–533. [CrossRef] [PubMed]
11. Seal, B.S.; Fouts, D.E.; Simmons, M.; Garrish, J.K.; Kuntz, R.L.; Woolsey, R.; Schegg, K.M.; Kropinski, A.M.; Ackermann, H.-W.; Siragusa, G.R. *Clostridium perfringens* bacteriophages ΦCP39O and ΦCP26F: Genomic organization and proteomic analysis of the virions. *Arch. Virol.* **2011**, *156*, 25–35. [CrossRef] [PubMed]
12. Kirby, K. A new method for the isolation of ribonucleic acids from mammalian tissues. *Biochem. J.* **1956**, *64*, 405. [CrossRef] [PubMed]
13. Zeugin, J.A.; Hartley, J.L. Ethanol precipitation of DNA. *Focus* **1985**, *7*, 1–2.
14. Altschul, S.F.; Gish, W.; Miller, W.; Myers, E.W.; Lipman, D.J. Basic local alignment search tool. *J. Mol. Biol.* **1990**, *215*, 403–410. [CrossRef]
15. Aziz, R.K.; Bartels, D.; Best, A.A.; DeJongh, M.; Disz, T.; Edwards, R.A.; Formsma, K.; Gerdes, S.; Glass, E.M.; Kubal, M. The RAST Server: Rapid annotations using subsystems technology. *BMC Genomics* **2008**, *9*, 75. [CrossRef] [PubMed]
16. Chang, Y.; Kim, M.; Ryu, S. Characterization of a novel endolysin LysSA11 and its utility as a potent biocontrol agent against *Staphylococcus aureus* on food and utensils. *Food Microbiol.* **2017**, *68*, 112–120. [CrossRef] [PubMed]

17. Walmagh, M.; Boczkowska, B.; Grymonprez, B.; Briers, Y.; Drulis-Kawa, Z.; Lavigne, R. Characterization of five novel endolysins from Gram-negative infecting bacteriophages. *Appl. Microbiol. Biotechnol.* **2013**, *97*, 4369–4375. [CrossRef] [PubMed]
18. Kong, M.; Shin, J.H.; Heu, S.; Park, J.-K.; Ryu, S. Lateral flow assay-based bacterial detection using engineered cell wall binding domains of a phage endolysin. *Biosens. Bioelectron.* **2017**, *96*, 173–177. [CrossRef] [PubMed]
19. Kong, M.; Sim, J.; Kang, T.; Nguyen, H.H.; Park, H.K.; Chung, B.H.; Ryu, S. A novel and highly specific phage endolysin cell wall binding domain for detection of *Bacillus cereus*. *Eur. Biophys. J.* **2015**, *44*, 437–446. [CrossRef] [PubMed]
20. Sheng, S.; Cherniak, R. Structure of the capsular polysaccharide of *Clostridium perfringens* Hobbs 10 determined by NMR spectroscopy. *Carbohydr. Res.* **1997**, *305*, 65–72. [CrossRef]
21. Kretzer, J.W.; Lehmann, R.; Schmelcher, M.; Banz, M.; Kim, K.-P.; Korn, C.; Loessner, M.J. Use of high-affinity cell wall-binding domains of bacteriophage endolysins for immobilization and separation of bacterial cells. *Appl. Environ. Microbiol.* **2007**, *73*, 1992–2000. [CrossRef] [PubMed]
22. Volozhantsev, N.V.; Oakley, B.B.; Morales, C.A.; Verevkin, V.V.; Bannov, V.A.; Krasilnikova, V.M.; Popova, A.V.; Zhilenkov, E.L.; Garrish, J.K.; Schegg, K.M. Molecular characterization of podoviral bacteriophages virulent for *Clostridium perfringens* and their comparison with members of the *Picovirinae*. *PLoS ONE* **2012**, *7*, e38283. [CrossRef] [PubMed]
23. Nelson, D.; Schuch, R.; Zhu, S.; Tscherne, D.M.; Fischetti, V.A. Genomic sequence of C1, the first streptococcal phage. *J. Bacteriol.* **2003**, *185*, 3325–3332. [CrossRef] [PubMed]
24. Mayer, M.J.; Narbad, A.; Gasson, M.J. Molecular characterization of a *Clostridium difficile* bacteriophage and its cloned biologically active endolysin. *J. Bacteriol.* **2008**, *190*, 6734–6740. [CrossRef] [PubMed]
25. Schleifer, K.H.; Kandler, O. Peptidoglycan types of bacterial cell walls and their taxonomic implications. *Bacteriol. Rev.* **1972**, *36*, 407–477. [PubMed]
26. Son, B.; Yun, J.; Lim, J.-A.; Shin, H.; Heu, S.; Ryu, S. Characterization of LysB4, an endolysin from the *Bacillus cereus*-infecting bacteriophage B4. *BMC Microbiol.* **2012**, *12*, 33. [CrossRef] [PubMed]
27. Pritchard, D.G.; Dong, S.; Baker, J.R.; Engler, J.A. The bifunctional peptidoglycan lysin of *Streptococcus agalactiae* bacteriophage B30. *Microbiology* **2004**, *150*, 2079–2087. [CrossRef] [PubMed]
28. Khakhum, N.; Yordpratum, U.; Boonmee, A.; Tattawasart, U.; Rodrigues, J.L.; Sermswan, R.W. Cloning, expression, and characterization of a peptidoglycan hydrolase from the *Burkholderia pseudomallei* phage ST79. *AMB Express* **2016**, *6*, 77. [CrossRef] [PubMed]
29. Mikoulinskaia, G.V.; Odinokova, I.V.; Zimin, A.A.; Lysanskaya, V.Y.; Feofanov, S.A.; Stepnaya, O.A. Identification and characterization of the metal ion-dependent L-alanoyl-D-glutamate peptidase encoded by bacteriophage T5. *FEBS J.* **2009**, *276*, 7329–7342. [CrossRef] [PubMed]
30. Lavigne, R.; Briers, Y.; Hertveldt, K.; Robben, J.; Volckaert, G. Identification and characterization of a highly thermostable bacteriophage lysozyme. *Cell. Mol. Life Sci. CMLS* **2004**, *61*, 2753–2759. [CrossRef] [PubMed]
31. Schmelcher, M.; Waldherr, F.; Loessner, M.J. *Listeria* bacteriophage peptidoglycan hydrolases feature high thermoresistance and reveal increased activity after divalent metal cation substitution. *Appl. Microbiol. Biotechnol.* **2012**, *93*, 633–643. [CrossRef] [PubMed]
32. Dong, H.; Zhu, C.; Chen, J.; Ye, X.; Huang, Y.-P. Antibacterial Activity of *Stenotrophomonas maltophilia* Endolysin P28 against both Gram-positive and Gram-negative Bacteria. *Front. Microbiol.* **2015**, *6*, 1299. [CrossRef] [PubMed]
33. Swift, S.M.; Seal, B.S.; Garrish, J.K.; Oakley, B.B.; Hiett, K.; Yeh, H.-Y.; Woolsey, R.; Schegg, K.M.; Line, J.E.; Donovan, D.M. A thermophilic phage endolysin fusion to a *Clostridium perfringens*-specific cell wall binding domain creates an anti-*Clostridium* antimicrobial with improved thermostability. *Viruses* **2015**, *7*, 3019–3034. [CrossRef] [PubMed]
34. Loessner, M.J. Bacteriophage endolysins—Current state of research and applications. *Curr. Opin. Microbiol.* **2005**, *8*, 480–487. [CrossRef] [PubMed]

5

The Robust Self-Assembling Tubular Nanostructures Formed by gp053 from Phage vB_EcoM_FV3

Eugenijus Šimoliūnas [1,*], Lidija Truncaitė [1], Rasa Rutkienė [1], Simona Povilonienė [1], Karolis Goda [2], Algirdas Kaupinis [3], Mindaugas Valius [3] and Rolandas Meškys [1]

1 Department of Molecular Microbiology and Biotechnology, Institute of Biochemistry, Life Sciences Centre, Vilnius University, Saulėtekio av. 7, LT-10257 Vilnius, Lithuania; lidija.truncaite@bchi.vu.lt (L.T.); rasa.rutkiene@bchi.vu.lt (R.R.); simona.poviloniene@bchi.vu.lt (S.P.); rolandas.meskys@bchi.vu.lt (R.M.)
2 Sector of Microtechnologies, Institute of Biotechnology, Life Sciences Centre, Vilnius University, Saulėtekio av. 7, LT-10257 Vilnius, Lithuania; karolis.goda@gmail.com
3 Proteomics Centre, Institute of Biochemistry, Life Sciences Centre, Vilnius University, Saulėtekio av. 7, LT-10257 Vilnius, Lithuania; algirdas.kaupinis@gf.vu.lt (A.K.); mindaugas.valius@bchi.vu.lt (M.V.)
* Correspondence: eugenijus.simoliunas@bchi.vu.lt

Abstract: The recombinant phage tail sheath protein, gp053, from *Escherichia coli* infecting myovirus vB_EcoM_FV3 (FV3) was able to self-assemble into long, ordered and extremely stable tubular structures (polysheaths) in the absence of other viral proteins. TEM observations revealed that those protein nanotubes varied in length (~10–1000 nm). Meanwhile, the width of the polysheaths (~28 nm) corresponded to the width of the contracted tail sheath of phage FV3. The formed protein nanotubes could withstand various extreme treatments including heating up to 100 °C and high concentrations of urea. To determine the shortest variant of gp053 capable of forming protein nanotubes, a set of N- or/and C-truncated as well as poly-His-tagged variants of gp053 were constructed. The TEM analysis of these mutants showed that up to 25 and 100 amino acid residues could be removed from the N and C termini, respectively, without disturbing the process of self-assembly. In addition, two to six copies of the gp053 encoding gene were fused into one open reading frame. All the constructed oligomers of gp053 self-assembled in vitro forming structures of different regularity. By using the modification of cysteines with biotin, the polysheaths were tested for exposed thiol groups. Polysheaths formed by the wild-type gp053 or its mutants possess physicochemical properties, which are very attractive for the construction of self-assembling nanostructures with potential applications in different fields of nanosciences.

Keywords: self-assembly; nanotubular structures; tail sheath protein; bacteriophage vB_EcoM_FV3

1. Introduction

Over the past decades, the construction of various nanostructures, based on self-assembling biomolecules, is of special interest for both an understanding of the fundamental point of view as well as due to the enormous potential for application in industry, medicine and many other areas [1–3]. Self-assembling structures open up exciting opportunities for the development of various tools, including biosensors, energy storage devices, drug delivery systems and nanobiopolymeric scaffolds [4–7]. Therefore, it is not surprising that viruses, including bacteriophages, have been used for the preparation of nanoscaled materials [8–15].

Filamentous bacteriophage M13 and its relatives are the phages of choice, most often used for the construction of genetically engineered viruses, which have been adapted to build phage-based nanosensors, liquid crystals and films [16,17], micro- and nanofibers [14,18], as well as having use in tissue regeneration and to build other functional materials [19–23], or to fabricate nanorings [24].

The construction of self-assembling structures, made from the genetically engineered recombinant proteins of tailed bacteriophages, including podoviruses P22, phi29 [25,26] and myovirus T4 [4,27,28], have also been reported.

During the morphogenesis process of the tailed bacteriophages, the assembly pathways of the structural components follow a strict order and consist of many steps where the subsequent attachment of proteins is controlled by different mechanisms [29–31]. Thus, tail sheath proteins alongside tail tube proteins constitute a major part of the contractile tail of myoviruses [31,32]. The polymerisation of the tail sheath proteins, which assemble around the tail tube, starts at the tail tip complex (the baseplate) and is propagated to the end of the tube. The sheath subunits arrange into a stack of hexameric rings, rotated relative to each other, thus creating a 6-start helix [33,34]. During the process of virion assembly inside the cell, the length of the tail sheath is determined by the length of the tail tape measure protein, which is used as a scaffold for the polymerisation of both tail tube and tail sheath proteins. When the sheath reaches the length of the tube, the tail terminator protein binds to the tail tube terminator protein and the last row of the tail sheath subunits to complete the tail, which then becomes available for attachment to the head [31,33].

It has been demonstrated that, in the absence of the baseplate or the tail tube, the recombinant tail sheath proteins of bacteriophages self-assemble both in vivo and in vitro into tubular structures of variable lengths called polysheaths which have the same helical parameters as the contracted tail sheath [35–38]. The polysheath is a stable structure, which withstands treatment with various chemical and physical factors [39–41]. A number of deletion mutants of tail sheath proteins have also been constructed and analysed [29,38,42,43].

In order to expand the current knowledge on biotechnologically attractive self-assembling structural proteins from phages, a tail sheath protein (gp053) of *E. coli* infecting myovirus vB_EcoM_FV3 (FV3) [44] was studied in detail. We showed that the recombinant gp053 self-assembled into long ordered and very stable polysheaths. In addition, the gp053 mutants harbouring the deletions at the N and/or C terminus, with or without the His-tag, as well as the genetically fused homooligomers of gp053 were constructed and their tendency to form nanostructures was analysed.

2. Materials and Methods

2.1. Cloning Procedures

The PCR fragment of gene *053* (GeneID: 14011712) from phage FV3 was obtained by amplification of the phage DNA using the primers presented in Supplementary Table S1. The purified PCR products were cleaved with NdeI and BamHI/XhoI (Thermo Fisher Scientific, Vilnius, Lithuania) and then inserted into the pET-16b or pET-21a (Novagene, Madison, WI, USA) vectors, digested with the appropriate restriction endonucleases. The resulting vectors were amplified in E. coli DH10B (Invitrogen, Dublin, Ireland) cells and verified by DNA sequencing. These vectors were used for the production of a recombinant full-length gp053 and the truncated gp053 mutants with various variations in length and His-tag (Supplementary Table S2). The construction scheme of the gp053 oligomers is presented in detail in Supplementary Figure S1.

2.2. Protein Expression

Protein production was carried out in the *E. coli* strain BL21 (DE3) (Novagene, Madison, WI, USA). The 15 ml of cells that were transformed with the appropriate plasmid were grown at 37 °C to an OD_{600} of 0.5, induced with 0.1 mM IPTG and then incubated at 30 °C for 3 hours. The cells were harvested by centrifugation at 4000× *g* for 5 min at 4 °C, resuspended in 1.5 mL TE (20 mM Tris–HCl (pH 7.8)) and 1 mM EDTA, 4 °C) buffer solution and then disrupted by sonication. The cells with expressed His-tagged recombinant gp053 proteins, which were further purified by using a metal-chelating sorbent, were resuspended in 1.5 mL His-Wash (50 mM sodium phosphate buffer (pH 7.7)), 300 mM NaCl, 50 mM imidazole and 0.03% Triton X-100, 4 °C) buffer solution. The cells were sonicated on ice, in 2.0 mL tubes at 30% amplitude for 5 min of total ON time (30 s on/30 s off) by using the Bandelin SonoPuls HD 2070 homogeniser (BANDELIN, Berlin, Germany). The crude extracts were centrifuged at 4 °C for 15 min at 21,000× *g* to remove cell debris. The cell-free extracts and pellets were directly analysed by SDS-PAGE as well as by TEM.

2.3. Protein Purification

The polysheaths formed by the recombinant gp053 were precipitated from the supernatant by the addition of ammonium sulphate to a final concentration of 10%. After incubation for 10 min on ice and centrifugation at 9000× *g* for 15 min at 4 °C, the supernatant was removed, and the pellet was suspended in TE buffer (1/10 of the initial volume) and stored at 4 °C. The shorter, less ordered polysheaths were purified by the addition of ammonium sulphate to a final concentration of 15–20% and/or by repeated centrifugation at 9000× *g* for 15 min at 4 °C. Alternatively, the recombinant His-tagged protein purification using the metal-chelating sorbent was performed by using the His-Spin Protein Miniprep kit (Zymo Research, Irvine, CA, USA) according to the manufacturer's recommendations. The concentration of the recombinant protein was determined by using a method described by Lowry et al. [45].

2.4. Limited Proteolysis of Recombinant Proteins with Trypsin

Trypsin (2 mg/mL) was added into the protein solution (~2 mg/mL) to form a final protease:protein ratio of 1:100 (*w*/*w*), and the solution was then incubated at 22 °C. Aliquots (10 μL) were withdrawn from the reaction mixture at different time intervals and mixed with 10 μL of SDS sample buffer (2× concentrated). The reaction was stopped by heating the solution in boiling water for 5 min. The samples were analysed by electrophoresis in SDS-polyacrylamide gel.

2.5. In-Gel Protein Digestion for Mass Spectrometry Analysis, Liquid Chromatography and Mass Spectrometry

In-gel trypsin digestion was performed according to the protocol described by Hellman et al. [46]. LC–MS data were collected as described previously [47]. Briefly, the liquid chromatographic analysis was performed in a Waters Acquity ultra performance LC system (Waters Corporation, Wilmslow, UK). The peptide separation was performed on an Acquity UPLC HSS T3 250 mm analytical column. The MS data were acquired using the Synapt G2 mass spectometer and Masslynx 4.1 software (Waters Corporation) in positive ion mode using data independent (DIA) acquisition (MSE).

2.6. Labelling of gp053 Polysheaths with Neutravidin-Conjugated Gold Nanoparticles

The labelling of polysheaths with gold nanoparticles was carried out according to the published procedure [48] with some modifications. The suspension of the purified recombinant gp053 was treated with 1 mM of tris(2-carboxyethyl)phosphine (TCEP) (AppliChem, Darmstadt, Germany) to reduce possible disulphide bonds. The reduced protein was dialysed against PBS (pH 7.5) for 1 h at 22 °C. After 1 hour of biotinylation at 22 °C, the mixture was dialysed against PBS (pH 7.5) buffer. The efficiency of biotinylation was evaluated and the biotinylated polysheaths were incubated with neutravidin-conjugated 10 nm gold nanoparticles (Nanopartz, Loveland, CO, USA) as described previously [48]. The mixtures of polysheaths and nanoparticles were incubated for 1 to 5 days at 4 °C. The incubation of non-biotinylated polysheaths with neutravidin-conjugated 10 nm gold nanoparticles was carried out under the same conditions.

2.7. Transmission Electron Microscopy (TEM)

The images of gp053 were obtained by transmission electron microscopy of the negatively-stained samples. Approximately 10 μL of the diluted sample solution was directly applied on the carbon-coated nickel grid (Agar Scientific, Essex, UK), and the excess liquid was drained with filter paper before staining with two successive drops of 2% uranyl acetate (pH 4.5). The prepared sample was dried and examined with a Morgagni 268(D) transmission electron microscope (FEI, Hillsboro, OR, USA).

2.8. Bioinformatics and Molecular Modeling

The bioinformatics analysis of gp053 was performed using the Fasta-Nucleotide, Fasta-Protein, BLASTP, Transeq [49] and Clustal Omega [50]. The phylogenetic analysis was conducted using MEGA version 5 [51]. The Protein Information Resource (PIR) server was used for calculating the predicted molecular mass of the recombinant protein [52]. The search for the gp053 fold was conducted using the HHpred server [53–55]. The I-TASSER server [56,57] was used to model the whole gp053. UCSF Chimera [58,59] was used for the visualisation and analysis of the predicted molecular structure of gp053.

3. Results

3.1. Bioinformatics Analysis

Based on the results of the bioinformatics analysis, the tail sheath protein encoded by gene *053* (GeneID: 14011712) of the enterobacteria phage FV3 [44] had the closest identity (96–99%) to the tail sheath proteins from the eleven *Escherichia* infecting bacteriophages, which belonged to the genus *V5virus* within the subfamily *Vequintavirinae* (Supplementary Table S3).

The HHpred analysis of the amino acid sequence of gp053 revealed that it corresponded to the fold of the two tail sheath proteins of bacteriophages. Hence, the residues 8 to 442 of gp053 were predicted to adopt the fold of the tail sheath protein from *Staphylococcus* phage phi812 (PDB ref 5LI4) with a probability of 97.05 (*E*-value, 0.047), whereas C-terminal fragment of gp053 (residues 258 to 442) was predicted to adopt the fold of the C-terminal fragment (residues 470 to 643) of the tail sheath protein, gp18, of the phage T4 (PDB ref 3J2M) with a probability of 97.2 (*E*-value, 3.4×10^{-4}). In addition, the HHpred analysis revealed that the predicted fold of gp053 was similar to the contractile phage-like structures from bacteria. The residues 1 to 441 of gp053 were predicted to adopt the fold of the R-type pyocins from *Clostridium difficile* (PDB ref 6GKW) and *Pseudomonas aeruginosa* (PDB ref 3J9Q) with a probability of 99.51 (*E*-value, 1.2×10^{-14}) and 98.6 (*E*-value, 9.9×10^{-7}), respectively. The residues 18 to 442 were aligned with the tail sheath protein encoded by gene *lin1278* from the prophage infecting *Listeria innocua* (PDB ref 3LML) with a probability of 98.5 (*E*-value, 1.3×10^{-5}). Furthermore, the residues 16 to 442 were aligned to the tail sheath protein encoded by gene *dsy3957* from the prophage of *Desulfitobacterium hafniense* (PDB ref 3HXL) with a probability of 98.5 (*E*-value, 6.3×10^{-6}).

The best 3D model of the full-length gp053 predicted using the I-TASSER server (Figure 1) showed a C-score of –1.77 and a TM-score of 0.50 ± 0.15. The top three proteins from the PDB that had the closest structural similarity to the predicted model were 3LML, 3HXL and 6GKW with a TM-score of 0.801, 0.659 and 0.602, respectively.

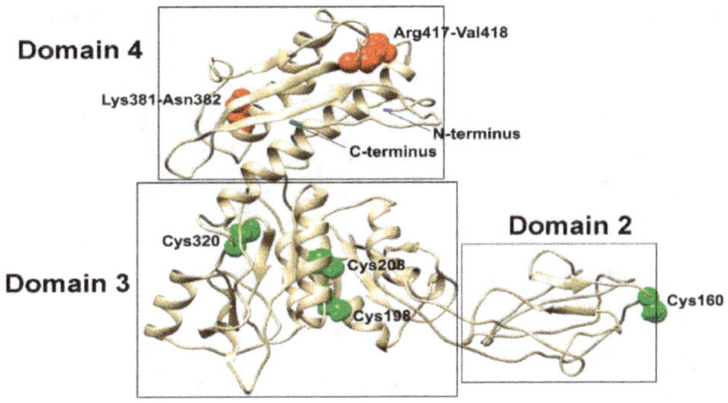

Figure 1. A model of gp053 generated by I-TASSER. The locations of the N and C termini are shown in blue; Cys160, Cys196, Cys208 and Cys320 are shown in green; Lys381-Asn382 and Arg417-Val418 are shown in red. The domain numbering was done according to Kurochkina et al. [38]. The model was visualised using UCSF Chimera.

3.2. Production and Analysis of the Recombinant gp053

Initially, for gene *053* expression in *E. coli* cells, three plasmids encoding wild-type, N- and C-His-tagged gp053 were designed (Supplementary Table S2). The soluble recombinant gp053 with a molecular mass of ~50 kDa (predicted mass—50.245 kDa) was produced after induction with IPTG and incubation at 30 °C for three hours. The TEM analysis of the cell-free extracts revealed that the overexpressed gp053 self-assembled into regular tubular structures—polysheaths (Figure 2C). The width of the gp053 polysheaths (27.13 ± 2.69 nm) corresponded to the width of the contracted tail (26.79 ± 1.78 nm) of the phage FV3 [44]. The length of the polysheaths varied from separate rings with an internal hole of 11.51 ± 0.78 nm and a diameter of 27.83 ± 2.68 nm to nanotubes up to 1000 nm long. The attempts to purify long ordered His-tagged gp053 polysheaths effectively by using the metal-chelating sorbent were unsuccessful (Figure 2). The SDS-PAGE analysis revealed that the vast majority of the recombinant proteins did not adsorb onto the sorbent. Moreover, according to the results of the TEM analysis, the affinity-purified recombinant gp053 formed noticeably shorter tubular structures (Figure 2D,E).

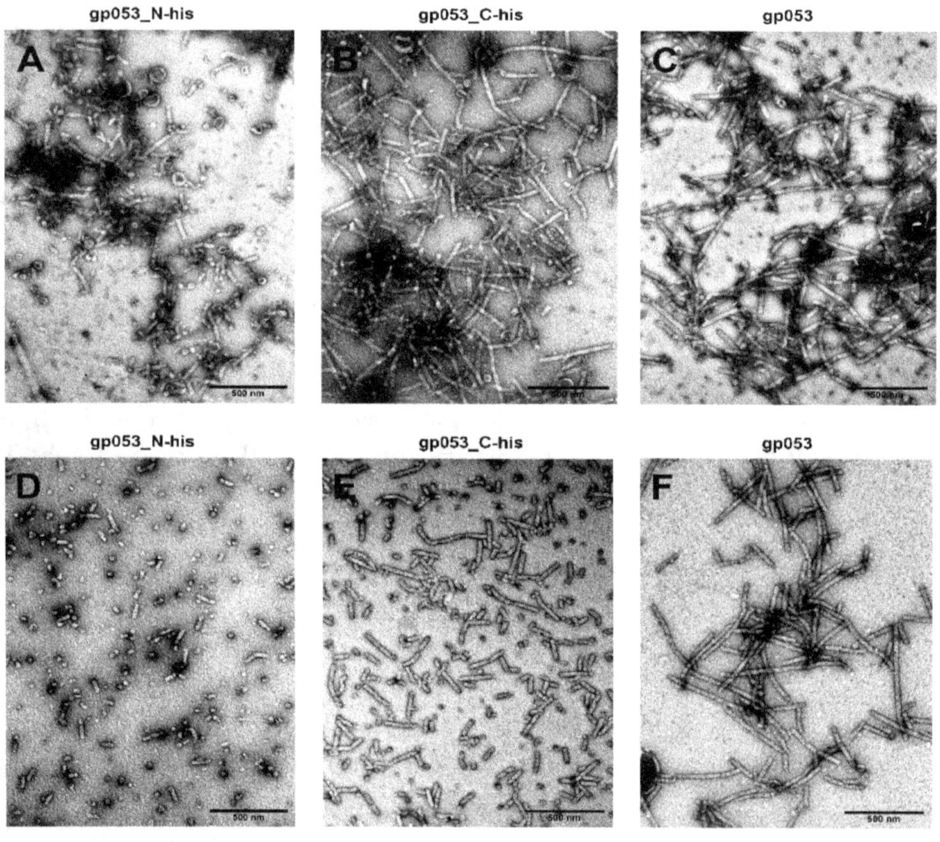

Figure 2. TEM analysis of polysheaths formed by recombinant gp053. The electron micrographs represent the structures before (above) and after (below) purification from the crude cell lysates. The polysheaths formed by recombinant gp053_N-his (**A**,**D**) and gp053_C-his (**B**,**E**) were purified using the His-Spin Protein Miniprep kit (Zymo Research) according to the manufacturer's recommendations. The polysheaths, formed by recombinant gp053 (**C**,**F**) were purified by precipitation using ammonium sulphate (10% final concentration).

Alternatively, we have demonstrated that the purification of the polysheaths could be performed quickly and effectively by using ammonium sulphate precipitation (Figures 2F and 3). The yield of the bacteria-derived and purified structures was 3.75 ± 0.35 mg/g of wet cells. The TEM analysis of the purified protein showed that the morphology of the polysheaths (Figure 3B) corresponded to the morphology of the tubular structures observed in the cell-free extracts.

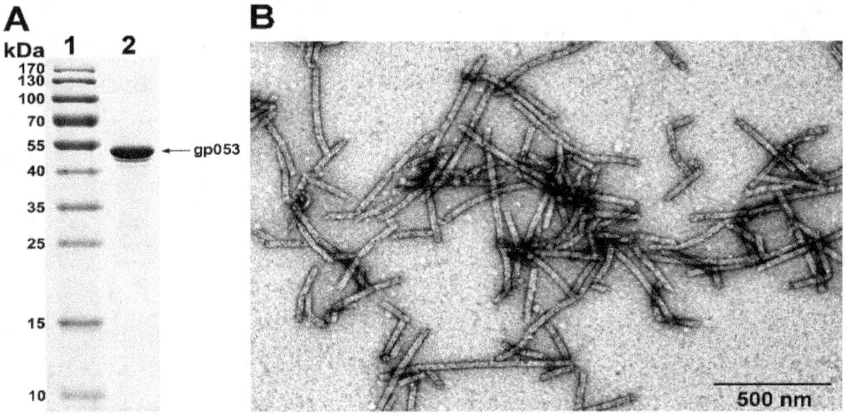

Figure 3. SDS-PAGE (**A**) and TEM (**B**) analysis of the purified wild-type recombinant gp053 produced in *E. coli* BL21 (DE3) cells. Protein synthesis was induced by 0.1 mM IPTG and the cells were incubated at 30 °C for 3 h. The recombinant gp053 was precipitated from the supernatant by the addition of ammonium sulphate to a final concentration of 10%. Lanes: 1—the molecular mass marker, Page Ruler^{TM} prestained Protein Ladder Plus (Thermo Fisher Scientific, Vilnius, Lithuania); 2—the purified wild-type recombinant gp053.

The purified polysheaths exhibited extraordinary stability. The structures withstood overnight incubation in 8 M urea. Moreover, even boiling them for 30 min or prolonged storage (more than one year) in TE buffer at 4 °C did not affect the morphology of the gp053 polysheaths (Figure 4C,E,F). Also, the polysheaths remained as ordered tubular structures after incubation with trypsin (Figure 4D).

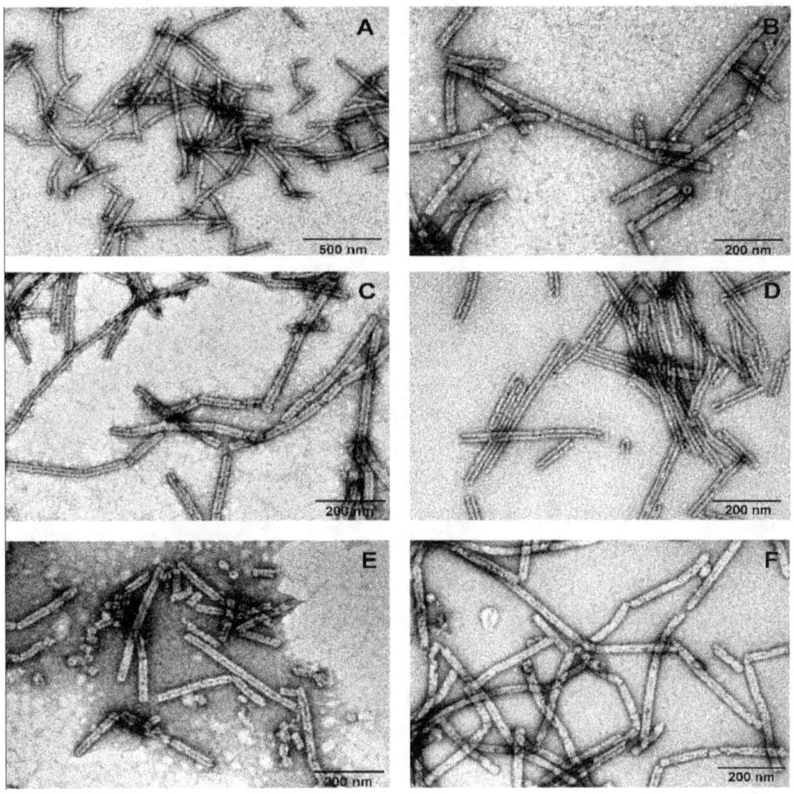

Figure 4. Electron micrographs of the polysheaths formed by wild-type recombinant gp053 from phage FV3. The samples of the purified polysheaths were analysed immediately after purification (**A**,**B**), following storage in TE buffer at 4 °C for 12 months (**C**), after treatment with trypsin (0.02 mg/mL) at 22 °C for 16 h (**D**), after incubation in the presence of 8 M urea at 22 °C for 16 h (**E**), and after incubation in boiling water for 30 min (**F**).

The detailed analysis of the trypsin-treated protein tubes by SDS-PAGE showed that fragments with molecular masses of approximately 43, 39 and 37 kDa appeared in addition to the full-length gp053 (Figure 5). Based on these results, it was concluded that only several trypsin digestion sites out of the 22 existing in the gp053 protein were available on the surface of the polysheaths. To identify those sites, the protein bands were excised from the gel before and after proteolysis by trypsin, purified according to the protocol described by Hellman et al. [46], and liquid chromatography coupled with mass spectrometry (LC–MS/MS) was performed. The protein sequence coverage map was obtained from the ProteinLynx Global Server (PLGS) processed MS^E data (three technical replicates) (Supplementary Figure S2). The results of the qualitative and quantitative MS analysis of the peptides showed that the vast majority of the proteins in the suspension of the purified polysheaths consisted of the potential full-length gp053 (Figure 5B, lane 1, arrow 1), whereas a minor protein band (Figure 5B, lane 1, arrow 2) contained gp053 with a potential 86 N-terminal amino acid truncation. This truncated protein was expected to occur due to the proteolytic activity of some of the proteases from the *E. coli* cells, since the protease inhibitors were not added during the purification of the polysheaths. After treatment with trypsin, the most intensive protein bands were identified as potentially full-length gp053 (Figure 5B, lane 2, arrow 3), gp053 with the C-terminal truncation of 41 amino acids (Figure 5B, lane 2, arrow 4) and gp053 with the C-terminal truncation of 77 amino acids (Figure 5B, lane 2, arrow 5). The minor protein band consisted of the gp053 with potential truncations of 86 N-terminal and 77 C-terminal amino acids (Figure 5B, lane 2, arrow 6).

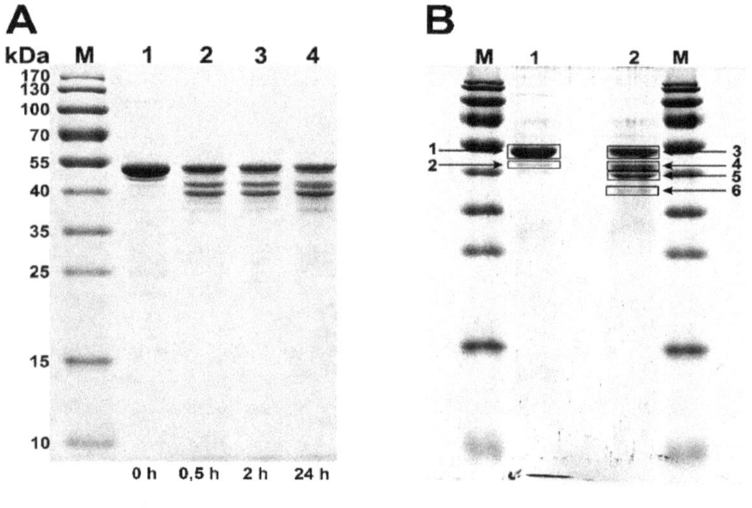

Figure 5. SDS-PAGE analysis of the trypsinisation products of the purified gp053 polysheaths. (**A**) Lanes: M—molecular mass marker, Page Ruler^{TM} Prestained Protein Ladder (Thermofisher), 1—recombinant gp053, 2–4 recombinant gp053 incubated with trypsin (the digestion time is indicated under the lanes). (**B**) Lanes: M—molecular mass marker, Page Ruler^{TM} Prestained Protein Ladder (Thermofisher); 1—recombinant gp053 incubated for 24 hours in TE buffer; 2—recombinant gp053 incubated for 24 hours in TE buffer with trypsin. The arrows indicate the full-length gp053 (1,3) and its fragments (2,4–6), excised from the gel for LC–MS/MS analysis.

According to this analysis, trypsin digestion of gp053 proceeded in a specific manner and resulted in the potential cleavage of the peptide bond between Lys381–Asn382 and Arg417–Val418 (Figure 1).

3.3. Construction of gp053 Mutants

In order to obtain truncated recombinant proteins still able to form regular, stable tubular structures and to understand the polymerisation properties of gp053, we constructed a set of the gp053 mutants truncated at the N or/and C terminus. Those recombinant proteins corresponded to the full-length gp053 from 98.0% (gp053_NΔ9) to 43.7% (gp053_CΔ200) (Supplementary Table S2, Figure 6).

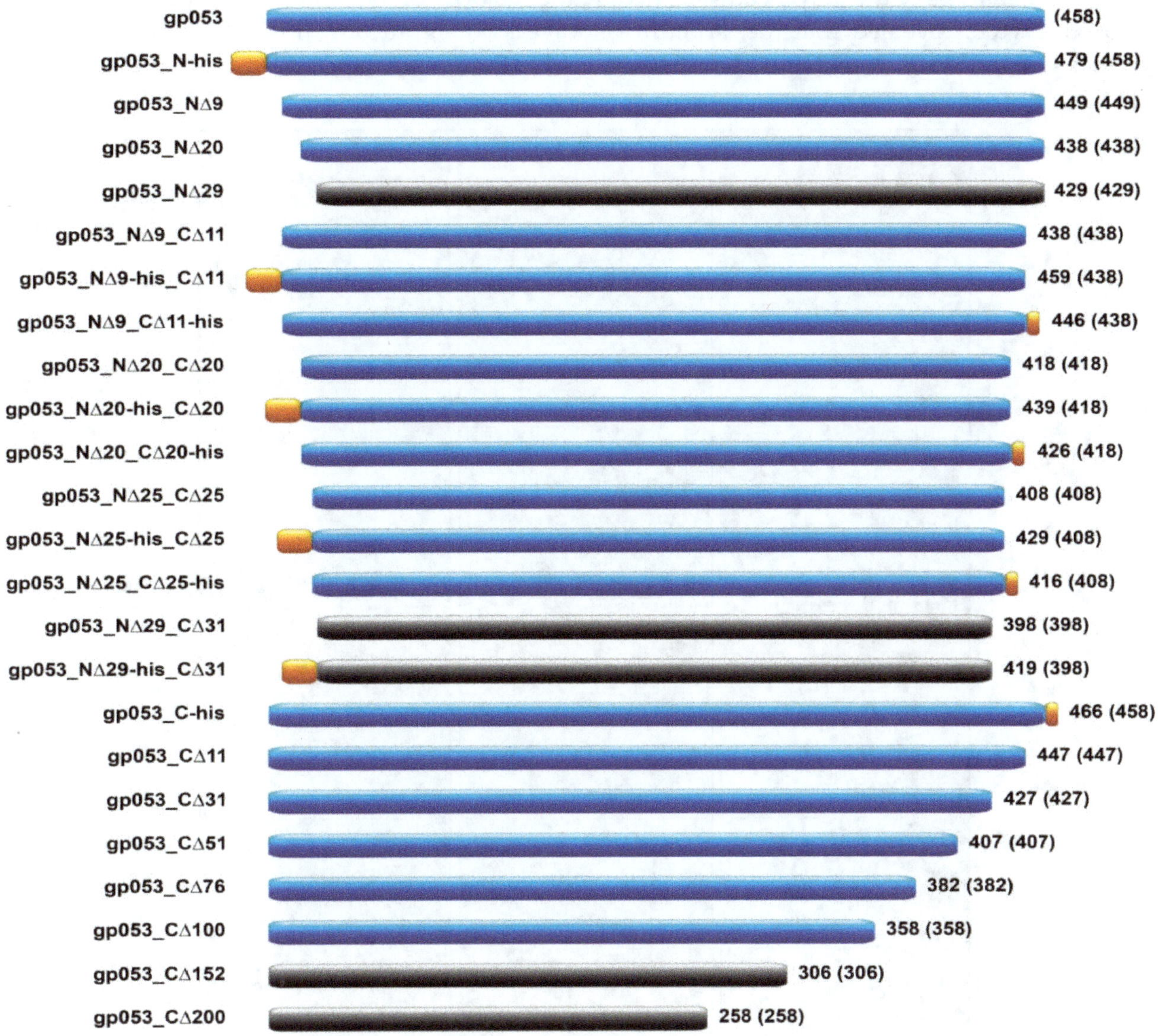

Figure 6. A schematic picture of the truncated mutants of the gp053 protein. The mutants are named (shown on the left) according to N-terminal or C-terminal mutations and the presence of His-tag. The proteins, able to polymerise into nanotubes, are shown in blue. The mutants, incapable of polymerisation, are shown in black. The His-tags are shown in orange. The size of a protein in amino acids is shown on the right, and the portion of wild-type gp053 within a construct is depicted in parenthesis.

The recombinant proteins were produced in *E. coli* BL21 (DE3) cells and the cell-free extracts were analysed by TEM. Whenever an assembly into the polysheaths was observed, the purification procedures using ammonium sulphate were performed. The TEM analysis revealed that deletions of 9, 20 and 25 amino acids from the N-terminal region did not disturb the protein assembly into polysheaths. Meanwhile, a deletion of 29 amino acids from the N-terminal region abolished the ability of the truncated gp053 to assemble into the typical polysheaths. Therefore, the gp053_NΔ29 was found in a soluble fraction, where it was folded into the unordered protein ribbons (Supplementary Table S2, Figure 6).

On the other hand, the deletions of 11, 31, 51, 76 and even 100 amino acids from the C terminus did not abolish the formation of the ordered tubular polysheaths. However, the gp053_CΔ150 mutant lacking 152 amino acids at the C terminus formed insoluble aggregates in *E. coli* cells (Supplementary Table S2, Figure 6).

The TEM analysis of the polysheaths formed by the majority of the mutants, except for gp053_N-his and gp053_CΔ100, revealed that the structures showed no significant differences compared to the polysheaths formed by the wild-type recombinant gp053 (Supplementary Table S2, Figure 7).

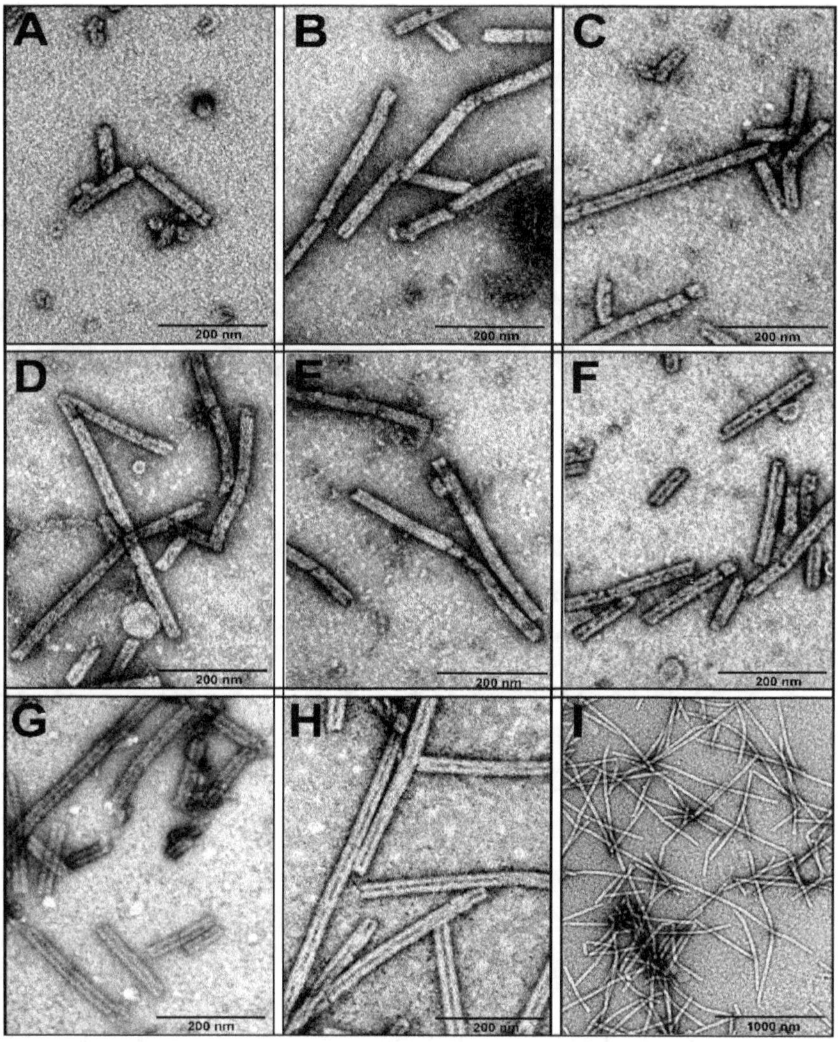

Figure 7. Electron micrographs of the polysheaths formed by the gp053 mutants. (**A**) gp053_N-his, (**B**) gp053_C-his, (**C**) gp053_NΔ9_CΔ11, (**D**) gp053_NΔ9-his_CΔ11, (**E**) gp053_NΔ25-his_CΔ25, (**F**) gp053_NΔ20, (**G**) gp053_CΔ76, and (**H**,**I**) gp053_CΔ100.

On the contrary, the polysheaths formed by the gp053_N-his mutant were less ordered and approximately 2–5-fold shorter in length (~100–300 nm) (Figure 7A). In the case of the gp053_CΔ100, the tubular structures were longer (up to ~3000 nm) and slightly larger in diameter (~32.47 nm) (Figure 7H,I) compared to the wild-type ones. In addition, it was observed that some of the nanotubes, for example, formed by gp053_NΔ20, gp053_CΔ76 and gp053_CΔ100 (Figure 7F–H, respectively), contained more clearly defined internal channels. A similar morphology was also detected in the case of the full-length gp053 after prolonged storage in TE buffer or treatment with trypsin (Figure 4C,D, respectively).

3.4. Investigation of the Oligomeric Constructs of gp053

According to the 6-start helix symmetry of the polysheath, the gp053 oligomers were genetically engineered (Supplementary Table S2, Figure 8A) and their ability to form the ordered structures was analysed. Initially, it was found that all the recombinant proteins (gp053_N_C_mon, gp053_N_C_dim, gp053_N_C_trim, gp053_N_C_tet, gp053_N_C_pent and gp053_N_C_hex) were soluble under the investigated conditions (Figure 8C).

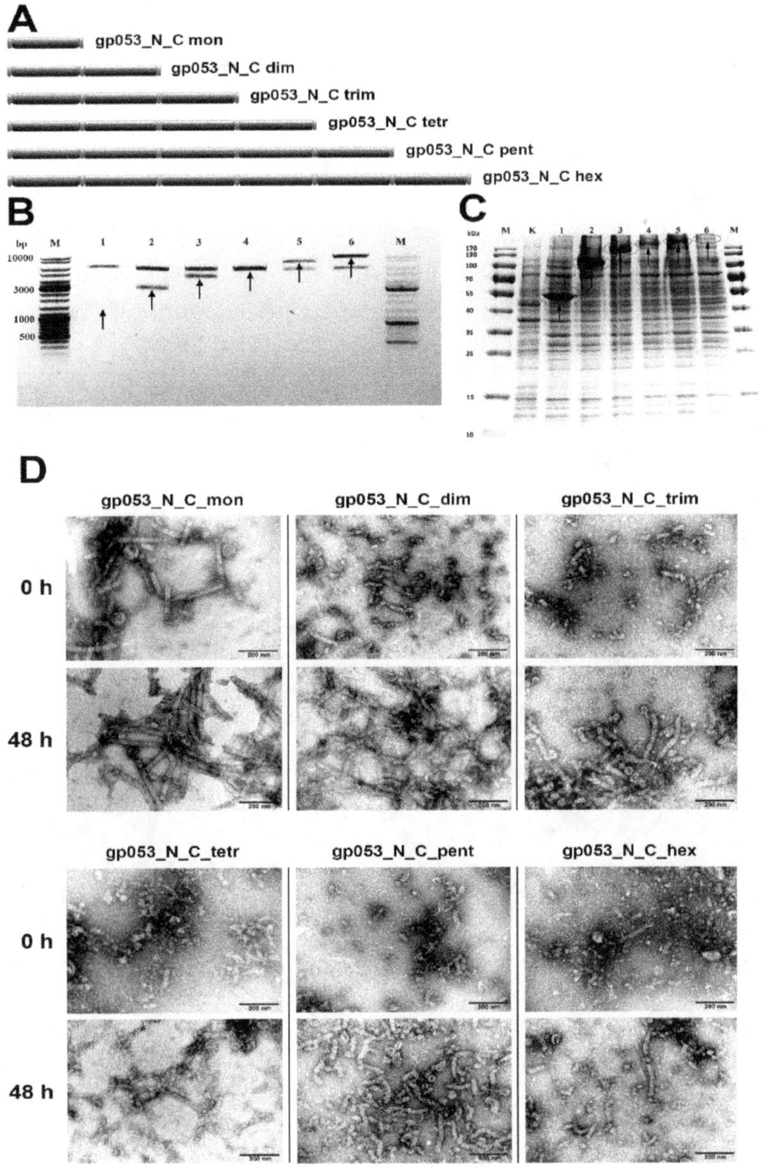

Figure 8. Assembling of the gp053 oligomers. (**A**) A schematic representation of the gp053 oligomeric constructs. The amino acids of the wild-type gp053 within the construct are shown in black, and the insertions with cloning sites are shown in orange. The constructs are named according to the number of copies of the gene *053*. (**B**) The agarose gel electrophoresis analysis of plasmids harbouring the fused *053* genes. The arrows indicate DNA fragments encoding gp053 after hydrolysis with NdeI and XhoI. Lanes: M—molecular mass marker, GeneRuler™ DNA Ladder Mix (Thermo Fisher Scientific, Vilnius, Lithuania); 1—gp053_N_C_mon; 2—gp053_N_C_dim; 3—gp053_N_C_trim; 4—gp053_N_C_tetr; 5—gp053_N_C_pent; and 6—gp053_N_C_hex. (**C**) The SDS-PAGE analysis of the cell-free extracts of the *E. coli* BL21 (DE3) cells producing the recombinant gp053 oligomers. Lanes: M—molecular mass marker, Page Ruler™ prestained Protein Ladder Plus (Thermo Fisher Scientific, Vilnius, Lithuania); K—pET21a (plasmid vector, control); 1—gp053_N_C_mon; 2—gp053_N_C_dim; 3—gp053_N_C_trim; 4—gp053_N_C_tetr; 5—gp053_N_C_pent; and 6—gp053_N_C_hex. (**D**) The TEM analysis of the structures formed by the recombinant gp053 oligomers. The cell-free extracts were analysed immediately after cell disruption and sample preparation or after incubation with periodical shaking at 22 °C. The incubation time is indicated on the left.

When the samples were analysed by TEM immediately after the disruption of cells, the ordered tubular structures up to ~400 nm in length were visible only in the case of the monomeric gp053. Some tubular structures up to ~200 nm in length were also visible in the case of the dimeric gp053. The other oligomers of gp053 formed indefinite structures (Figure 8D). However, the tubular structures, although of a less regular structure, were observed in the case of all oligomers after an incubation of the samples in TE buffer with periodic shaking at 22 °C for 48 h (Figure 8D).

3.5. Labelling of gp053 Polysheaths with Neutravidin-Conjugated Gold Nanoparticles

Based on the model of the structure of gp053, one of the four cysteines, Cys160, was located on putative domain 2 (Figure 1), which was found on the outer surface of the polysheaths formed by LIN1278, DSY3957 and gp18 [38]. In order to determine whether this amino acid was on the surface of the nanotubes formed by recombinant gp053, the purified polysheaths were modified with biotin and incubated with neutravidin-conjugated gold nanoparticles. The TEM analysis revealed that the gold nanoparticles in most cases were visible on the ends or breakpoints of the polysheaths, single rings or short gp053 aggregates, and the rest of the outer surface of the protein nanotubes was not decorated with gold nanoparticles (Figure 9B).

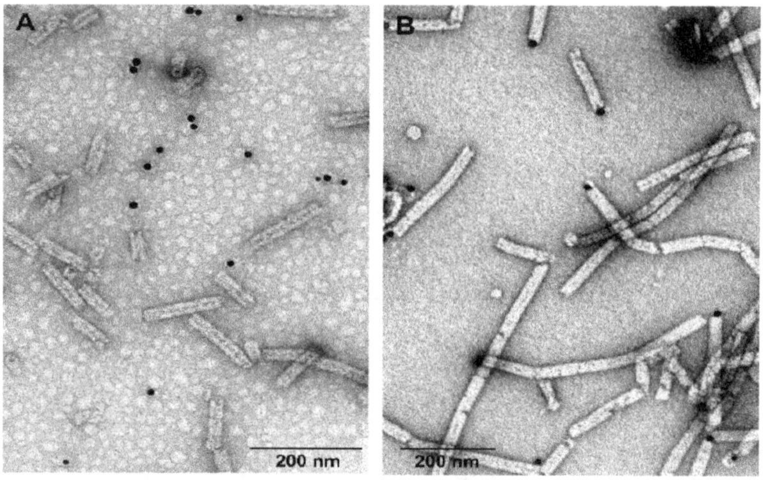

Figure 9. Electron micrographs of the polysheaths formed by the purified recombinant gp053 incubated with the neutravidin-conjugated gold nanoparticles. (**A**) Non-modified gp053, and (**B**) gp053-harbouring cysteines modified with biotin.

Thus, we can conclude that the cysteine residues of gp053 were buried inside of the protein nanotubes, and only could be accessible for labelling with neutravidin-conjugated gold nanoparticles at the termini of the polysheaths.

4. Discussion

The construction of various nanostructures based on self-assembling biomolecules is currently of special interest due to the potential for diverse application in different fields of nanobiosciences. However, only a limited number of self-assembling proteins from bacteriophages have been studied in detail to date. Therefore, in this study, we aimed to shed more light on the diversity of self-assembling tubular nanostructures and chose to investigate the tail sheath protein (gp053) from enterobacteria phage vB_EcoM_FV3.

Despite 98–99% sequence identity of the tail sheath proteins from phages of the genus *V5virus*, the molecular architecture of FV3 gp053 is largely undetermined. To our knowledge, none of the crystal structures of the tail sheath proteins from the genus *V5virus* or their close relatives have been resolved to date. On the other hand, it has been reported that the structure of the domains constituting the tail sheath proteins from phages T4, phiKZ, and phi812 are similar despite lower than

15% sequence identity between the aforementioned proteins [29,60,61]. It has also been shown that contractile structures from phages have very similar architecture to contractile molecular machines found in many prokaryotes: R-type pyocins, the Type VI secretion system (T6SS) and phage tail-like protein translocation structures (PLTS) [32,62–65]. Moreover, phylogenetic analysis of the tail sheath proteins DSY3957 (PDB ref: 3HXL) and LIN1728 (PDB ref: 3LML) from the prophages revealed that, phylogenetically, they are more closely related to the R-type pyocin from *Clostridium difficile,* than to the tail sheath proteins from other phages [63]. Therefore, it is unsurprising that the HHpred analysis showed the predicted fold of gp053 from phage FV3 to be more similar to the R-type pyocins from *Clostridium difficile* (PDB ref 6GKW) and *Pseudomonas aeruginosa* (PDB ref 3J9Q) than to the fold of the tail sheath proteins from prophages (PDB ref 3LML, PDB ref 3HXL) or phages (PDB ref 5LI4, PDB ref 3J2M). On the other hand, bioinformatics analysis revealed that the gp053 identity compared to the previously mentioned proteins at the amino acid level ranged only from 14.6% (3HXL) to 18.8% (6GKW). Therefore, it has been demonstrated that despite the fact that the tail sheath proteins of different *Myoviridae* phages appeared to have similar helical parameters and functioned in a similar manner [29,31,32,38,66–70], completely different sequences of amino acids forming the tail sheath proteins could exist.

Here, we demonstrated that it was possible to obtain comparatively high amounts of recombinant gp053 protein quickly and efficiently through the use of *E. coli* expression systems. It has been shown that recombinant gp053 formed soluble high molecular weight polysheaths during production in the *E. coli* cells, and the purification of these nanostructures could be performed very quickly, cheaply and efficiently using an ammonium sulphate precipitation from the cell-free extracts only (Figure 2). In contrast, the precipitation of the polysheaths of phage PaBG by using ammonium sulphate was not as effective. Therefore, additional fractionation by ultracentrifugation was needed [71]. Additional purification procedures, including ultracentrifugation, hydroxyapatite and anion-exchange chromatography, have been used to purify nanostructures formed by the recombinant tail sheath proteins of phages T4 and phiKZ [38,42].

Another important characteristic of the polysheaths formed by gp053 is their extreme stability towards various physical and chemical treatments. Similarly, the polysheaths consisting of the tail sheath proteins of other phages have also been shown to be very stable structures. It has been demonstrated that the polysheaths of phage T4 retained their structure after incubation in 8 M urea (pH 7.6–12.2), 6 M guanidine hydrochloride (pH 5.6) and 7.8 M acetic acid (pH 1.8–6.1) [39]. Moreover, the polysheaths of phages T4 and phiKZ were also found to be extremely resistant to proteolysis by trypsin [38,41]. However, although the polysheaths of phage phiKZ remained uncrumbled after incubation with trypsin, the protein was completely cleaved into two fragments, as indicated by the disappearance of the full-length protein band and the appearance of two smaller fragments with molecular masses of about 60 and 15 kDa [38]. In contrast, a significant amount of the recombinant gp053 from phage FV3 remained undigested even after 24 hours of incubation with trypsin (Figure 5). It is likely that different trypsinisation profiles can be concerned with the morphological differences of polysheaths of the previously mentioned phages. In the case of the recombinant tail sheath protein of phage phiKZ, polysheaths with more visible internal channels have been observed. It has been suggested that these structures were caused by less compact packing of protein subunits that facilitates stain penetration into the internal channel of the polysheaths [38]. Meanwhile, the polysheaths composed of recombinant gp053 were different. It is possible that these structures are formed of highly compact protein subunits which are difficult for trypsin to reach. On the other hand, polysheaths with a better defined internal channel were observed in the cases of gp053 mutants such as gp053_NΔ20, gp053_CΔ76 and gp053_CΔ100 as well as a full-length gp053 after prolonged storage in TE buffer or after treatment with trypsin (Figures 4 and 7). Therefore, it is likely that less compact structures were formed in these cases.

LC–MS^E analysis of the trypsinisation products of the recombinant wild-type gp053 of FV3 revealed the potential trypsin-accessible cleavage sites between Lys381–Asn382 and Arg417–Val418

(Supplementary Figure S2). Based on the I-TASSER generated model of gp053, Lys381–Asn382 and Arg417–Val418, together with its N and C termini, are found in domain IV (Figure 1), which is structurally similar to the domains of the prophage tail sheath proteins DSY3957 and LIN1278, which have the same fold as the corresponding region (domain IV) of bacteriophage T4 [38]. It was demonstrated that the N and C termini of the tail sheath proteins are found close to each other and form a domain, which is located in the inner part of the contracted sheath [29–32,38]. Thus, with reference to the conserved structural model of the tail sheath proteins of bacteriophages and contractile molecular machines found in prokaryotes [65], it is the most likely that the N and C termini, as well as Lys381–Asn382 and Arg417–Val418, are located in the inner part of the polysheaths formed by gp053 of FV3. The results of the purification procedures of recombinant gp053 maintained these speculations—attempts to purify long tubular His-tagged gp053 polysheaths effectively by using the metal-chelating sorbent were unsuccessful (Figure 2).

The results of the experiments with FV3 gp053 deletion mutants were in accordance with previous observations that the elimination of the C-terminal residues of tail sheath proteins (gp18) from phage T4 had a less negative effect on the polymerisation properties, compared to deletions of the N terminus [42,43,72]. On the other hand, it has been demonstrated that some T4 gp18 mutants assembled into thinner filaments called "noncontracted polysheaths" (NCPs) [42], which might correspond to the transitional helices described previously [73]. However, NCPs in the case of gp053 mutants were not observed. In contrast, it has been demonstrated that the tubular structures, formed by gp053_CΔ100, were even longer and of a larger diameter than the polysheaths formed by wild-type recombinant gp053 (Figure 7).

In this study, for the first time, to our knowledge, the genetically engineered homooligomers of the tail sheath proteins from bacteriophages have been constructed. It was demonstrated that all of the recombinant gp053 duplicates were soluble under the investigated conditions. Moreover, a self-assembling process of a number of these proteins *in vitro*, in the absence of other viral proteins, was observed (Figure 8). Hence, the tubular, although irregular structures, were visible not only in the case of recombinant gp053 monomer (gp053_N_C_mon) but also and in the cases of other gp053 oligomers including gp053 pentamer (gp053_N_C_pent). These results are not only promising for the construction of self-assembling nanostructures with various materials exhibited in specific locations on their surfaces but they also raise a number of issues about the polymerisation processes of the tail sheath proteins of bacteriophages.

The I-TASSER-generated model of gp053 (Figure 1) showed that one out of four cysteine residues found in gp053, Cys160, was located in the predicted domain 2, which was structurally similar to the protruding domains forming the outer surface of contractile tail machines [29–32,38]. On the other hand, the results of the labelling of gp053 polysheaths with neutravidin-conjugated gold nanoparticles revealed that maleimide-based biotinylation occurred only at the termini of the polysheaths and none of the four cysteine residues (including Cys160) were located on the outer surface of the polysheaths freely available for biotinylation (Figure 9). Despite the fact that the specific cysteine residue (or several of them) that was biotinylated within gp053 is still unknown, a simple method to adhere gold nanoparticles to the termini of the polysheaths was shown. The attachment of gold nanoparticles could be very beneficial for the immobilisation of uniformly oriented nanostructures [74]. Thus, the results of this study suggest novel ways to construct hybrid self-assembling nanostructures, particularly ones appropriately prearranged on surfaces.

It has been demonstrated that self-assembling tubular nanostructures made of peptides, proteins or filamentous bacteriophages have been used for both fundamental studies and practical applications, including in the construction of biosensors, energy storage devices, drug delivery systems or tissue engineering [12–14,75,76]. Given the fact that the polysheaths, formed of the wild-type gp053 or its mutants, possess physicochemical properties which are very similar to the properties of self-assembling nanostructures formed of biomolecules mentioned above (well-defined shape and dimensions in the nanoscale, robust, ordered and intrinsically monodisperse structures), it is likely that gp053 polysheaths can be used for the same practical applications as tubular nanostructures made of peptides, proteins or filamentous bacteriophages. For example, the gp053 polysheaths, which are up to 1000 nm in length, ~28 nm in diameter and with an internal hole of ~12 nm, are very attractive tools for the synthesis of inorganic matter to produce nanowires with properties of interest to energy sciences and the electronic industry. Polysheaths can potentially be used for mineralisation, both in their interior channel and around their exterior surface as has been demonstrated in the case of the tobacco mosaic virus [77,78]. On the other hand, the polymerisation of recombinant tail sheath proteins is not a precisely controlled process and this still remains the main drawback associated with these nanostructures, hence additional studies are needed to tackle this issue.

Nevertheless, the production and purification of polysheaths, formed from recombinant gp053, is a relatively short, simple and cheap process: only after three hours following induction and after a few hours of purification procedures, it is possible to produce a high yield of purified nanostructures. In contrast, the in vivo or in vivo/in vitro propagation of phage M13 and plant viruses takes several days or several weeks, respectively [79–81]. In addition, the polysheaths are released from bacteria by cell lysis, thus, these structures are not strictly limited in either the size of the displayed (poly)peptides or in terms of the hydrophobicity of the polypeptide chain, which was demonstrated as a serious limitation of nascent rod-shaped virions to passage through the host's exit pore [81].

To summarise, gp053 from phage FV3 self-assembles into ordered tubular structures—polysheaths. The assembly proceeds in vivo and in the absence of other viral proteins. Polysheaths formed of the wild-type gp053 or its mutants possess physicochemical properties which are very attractive for the construction of self-assembling nanostructures with potential applications in different fields of nanosciences.

Author Contributions: R.M., L.T. and E.Š. conceived and designed the experiments; E.Š., L.T., R.R., S.P., K.G. and A.K. performed the experiments; R.M., E.Š. and A.K. analysed the data; R.M. and M.V. contributed reagents/materials/analysis tools; E.Š. and R.M. wrote the paper.

References

1. Busseron, E.; Ruff, Y.; Moulin, E.; Giuseppone, N. Supramolecular self-assemblies as functional nanomaterials. *Nanoscale* **2013**, *5*, 7098–7140. [CrossRef] [PubMed]
2. Li, F.; Wang, Q. Fabrication of nanoarchitectures templated by virus-based nanoparticles: Strategies and applications. *Small* **2014**, *10*, 230–245. [CrossRef]
3. Lee, E.J.; Lee, N.K.; Kim, I.S. Bioengineered protein-based nanocage for drug delivery. *Adv. Drug Deliv. Rev.* **2016**, *106*, 157–171. [CrossRef] [PubMed]
4. Yokoi, N.; Inaba, H.; Terauchi, M.; Stieg, A.Z.; Sanghamitra, N.J.; Koshiyama, T.; Yutani, K.; Kanamaru, S.; Arisaka, F.; Hikage, T.; et al. Construction of robust bio-nanotubes using the controlled self-assembly of component proteins of bacteriophage T4. *Small* **2010**, *6*, 1873–1879. [CrossRef] [PubMed]
5. Zhou, J.C.; Soto, C.M.; Chen, M.S.; Bruckman, M.A.; Moore, M.H.; Barry, E.; Ratna, B.R.; Pehrsson, P.E.; Spies, B.R.; Confer, T.S. Biotemplating rod-like viruses for the synthesis of copper nanorods and nanowires. *J. Nanobiotechnol.* **2012**, *10*, 18. [CrossRef]
6. Drygin, Y.; Kondakova, O.; Atabekov, J. Production of platinum atom nanoclusters at one end of helical plant viruses. *Adv. Virol.* **2013**, *2013*, 746796. [CrossRef]
7. Swaminathan, S.; Cui, Y. Biochemical functionalization of peptide nanotubes with phage displayed peptides. *Nanotechnology* **2016**, *27*, 365703. [CrossRef]
8. Huang, Y.; Chiang, C.Y.; Lee, S.K.; Gao, Y.; Hu, E.L.; de Yoreo, J.; Belcher, A.M. Programmable assembly of nanoarchitectures using genetically engineered viruses. *Nano Lett.* **2005**, *5*, 1429–1434. [CrossRef]
9. Singh, P.; Gonzalez, M.J.; Manchester, M. Viruses and their uses in nanotechnology. *Drug Dev. Res.* **2006**, *67*, 23–41. [CrossRef]
10. Young, M.; Willits, D.; Uchida, M.; Douglas, T. Plant viruses as biotemplates for materials and their use in nanotechnology. *Annu. Rev. Phytopathol.* **2008**, *46*, 361–384. [CrossRef]
11. Lee, S.Y.; Lim, J.S.; Harris, M.T. Synthesis and application of virus-based hybrid nanomaterials. *Biotechnol. Bioeng.* **2012**, *109*, 16–30. [CrossRef]

12. Hyman, P. Bacteriophages and nanostructured materials. *Adv. Appl. Microbiol.* **2012**, *78*, 55–73. [CrossRef] [PubMed]
13. Molek, P.; Bratkovič, T. Bacteriophages as scaffolds for bipartite display: Designing swiss army knives on a nanoscale. *Bioconjug. Chem.* **2015**, *3*, 367–378. [CrossRef] [PubMed]
14. Pires, D.P.; Cleto, S.; Sillankorva, S.; Azeredo, J.; Lu, T.K. Genetically engineered phages: A review of advances over the last decade. *Microbiol. Mol. Biol. Rev.* **2016**, *80*, 523–543. [CrossRef] [PubMed]
15. Ju, Z.; Sun, W. Drug delivery vectors based on filamentous bacteriophages and phage-mimetic nanoparticles. *Drug Deliv.* **2017**, *24*, 1898–1908. [CrossRef]
16. Kim, I.; Moon, J.S.; Oh, J.W. Recent advances in M13 bacteriophage-based optical sensing applications. *Nano Converg.* **2016**, *3*, 27. [CrossRef] [PubMed]
17. Lee, J.H.; Fan, B.; Samdin, T.D.; Monteiro, D.A.; Desai, M.S.; Scheideler, O.; Jin, H.E.; Kim, S.; Lee, S.W. Phage-based structural color sensors and their pattern recognition sensing system. *ACS Nano* **2017**, *11*, 3632–3641. [CrossRef]
18. Lee, S.W.; Belcher, A.M. Virus-based fabrication of micro- and nanofibers using electrospinning. *Nano Lett.* **2004**, *4*, 387–390. [CrossRef]
19. Merzlyak, A.; Indrakanti, S.; Lee, S.W. Genetically engineered nanofiber-like viruses for tissue regenerating materials. *Nano Lett.* **2009**, *9*, 846–852. [CrossRef]
20. Yang, S.H.; Chung, W.J.; McFarland, S.; Lee, S.W. Assembly of bacteriophage into functional materials. *Chem. Rec.* **2013**, *13*, 43–59. [CrossRef]
21. Lee, J.H.; Warner, C.M.; Jin, H.E.; Barnes, E.; Poda, A.R.; Perkins, E.J.; Lee, S.W. Production of tunable nanomaterials using hierarchically assembled bacteriophages. *Nat. Protoc.* **2017**, *12*, 1999–2013. [CrossRef] [PubMed]
22. Devaraj, V.; Han, J.; Kim, C.; Kang, Y.C.; Oh, J.W. Self-assembled nanoporous biofilms from functionalized nanofibrous M13 bacteriophage. *Viruses* **2018**, *10*, 322. [CrossRef] [PubMed]
23. Sawada, T.; Serizawa, T. Filamentous viruses as building blocks for hierarchical self-assembly toward functional soft materials. *Bull. Chem. Soc. Jpn.* **2018**, *91*, 455–466. [CrossRef]
24. Nam, K.T.; Peelle, B.R.; Lee, S.W.; Belcher, A.M. Genetically driven assembly of nanorings based on the M13 virus. *Nano Lett.* **2004**, *4*, 23–27. [CrossRef]
25. Bhardwaj, A.; Walker-Kopp, N.; Wilkens, S.; Cingolani, G. Foldon-guided self-assembly of ultra-stable protein fibers. *Protein Sci.* **2008**, *17*, 1475–1485. [CrossRef] [PubMed]
26. Guo, P. Bacterial virus phi29 DNA-packaging motor and its potential applications in gene therapy and nanotechnology. *Methods Mol. Biol.* **2005**, *300*, 285–324. [CrossRef] [PubMed]
27. Hyman, P.; Valluzzi, R.; Goldberg, E. Design of protein struts for self-assembling nanoconstructs. *Proc. Natl. Acad. Sci. USA* **2002**, *99*, 8488–8493. [CrossRef] [PubMed]
28. Daube, S.S.; Arad, T.; Bar-Ziv, R. Cell-free co-synthesis of protein nanoassemblies: Tubes, rings, and doughnuts. *Nano Lett.* **2007**, *7*, 638–641. [CrossRef]
29. Aksyuk, A.A.; Leiman, P.G.; Kurochkina, L.P.; Shneider, M.M.; Kostyuchenko, V.A.; Mesyanzhinov, V.V.; Rossmann, M.G. The tail sheath structure of bacteriophage T4: A molecular machine for infecting bacteria. *EMBO J.* **2009**, *28*, 821–829. [CrossRef]
30. Arisaka, F.; Yap, M.L.; Kanamaru, S.; Rossmann, M.G. Molecular assembly and structure of the bacteriophage T4 tail. *Biophys. Rev.* **2016**, *8*, 385–396. [CrossRef]
31. Fokine, A.; Rossmann, M.G. Molecular architecture of tailed double-stranded DNA phages. *Bacteriophage* **2014**, *4*, e28281. [CrossRef] [PubMed]
32. Leiman, P.G.; Shneider, M.M. Contractile tail machines of bacteriophages. *Adv. Exp. Med. Biol.* **2012**, *726*, 93–114. [CrossRef] [PubMed]
33. Leiman, P.G.; Arisaka, F.; van Raaij, M.J.; Kostyuchenko, V.A.; Aksyuk, A.A.; Kanamaru, S.; Rossmann, M.G. Morphogenesis of the T4 tail and tail fibers. *Virol. J.* **2010**, *7*, 355. [CrossRef] [PubMed]
34. Kostyuchenko, V.A.; Chipman, P.R.; Leiman, P.G.; Arisaka, F.; Mesyanzhinov, V.V.; Rossmann, M.G. The tail structure of bacteriophage T4 and its mechanism of contraction. *Nat. Struct. Mol. Biol.* **2005**, *12*, 810–813. [CrossRef] [PubMed]
35. Kellenberger, E.; Boy de la Tour, E. On the fine structure of normal and "polymerized" tail sheath of phage T4. *J. Ultrastruct. Res.* **1964**, *11*, 545–563. [CrossRef]

36. Moody, M.F. Structure of the sheath of bacteriophage T4. I. Structure of the contracted sheath and polysheath. *J. Mol. Biol.* **1967**, *25*, 167–200. [CrossRef]
37. Tschopp, J.; Arisaka, F.; van Driel, R.; Engel, J. Purification, characterization and reassembly of the bacteriophage T4D tail sheath protein P18. *J. Mol. Biol.* **1979**, *128*, 247–258. [CrossRef]
38. Kurochkina, L.P.; Aksyuk, A.A.; Sachkova, M.Y.; Sykilinda, N.N.; Mesyanzhinov, V.V. Characterization of tail sheath protein of giant bacteriophage φKZ *Pseudomonas aeruginosa*. *Virology* **2009**, *395*, 312–317. [CrossRef]
39. To, C.M.; Kellenberger, Y.; Eisenstark, A. Disassembly of T-even bacteriophage into structural parts and subunits. *J. Mol. Biol.* **1969**, *46*, 493–511. [CrossRef]
40. Arisaka, F.; Engel, J.; Horst, K. Contraction and dissociation of the bacteriophage T4 tail sheath induced by heat and urea. *Prog. Clin. Biol. Res.* **1981**, *64*, 365–379.
41. Arisaka, F.; Takeda, S.; Funane, K.; Nashijima, N.; Ishii, S. Structural studies of the contractile tail sheath protein of bacteriophage T4. Structural analyses of the tail sheath protein, gp18, by limited proteolysis, immunoblotting, and immunoelectron microscopy. *Biochemistry* **1990**, *29*, 5057–5062. [CrossRef] [PubMed]
42. Poglazov, B.F.; Efimov, A.V.; Marco, S.; Carrascosa, J.; Kuznetsova, T.A.; Aijrich, L.G.; Kurochkina, L.P.; Mesyanzhinov, V.V. Polymerization of bacteriophage T4 tail sheath protein mutants truncated at the C-termini. *J. Struct. Biol.* **1999**, *127*, 224–230. [CrossRef] [PubMed]
43. Efimov, V.P.; Kurochkina, L.P.; Mesyanzhinov, V.V. Engineering of bacteriophage T4 tail sheath protein. *Biochemistry* **2002**, *67*, 1366–1370. [CrossRef] [PubMed]
44. Truncaite, L.; Šimoliūnas, E.; Zajanckauskaite, A.; Kaliniene, L.; Mankeviciute, R.; Staniulis, J.; Klausa, V.; Meskys, R. Bacteriophage vB_EcoM_FV3: A new member of "rV5-like viruses". *Arch. Virol.* **2012**, *157*, 2431–2435. [CrossRef] [PubMed]
45. Lowry, O.H.; Rosebrough, N.J.; Farr, A.L.; Randall, R.J. Protein measurement with the Folin phenol reagent. *J. Biol. Chem.* **1951**, *193*, 265–275.
46. Hellman, U.; Wernstedt, C.; Gonez, J.; Heldin, C.H. Improvement of an "in-gel" digestion procedure for the micro preparation of internal protein fragments for amino acid sequencing. *Anal. Biochem.* **1995**, *224*, 451–455. [CrossRef]
47. Ger, M.; Kaupinis, A.; Nemeikaite-Ceniene, A.; Sarlauskas, J.; Cicenas, J.; Cenas, N.; Valius, M. Quantitative proteomic analysis of anticancer drug RH1 resistance in liver carcinoma. *Biochim. Biophys. Acta* **2016**, *1864*, 219–232. [CrossRef]
48. Povilonienė, S.; Časaitė, V.; Bukauskas, V.; Šetkus, A.; Staniulis, J.; Meškys, R. Functionalization of α-synuclein fibrils. *Beilstein J. Nanotechnol.* **2015**, *6*, 124–133. [CrossRef]
49. Transeq. Available online: http://www.ebi.ac.uk/Tools/st/emboss_transeq (accessed on 10 January 2019).
50. Clustal Omega. Available online: http://www.ebi.ac.uk/Tools/msa/clustalo (accessed on 10 January 2019).
51. Tamura, K.; Peterson, D.; Peterson, N.; Stecher, G.; Nei, M.; Kumar, S. MEGA5: Molecular evolutionary genetics analysis using maximum likelihood, evolutionary distance, and maximum parsimony methods. *Mol. Biol. Evol.* **2011**, *28*, 2731–2739. [CrossRef]
52. PIR. Available online: http://pir.georgetown.edu/pirwww/search/comp_mw.shtml (accessed on 10 January 2019).
53. Söding, J.; Biegert, A.; Lupas, A.N. The HHpred interactive server for protein homology detection and structure prediction. *Nucleic Acids Res.* **2005**, *33*, W244–W248. [CrossRef]
54. Zimmermann, L.; Stephens, A.; Nam, S.Z.; Rau, D.; Kubler, J.; Lozajic, M.; Gabler, F.; Söding, J.; Lupas, A.N.; Alva, V. A completely reimplemented mpi bioinformatics toolkit with a new HHpred server at its core. *J. Mol. Biol.* **2018**, *430*, 2237–2243. [CrossRef] [PubMed]
55. HHpred. Available online: https://toolkit.tuebingen.mpg.de/hhpred (accessed on 10 January 2019).
56. Zhang, Y. I-TASSER server for protein 3D structure prediction. *BMC Bioinform.* **2008**, *9*, 40. [CrossRef] [PubMed]
57. I-TASSER. Available online: https://zhanglab.ccmb.med.umich.edu/I-TASSER (accessed on 10 January 2019).
58. Pettersen, E.F.; Goddard, T.D.; Huang, C.C.; Couch, G.S.; Greenblatt, D.M.; Meng, E.C.; Ferrin, T.E. UCSF Chimera–a visualization system for exploratory research and analysis. *J. Comput. Chem.* **2004**, *13*, 1605–1612. [CrossRef] [PubMed]
59. UCSF Chimera. Available online: http://www.rbvi.ucsf.edu/chimera (accessed on 10 January 2019).
60. Aksyuk, A.A.; Kurochkina, L.P.; Fokine, A.; Forouhar, F.; Mesyanzhinov, V.V.; Tong, L.; Rossmann, M.G. Structural conservation of the myoviridae phage tail sheath protein fold. *Structure* **2011**, *19*, 1885–1894. [CrossRef] [PubMed]

61. Novacek, J.; Siborova, M.; Benesik, M.; Pantucek, R.; Doskar, J.; Plevka, P. Structure and genome release of Twort-like Myoviridae phage with a double-layered baseplate. *Proc. Natl. Acad. Sci. USA* **2016**, *113*, 9351–9356. [CrossRef] [PubMed]
62. Sarris, P.F.; Ladoukakis, E.D.; Panopoulos, N.J.; Scoulica, E.V. A phage tail-derived element with wide distribution among both prokaryotic domains: A comparative genomic and phylogenetic study. *Genome Biol. Evol.* **2014**, *6*, 1739–1747. [CrossRef] [PubMed]
63. Kube, S.; Wendler, P. Structural comparison of contractile nanomachines. *AIMS Biophys.* **2015**, *2*, 88–115. [CrossRef]
64. Kudryashev, M.; Wang, R.Y.; Brackmann, M.; Scherer, S.; Maier, T.; Baker, D.; DiMaio, F.; Stahlberg, H.; Egelman, E.H.; Basler, M. Structure of the type VI secretion system contractile sheath. *Cell* **2015**, *160*, 952–962. [CrossRef]
65. Taylor, N.M.I.; van Raaij, M.J.; Leiman, P.G. Contractile injection systems of bacteriophages and related systems. *Mol. Microbiol.* **2018**, *108*, 6–15. [CrossRef]
66. Donelli, G.; Guglielmi, F.; Paoletti, L. Structure and physico-chemical properties of bacteriophage G. I. Arrangement of protein subunits and contraction process of tail sheath. *J. Mol. Biol.* **1972**, *71*, 113–125. [CrossRef]
67. Cremers, A.F.M.; Schepman, A.M.H.; Visser, M.P.; Mellema, J.E. An analysis of the contracted sheath structure of bacteriophage Mu. *Eur. J. Biochem.* **1977**, *80*, 393–400. [CrossRef] [PubMed]
68. Parker, M.L.; Eiserling, F.A. Bacteriophage SPO1 structure and morphogenesis. I. Tail structure and length regulation. *J. Virol.* **1983**, *46*, 239–249.
69. Muller, M.; Engel, A.; Aebi, U. Structural and physicochemical analysis of the contractive MM phage tail and comparison with the bacteriophage T4 tail. *J. Struct. Biol.* **1994**, *112*, 11–31. [CrossRef]
70. Fokine, A.; Battisti, A.J.; Bowman, V.D.; Efimov, A.V.; Kurochkina, L.P.; Chipman, P.R.; Mesyanzhinov, V.V.; Rossmann, M.G. Cryo-EM study of the *Pseudomonas* bacteriophage phiKZ. *Structure* **2007**, *15*, 1099–1104. [CrossRef] [PubMed]
71. Kurochkina, L.P.; Semenyuk, P.I.; Sykilinda, N.N.; Miroshnikov, K.A. The unique two-component tail sheath of giant *Pseudomonas* phage PaBG. *Virology* **2018**, *515*, 46–51. [CrossRef] [PubMed]
72. Kuznetsova, T.A.; Efimov, A.V.; Aijrich, L.G.; Kireeva, I.Y.; Marusich, E.I.; Cappuccinelli, P.; Fiori, P.; Rappelli, P.; Kurochkina, L.P.; Poglazov, B.F.; et al. Properties of recombinant bacteriophage T4 tail sheath protein and its deletion fragments. *Biokhimia* **1998**, *63*, 702–709.
73. Moody, M.F. Sheath of bacteriophage T4. III. Contraction mechanism deduced from partially contracted sheaths. *J. Mol. Biol.* **1973**, *80*, 613–635. [CrossRef]
74. Peissker, T.; Deschaume, O.; Rand, D.R.; Boyen, H.G.; Conard, T.; Van Bael, M.J.; Bartic, C. Selective protein immobilization onto gold nanoparticles deposited under vacuum on a protein-repellent self-sssembled monolayer. *Langmuir* **2013**, *29*, 15328–15335. [CrossRef]
75. Sharma, P.; Rathi, B.; Rodrigues, J.Y.; Gorobets, N. Self-assembled peptide nanoarchitectures: Applications and future aspects. *Curr. Top. Med. Chem.* **2015**, *15*, 1268–1289. [CrossRef]
76. Glover, D.J.; Giger, L.; Kim, S.S.; Naik, R.R.; Clark, D.S. Geometrical assembly of ultrastable protein templates for nanomaterials. *Nat. Commun.* **2016**, *7*, 11771. [CrossRef]
77. Knez, M.; Bittner, A.M.; Boes, F.; Wege, C.; Jeske, H.; Maiß, E.; Kern, K. Biotemplate synthesis of 3-nm nickel and cobalt nanowires. *Nano Lett.* **2003**, *3*, 1079–1082. [CrossRef]
78. Bromley, K.M.; Patil, A.J.; Perriman, A.W.; Stubbs, G.; Mann, S. Preparation of high quality nanowires by tobacco mosaic virus templating of gold nanoparticles. *J. Mater. Chem.* **2008**, *18*, 4796–4801. [CrossRef]
79. Warner, C.M.; Barker, N.; Lee, S.W.; Perkins, E.J. M13 bacteriophage production for large-scale applications. *Bioprocess Biosyst. Eng.* **2014**, *37*, 2067–2072. [CrossRef] [PubMed]
80. Lane, L.C. Propagation and purification of RNA plant viruses. *Methods Enzymol.* **1986**, *118C*, 687–696. [CrossRef]
81. Shih, S.M.H.; Doran, P.M. In vitro propagation of plant virus using different forms of plant tissue culture and modes of culture operation. *J. Biotechnol.* **2009**, *143*, 198–206. [CrossRef] [PubMed]

6

Landscape Phage: Evolution from Phage Display to Nanobiotechnology

Valery A. Petrenko

Department of Pathobiology, Auburn University, Auburn, AL 36849-5519, USA; petreva@auburn.edu

Abstract: The development of phage engineering technology has led to the construction of a novel type of phage display library—a collection of nanofiber materials with diverse molecular landscapes accommodated on the surface of phage particles. These new nanomaterials, called the "landscape phage", serve as a huge resource of diagnostic/detection probes and versatile construction materials for the preparation of phage-functionalized biosensors and phage-targeted nanomedicines. Landscape-phage-derived probes interact with biological threat agents and generate detectable signals as a part of robust and inexpensive molecular recognition interfaces introduced in mobile detection devices. The use of landscape-phage-based interfaces may greatly improve the sensitivity, selectivity, robustness, and longevity of these devices. In another area of bioengineering, landscape-phage technology has facilitated the development and testing of targeted nanomedicines. The development of high-throughput phage selection methods resulted in the discovery of a variety of cancer cell-associated phages and phage proteins demonstrating natural proficiency to self-assemble into various drug- and gene-targeting nanovehicles. The application of this new "phage-programmed-nanomedicines" concept led to the development of a number of cancer cell-targeting nanomedicine platforms, which demonstrated anticancer efficacy in both in vitro and in vivo experiments. This review was prepared to attract the attention of chemical scientists and bioengineers seeking to develop functionalized nanomaterials and use them in different areas of bioscience, medicine, and engineering.

Keywords: phage display; landscape phage; major coat protein; nanomedicine; diagnostics; biosensors

---

1. Introduction

Filamentous bacteriophages (shortly phages), remarkably tolerant of dramatic structural alteration, are in every respect the perfect material for bioengineering. The structure of phage capsids is encoded in small single-stranded circular DNA and can be reengineered using common DNA techniques. Their capsids have a well-defined architecture, consistently adaptable for nanofabrication. The precise structures of filamentous phages and their coat proteins have been determined, allowing for the engineering of variants with precise arrangements of fused functional peptides and desired shapes and functions of phage-originated nanocomposites. The development of phage-functionalized biosensors and phage-targeted nanomedicines, which is the focus of this review, has been closely linked to the advances in structural phage biology and phage display technology. For a complete summary of Ff phage structure and biology, one can be referred to comprehensive reviews [1,2]. This review focuses on Ff phage display vectors, as precursors of landscape phages and as major workhorses in phage nanobiotechnology. It considers also those aspects of the structure and life cycle of Ff phages that are of direct practical interest to chemical scientists and bioengineers seeking to develop phage virions as specifically functionalized nanoparticles and use them in different areas of bioscience, medicine, material science, and engineering, as summarized in [3].

2. The History of Filamentous Phages

The story of bacteriophage discovery is told by Stent in his book *Molecular Biology of Bacterial Viruses* [4]. After French microbiologist Félix d'Hérelle introduced phages as antibacterial therapeutic agents, phage therapy soon became a boom and brought about great optimism in biomedical scientists [5]. After the discovery of penicillin in 1928 [6] and its triumphal introduction in the 1940s, the interest in the practical applications of phages as antibacterial agents faded, along with increasing interest in phages as tools for fundamental biological studies [7]. On this new wave of "phage renaissance", in the early 1960s, a group of filamentous phages (M13, fd, and fl) infecting conjugative F-pilus-positive *Escherichia coli* strains were isolated from sewage systems in the United States and Europe [1]. These small male-specific filamentous coliphages incorporate a single-stranded circular DNA encapsulated in a rigid tubular capsid (Figure 1). Because genomes of these phages code for a small number of proteins, reproduction of the phages is mostly supported by host bacteria. For this and other reasons, the small DNA phages became very popular model systems in molecular biology and versatile tools in genetic engineering and biotechnology.

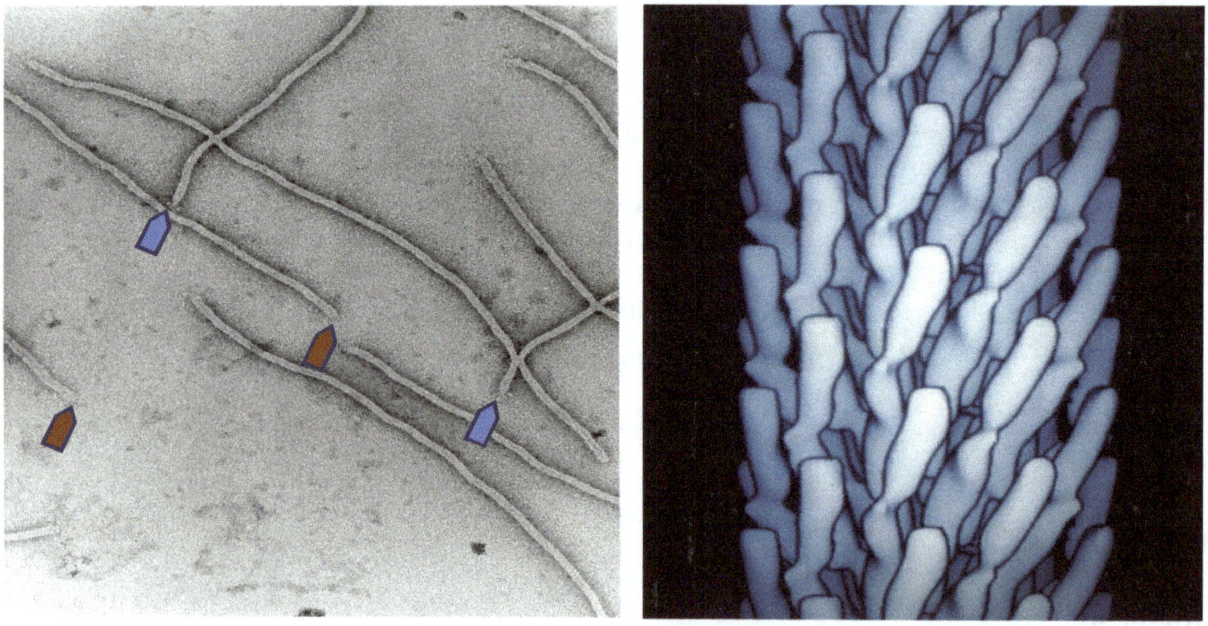

Figure 1. Electron microscopy image of filamentous phage (**left**) and electron density model (**right**) of filamentous phage M13 (Courtesy of Lee Makowski and Gregory Kishchenko. Adapted from [8]). Blue and red arrows depict the sharp and blunt ends of the phage capsid with attached minor coat proteins pIII/pIV and pVII/pIX, respectively (five copies each). Major coat protein (~2700 copies) forms the tubular capsid around viral single-stranded DNA.

3. Phage Engineering

In the mid-1970s, as a result of the development of the gene-splicing technique, the creation of molecular chimeras became a novel stirring area in molecular biology and biotechnology [9]. For a comprehensive presentation of the most commonly used molecular cloning systems, the reader is referred to [10]. As oligonucleotide-directed mutagenesis and DNA sequencing grew into routine genetic engineering techniques in the 1970s [9], filamentous phages emerged as unique DNA cloning vectors, very convenient for gene reconstruction and protein engineering (reviewed in [11]). These vectors provide a researcher with one of the two strands of vector DNA in a very useful form—easily purified phage particles. The extracted single-stranded DNA serves then as a template for sequencing and site-directed mutagenesis. The extendable tubular capsid of filamentous phages, oppositely to spherical capsids, readily accommodate enclosed recombinant DNA of any length [2].

In 1985, the emerged filamentous-phage recombinant DNA technique was adapted to design a novel type of viral chimera that inspired the development of phage display technology [12]. To construct this type of viral chimera, an alien DNA fragment is inserted into a gene encoding one of the phage coat proteins in such a way that the encoded "foreign" peptide is genetically fused to a coat protein and, by these means, exposed on the surface of the phage particle. A phage display library is a collection of fusion phage particles, containing diverse foreign DNA sequences and displaying diverse fusion peptides on the surface. The inserted foreign DNA can be derived from a natural source or it can be chemically synthesized. This technique has allowed for the constructing of phage libraries displaying billions of random peptides on the phage surface, as reviewed in [13].

Surface exposure of guest peptides presents a possibility of affinity selection, an essential aspect of phage display technology [12]. Affinity selection of phages exemplifies a kind of natural selection or molecular evolution in vitro, in which the phage library is comparable to a population of biological species and affinity for the receptor is equivalent to the "fitness" that aids an individual organism to survive and produce more offspring in the next generation. General principles and numerous applications of phage display technology are covered in several books and reviews, for example [12–16]. Here, the focus is on the logic and advances of the landscape phages—multivalent p8-type filamentous-phage constructs, which stand out from traditional phage display systems as a result of their unique structure and emerging properties [17]. Because of these characteristics, landscape phages can be of major interest as a foundation of phage nanobiotechnology—a new area of material science and nanotechnology. Many other filamentous phage vectors, suitable for other types of phage display systems (see below), are presented in [15,18].

4. Classification of Filamentous Phage Display Systems

To compare the landscape phage with other types of filamentous phage display systems, it is worth introducing a classification of phage display constructs [12,18]. Some display systems, mainly pIII-type, are mostly monovalent, with one or even fewer peptides displayed per phage particle. The monovalent type of display is advantageous when the goal is a high monovalent affinity of displayed peptides to the target. Multivalent phage display vectors belong to the group *n* that includes type 3, 8, 6, 7, and 9 systems, for which the foreign peptide is fused to all copies of the pIII, pVIII, pVI, pVII or pIX protein. The phage vector genome harbors a single recombinant coat protein gene (*III*, *VIII*, *VI*, *VII*, or *IX*). Landscape phages, which are constructed using type 8 vectors (Figure 2), are an extreme variant of multivalent display, in which thousands of fusion peptides are arranged in an ultra-high-density array on the surface of the virion (Figure 3). The dramatic alteration of the phage architecture can lead to emergent properties of the whole particle that are not directly linked to the properties of the individual fusion phage proteins and inserted peptides, as was observed in [19]. Multivalent display in p8-type phage vectors (named the landscape phage) [17] undermines selection for high affinity by disregarding individually weak and strong peptide ligands, which cannot be distinguished between because of greatly increased avidity (effective affinity). There are many applications, however, in which multivalent display is a distinctive advantage. As exemplified below, polyvalent display in landscape phages is particularly effective in the engineering of dense, ideally organized, robust interfaces in biosensors [20,21]; as diagnostic and therapeutic probes for targeting multiple cellular receptors [22–25]; as leads for vaccine development; and for the assembly of a broad spectrum of new materials [26].

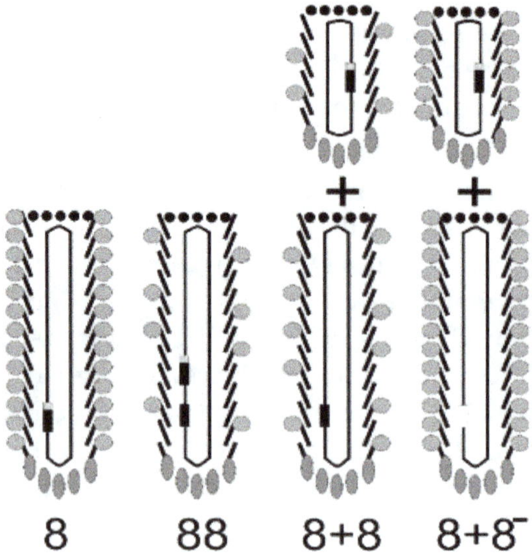

Figure 2. A type 8 vector (landscape-phage vector) contains multiple cloning sites in a single gene *gpVIII*. A type 88 vector harbors two genes: *VIII*, the wild-type *gpVIII*, and a recombinant rec-*gpVIII* that has the cloning sites. Type 88 phage capsids are composed of a mosaic of wild-type and recombinant pVIII proteins. In type 8 + 8 systems, the recombinant version and the wild-type version of gene *VIII* are on separate genomes: on a phagemid and on a helper phage ([11,27]). As for type 88 phage particles, the helper and phagemid virions in the type 8 + 8 system have a mosaic composition of recombinant and wild-type pVIII. In the type 8 + 8^- system, the helper phage lacks *gpVIII*, and all pVIII proteins are recombinant, as in the type 8 vector.

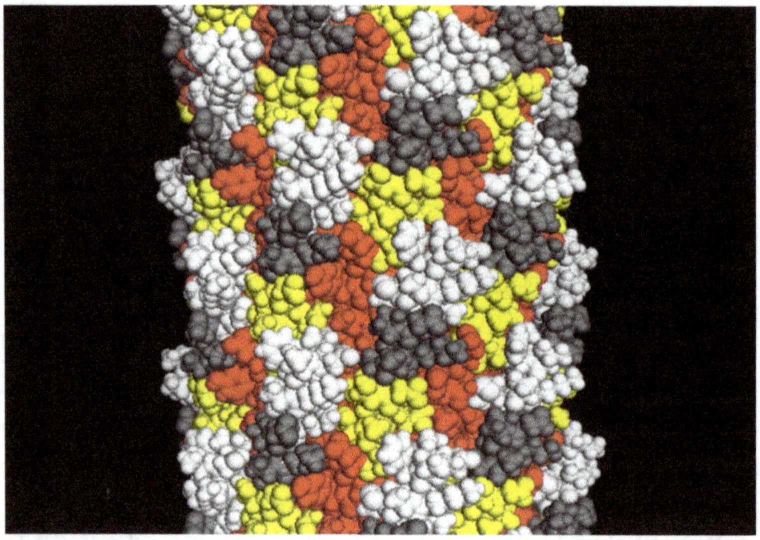

Figure 3. The landscape phage model (~10 nm segment of the full length). White: depicted atoms belong to the foreign peptides; yellow: pictured random amino acids in the phage α-library [28]; red: depicted amino acids belong to a small displayed segment of the hydrophobic domain predominantly buried in the capsid; grey: depicted amino acids belong to a "conservative" small segment of amphipathic domain. Adapted from [20].

The first *n*-type pVIII phage display vectors [29] originated from the recombinant phage fd-tet [30]. They were the basis of the first peptide and antibody libraries [31–34] and were employed in many different projects (for a review, see [12]). The remarkable feature of the fd-tet vectors is a small number of double-stranded phage replicative form (RF) DNA molecules per cell, because of a defect in the origin of replication that forces RNA polymerase to initiate the synthesis of a DNA replicative form using inefficient alternative initiation sites. Because of its replication defectiveness, the fd-tet vector produces in infected bacteria just several copies of RF DNA, which is hardly sufficient to support the phage assembly but allows the prevention of "cell killing" caused by toxic effects of excess fusion proteins.

Table 1 exemplifies the type 8 phage display vectors. The most advanced vectors f8-5 and f8-6 are suitable for the insertion of random peptides into the pVIII protein using *Pst*I, *Bam*HI, *Nhe*I, and *Mlu*I sites [27,35] (Figure 4). The presence of TAG amber stop codons between the cloning sites *Pst*I and *Bam*HI in the f8-6 vector precludes the synthesis of the wild-type vector in a non-suppressor strain, avoiding contamination of the library with vector phages.

Table 1. Type 8 vectors.

Name	Parent Phage	Antibiotic Resistance	Applications	Reference
f8-1	fd-tet	Tet	Billion-clone 8-mer peptide library	[17]
f8-5	fd-tet	Tet	Hundred-million-clone α-helical peptide library	[27]
f8-6	fd-tet	Tet	Billion-clone 9-mer peptide library	[35]
PM54	fd-tet	Tet	Small 6–16-mer peptide libraries	[36]
PM52	fd-tet	Tet	Small 6–16-mer peptide libraries	[36]
fdAMPLAY8	fd	Amp	Cloning of peptides	[37]
fdH	fd	None	Cloning of 4- and 6-mer peptides	[38]
fdISPLAY	fd	None	Cloning of peptides	[39,40]
PM48	f1	None	Ten-million-clone 8-mer peptide library; small 9-mer library	[36,41]
M13B	M13mp10	Amp	Cloning of 5-mer peptides	[42,43]

Tet: Tetracycline; Amp: ampicillin.

```
        ... -1  1  2  3  4  5  6  7  8  9 10 11 12 13 14 15 16 17 18 19 20 21 22 23 24 25 26
        ...  A  A  E  G  E  D  P  A  K  A  A  F  D  S  L  Q  A  S  A  T  E  Y  I  G  Y  A  W
fd-tet  ...GCTGCTGAGGGTGACGATCCCGCAAAAGCGGCCTTTGACTCCCTGCAAGCCTCAGCGACCGAATATATCGGTTATGCGTG

f8-5    ...GCTGCAGAGGGTGAGGATCCCGCAAAAGCGGCCTTTGACTCCCTGCAAGCTAGCGCGACCGAATATATCGGTTACGCGTG...
            PstI         BamHI                               NheI                   MluI

f8-6    ...GCTGCAGAGTAGTAGGATCCCGCAAAAGCGGCCTTTGACTCCCTGCAAGCTAGCGCGACCGAATATATCGGTTACGCGTG...
            PstI         BamHI                               NheI                   MluI

f8-1, PM52

        ...GCTGCAGAGGGTGAGGATCCCGCAAAAGCGG
            PstI         BamHI

M13B    ...GCTGCTGAGGGTGAGGATCCCGCAAAAGCGG...
                         BamHI

                 V* N*
fdH     ...GCTGCTGAGGTTAACGATCCCGCAAAAGCGG...
                      HpaI

fdISPLAY, fdAMPLAY8

        ...GCCGCGGAGGGTGACGATCCCGCCAAGGCGG...
            SacII                StyI

PM48, PM54
... -1  1  2  3  4        5  6  7  8  9 10 11
...  A  A  E  G  E  F* N* D  P  A  K  A  A  F
...GCTGCTGAGGGTGAATTCTAGGATCCCGCAAAAGCGGCCTTT
               EcoRI   BamHI
```

Figure 4. Cloning sites in p8-type filamentous phage vectors. Stop codons are underlined; asterisks mark amino acids synthesized in repressor strains.

5. Development of Landscape-Phage Libraries

"To the surprise of most phage biologists" [2], short guest peptides were displayed on the N-terminus of the major coat protein pVIII, shortly after the phage display on the coat protein pIII was developed [42–44]. The design of the first type 8 phage display constructs was motivated mostly by the interest of researchers in universal polyvalent vaccines and artificial immunodiagnostics; the foreign peptide was thus an epitope mimic interacting with an antibody [38,43,45,46]. They were constructed using the p8-type phage vector M13B [47] (Table 1; Figure 4). Phage particles displayed the foreign 5-mer peptide on every pVIII subunit [38,43], which led to the increasing of the phage mass by 10% (Figure 5).

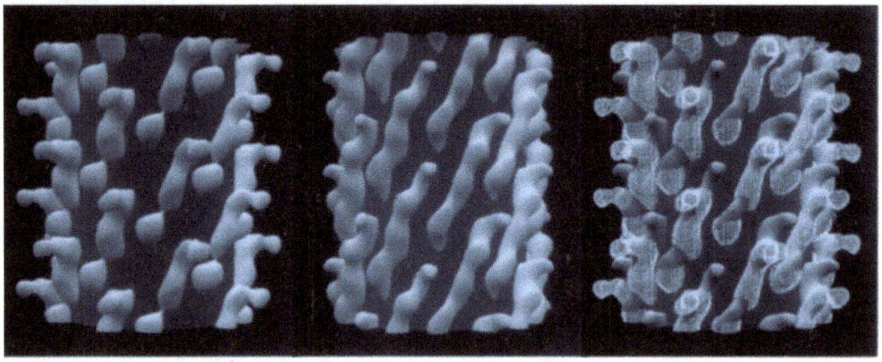

Figure 5. Configuration of a 5-mer peptide displayed on bacteriophage (phage) M13. Computer rendering of a ~10 nm length of the surface of electron density maps of M13 (**left**), fusion phage with 5-mer peptide inserted in all copies of p8 proteins (**center**), and a rendering of the differences between images (**right**). A cylinder of 2.5 nm radius was added to images to mask essentially identical interior features of the phages. About half of each coat protein is visible in phage surface images. Adapted from [48].

Surprisingly, such phage particles, with a dramatic distortion of their surface landscape, can infect *E. coli* and replicate in bacteria to produce plentiful progeny phage particles. Such phage particles were named the landscape phage to highlight the uniqueness of their architecture formed by thousands of the guest peptides displayed in a compact, reiterating pattern over the whole length of the phage capsid [17,25,27,41,49], as illustrated in Figures 3 and 5. The structures of these chimeric phages and fusion viral p8 proteins were studied, in comparison with the wild-type phage fd, by low- and high-resolution physical methods [44,48,50]. It was found that the inserted 5-mer peptides are fused to each p8 subunit and agglomerate on the surface of the phage. It was observed, however, that the size and structure of alien peptides, fused to the N-terminus of each phage protein p8, are restricted [36]. For example, the filamentous phage f1 readily tolerates a high diversity of 6-mer peptide fusion (resulting in an increase in the length of the major coat protein pVIII from 50 to 57 amino acids), but only a portion of phages are infective with longer fusions. As was exemplified, 40% of random 8-mer (59 amino acid long pVIII), 20% of 10-mer (61 amino acid long pVIII), and 1% of 16-mer (67 amino acid long pVIII) survived. The replacement of the M13 vector for the low-copy-number vector f8-1 (Table 1) allowed for the construction of the first multibillion-clone 8-mer landscape-phage display library f8/8 [11,17] (Figure 4; Table 1). To overcome limitations in the sizes of foreign peptides, we used a novel *replacement* strategy that combines the insertion of new peptides with the deletion of peptides that are not essential for phage viability. Thus, in constructing the first landscape library (named the f8/8 library) using the f8-1 vector, amino acids 2–4 (EGE) at the N-terminus of the mature pVIII protein were deleted and replaced by random 8-mers. As a result, the length of the phage fusion pVIII protein expanded only by five amino acids. In the f8/9 library, prepared using vector f8-6 (Figure 6), the amino acids EGED were substituted by random 9-mers. During this reconstruction of the pVIII protein, its length expanded to 55 amino acids, as in phage particles displaying foreign 5-mers [38,43] and in the f8/8 library [17] (Figure 7). As could be expected, the replacement of the negatively charged peptide EGDD for random peptides led to dramatic changes in the phage's physico-chemical and biological properties and, accordingly, to the censoring of the phage libraries' diversity, as demonstrated in [35].

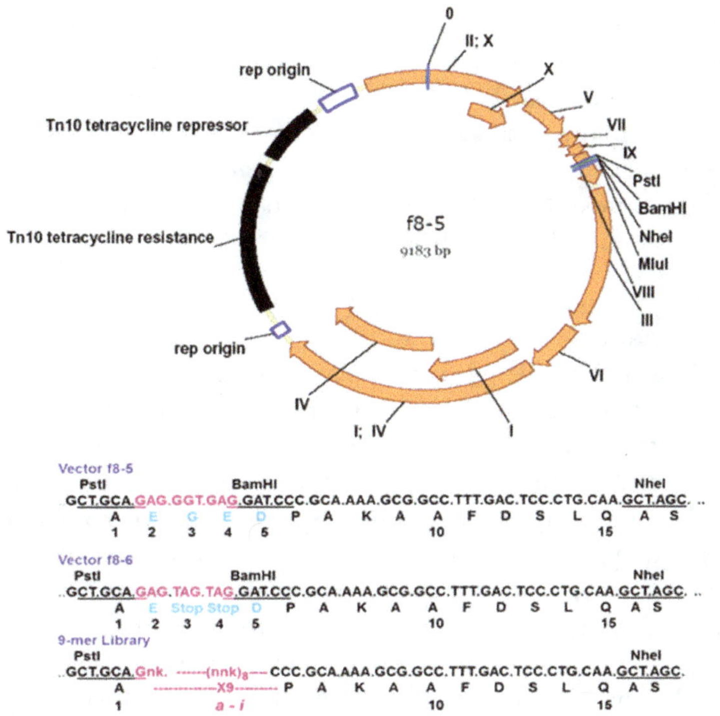

Figure 6. Phage display peptide libraries for vectors f8-5/f8-6 and f8-6. Characteristics of f8-6 vector: Restriction sites *Pst*I, *Bam*HI, *Nhe*I, and *Mlu*I in *pVIII* gene; Tet resistance gene spliced in origin of replication; low copy number of RF DNA and phages/cell; formation of tiny plaques on Tet-negative and colonies on Tet-positive agar plates.

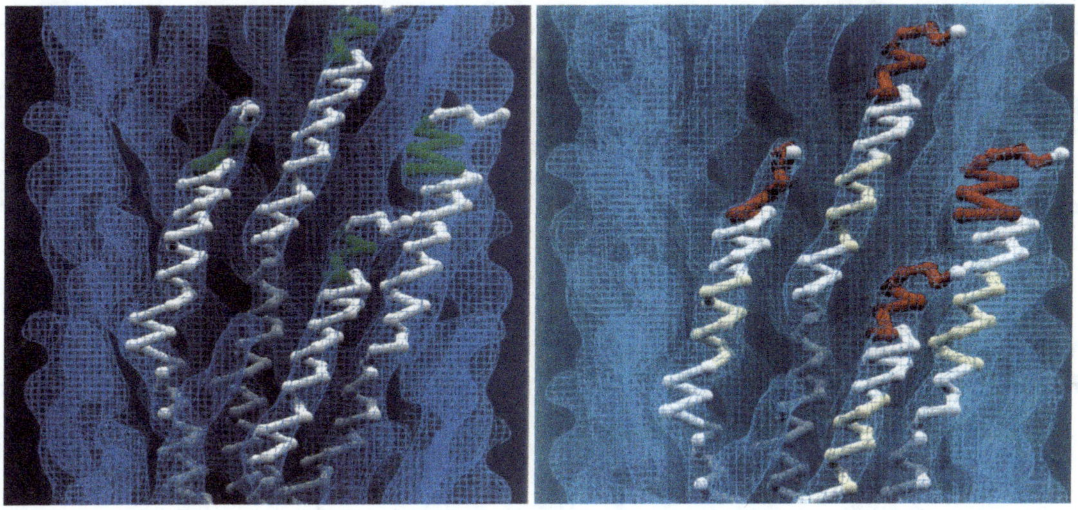

Figure 7. Models of landscape-phage libraries. **Left**: Ball-and-stick model of four neighboring pVIII proteins with inserted 5-mer peptides (green). About 1% of the phage length is shown. **Right**: Ball-and-stick model of three neighboring pVIII proteins with N-terminal random foreign 8-mer peptides (red) and mutagenized central segments (yellow) on the surface of the phage capsid (blue contour). The courtesy of Lee Makowski and Gregory Kishchenko.

Taking into account that the filamentous phage fd-tet can tolerate extension of the length of the major coat protein to 59–61 amino acids [36], one can propose that landscape-phage libraries with random 12–15-mer guest peptides can be designed using our replacement strategy. In particular, the performance of this strategy was confirmed by the construction of a conformationally constrained half-billion landscape-phage library, which displayed mutagenized amino acids 12, 13, 15–17, and 19 of the α-helical amphipathic domain of the pVIII protein [28], as portrayed in Figure 7. Such a combination of constrained and flexible random peptides grafted into the phage scaffold can mimic the architecture of antigen binding sites—regions in antibodies containing a number of highly mutable peptide loops, about 10 amino acids in length each, called *hypervariable regions* or *complementarity-determining regions* (CDRs) [51]. Indeed, in experiments with representative antigens, it was shown that landscape phages can serve as a new kind of substitute antibody that binds target antigens with a high affinity, specificity, and selectivity [49]. As substitute antibodies, landscape phages recommended themselves very well in biosensors and other analytical devices, as described below.

The multivalent display of peptides in p8-type landscape phages provides abundant freedom in engineering robust phage nanomaterials with rationally designed or selectable properties, as well as platforms for biospecific interactions. The phage-decorating foreign peptides constitute up to 25% of a phage particle's weight and cover about 50% of its surface (Figure 3). Because of the strong avidity revealed by the interaction of landscape phages with various biomolecules and nanomaterials, they are commonly used as substitute antibodies with subnanomolar effective affinities and high specificity for counterpart antigens. As antibody mimics, landscape phages overcome some intrinsic restrictions of antibody technology. Landscape phages are easily obtained from discontinuously growing bacteria—an effective recombinant protein manufacture system. The robust landscape phage shows an indefinite shelf life, without losing infectivity or specific functionality. These and other characteristics attest that landscape-phage libraries are an ideal rich source for specifically functionalized filamentous nanomaterials and recombinant peptide-fusion phage proteins. Landscape-phage probes have been discovered for a range of organic and inorganic materials that comprise widespread complexity [8,20]. They efficiently serve as substitutes for antibodies against bacterial and cancer cells in biosensors and targeted drug/gene delivery vehicles [49,52–56], as is exemplified below.

6. Landscape-Phage-Based Biosensors for Detection Monitoring of Biological Threats

Medical manifestations of disease, confirmed by biochemical, microbiological, and animal tests, remain the gold standard in clinical diagnostic laboratories. Currently, new requirements for fast, sensitive, accurate, and inexpensive detection platforms devalue the traditional detection methods. Modern immunoassays and biosensors require a biorecognition probe, which is attached to the interface of the analytical device, binding the target biological or chemical threat agent and generating a measurable signal. Most analytic platforms rely on the use of monoclonal antibodies as biorecognition probes. However, their broad application is limited by high cost, low specificity, less-than-optimal affinity, and sensitivity to environmental stresses. Landscape-phage display is a novel concept that allows for the development of diagnostic and detection interfaces that meet modern criteria for biological detection and monitoring [20,21,49,52,57]. It was proved that the landscape libraries represent an inexhaustible rich source of substitute antibodies—filaments that bind protein and glycoprotein antigens with nanomolar affinities and high specificity [49]. Landscape-phage-based interfaces performed in many ways much more effectively than their natural immunoglobulin counterparts in the same detector platforms. Target-specific landscape-phage probes can be prepared as described in commonly available protocols [58,59]. In my laboratory, *Bacillus anthracis* spore-specific landscape phages were selected by incubation of the landscape-phage library with spores immobilized in wells of enzyme-linked immunosorbent assay (ELISA) plates. Non-related phage particles were discarded, and spore-bound particles were released with acid buffer and collected [60]. To discover phage probes against the bacteria *Salmonella typhimurium*, a similar panning procedure was used, along with a coprecipitation procedure, in which complexes of bacterial cells and phage particles were isolated by centrifugation [61]. After titering of the phage in the host *E. coli* strain, random phage clones were used for the identification of selected displayed peptides by sequencing of recombinant *gpVIII* DNA [61]. The specificity of the phage probes selected against the *B. anthracis* spores was confirmed by ELISA, in which a phage-coated microtiter plate was incubated with spores and then treated with *B. anthracis* specific antibody. Similarly, the specificity and selectivity of phage probes against *S. typimurium* were confirmed by phage ELISA [61]. The established selection methods were successfully used for the discovery of diagnostic probes against classical swine fever virus [62], *Vibrio parahaemolyticus* [63], and *Staphylococcus aureus* [64]. For comprehensive reviews on phage-based pathogen biosensors, one can be referred to [21,58]. Data on the performance of landscape-based interfaces in different biosensor platforms are summarized in Table 2.

Table 2. Performance of landscape-phage-based interfaces.

Biosensor	Interface	Analyte	Sensitivity, Detection Range	Reference
Quartz crystal microbalance (QCM)	Phage coupled with phospholipid via biotin-streptavidin	β-galactosidase from *Escherichia coli* (β-gal)	Kd = 0.6 nM	[57]
	Phage immobilized by physical adsorption	β-gal	Kd = 1.7 nM	[65]
		Salmonella typhimurium	100 cells/mL	[66]
Surface plasmon resonance (SPR) spectroscopy	Phage immobilized by physical adsorption	β-gal	1 pM to 1 nM	[67]
	Enhanced green fluorescent protein (eGFP)	Phage in solution used in competition assay	1.2×10^{-14} M (in competiton assay)	[68]
Electrochemical impedance cytosensor	Phage immobilized on the electrode surface by physical adsorption	Colorectal carcinoma cells	79 cells/mL, 2×10^2–2×10^8 cells/mL	[69,70]
	Conjugate of the hybrid (8 + 8)-type M13 phage and electronically conductive polymer	Human serum albumin (HSA)	100 nM to 5 μM	[71]
Magnetoelastic particle resonators	Phage immobilized by physical adsorption	*Bacillus anthracis* spores	10^2–10^3 cfu/mL	[21,72,73]
		S. typhimurium	10^2–10^4 cfu/mL	[74]
Magnetoelastic microcantilever	Phage immobilized by physical adsorption	*S. typhimurium*	Not determined	[75]
Colorimetric immunoassay	Conjugate of pVIII fusion protein and cysteamine (CS)–gold nanoparticles (CS–AuNPs)	*Staphylococcus aureus* (*S. aureus*)	19 cfu/mL/mL	[64]
	Conjugate of pVIII fusion protein and protein–MnO_2 nanosheets	*Vibrio parahaemolyticus*	15 cfu/mL, 20–10^4 cfu/mL	[63]
Enzyme-linked immunosorbent assay (ELISA)	Phage immobilized by physical adsorption	*B. anthracis* spores	Not determined	[60]
		β-gal	Kd = 30 nM	[49,57]
		Neutravidin	Not determined	[49]
		Streptavidin		[49]
		Antibodies against gonadotropin-releasing hormone (GnRH) in patient sera		[76]
		Lyme disease patient sera		[77]
		Free prostate-specific antigen (f-PSA)	0.16 ng/mL, 0.825–165 ng/mL	[78]
		Total prostate-specific antigen (t-PSA)	1.6 ng/mL, 3–330 ng/mL	[79]
Differential pulse voltammetry (DPV) analytical platform	Phage conjugated to the gold electrochemical immunosensor		3 pg/mL, 0.01–100 ng/mL	
Phage microarray	Phage conjugated with NHS-functionalized slide	Cellulytic endoglucanase I (EG I)	5–500 nM	[80]

From these summarized data, one can conclude that landscape-phage display technology allows for the construction of libraries of diverse nanostructures boarding on the phage capsid—an enormous reserve of probes for the preparation of diagnostic and detection interfaces in analytical and biosensor platforms. Landscape phages may also serve as unique robust and inexpensive molecular recognition interfaces for field-use detectors and real-time monitoring devices for the control of biological and chemical threats. The commercialization of landscape-phage-based analytical interfaces may significantly enhance the performance of commercially valuable biosensors.

7. Diagnostic-Therapeutic Cancer Cell-Targeted Landscape

Targeted nanomedicines suggest a less toxic variant for cancer patients by reducing the drug supply to non-cancer tissues and enhancing their accumulation in tumors. The development and screening of targeted nanomedicines were recently facilitated by introducing the methods of high-throughput selection of cancer cell-binding landscape-phage proteins and their self-assembly into drug-loaded nanovehicles. These applications of the landscape-phage technology resulted in the development of diagnostic probes and targeted nanomedicines towards human prostate, breast, lung, pancreatic, and colorectal carcinoma cells [69,70,81–83]. The major paradigm that preceded the development of the landscape-phage technology was that phage-derived analytical and medical devices have to recognize the same antigens and receptors, which serve as targets in the course of phage selection [53,55,84,85]. Indeed, the cell-interacting specificity of landscape-phage particles and their isolated proteins was adequately translated to the phage protein-targeted nanomedicines, improving their cytotoxic efficacy towards cancer cells both in vitro and in vivo [8,59,86–98].

It was presumed that not only the tumor-targeting specificity of landscape phages but also their specific immunogenicity could be translated to the phage-driven nanomedicines. While intrinsic immunogenicity of landscape phages can be favorable, for example, in the construction of phage-based vaccines [12,14,43,99], it can create serious concerns about the efficacy and safety of phage-derived nanomedicines, which is equally true for many biologics [100]. To reduce the potential immunogenicity of phage-targeting micelles and liposomes, PEGylation was used to prevent nanoparticle aggregation and thereby mask antigenic epitopes, as illustrated in Figure 8. Interestingly, the phage protein-fusion peptides sheltered in the poly(ethylene glycol) (PEG) corona demonstrated expected targeting effects and profound anticancer cell activity both in vitro and in vivo [82,83,85,87,89,92–96,98].

The fusion major coat protein pVIII is the universal construction material that can be used in the preparation of cancer cell-targeted nanomedicines and diagnostics. The protein contributes ~90% to the virion mass [55] and can be easily isolated from phage particles in pure form using fast and simple procedures [55,85]. As a natural multifunctional membrane protein, pVIII readily associates with lipid nanoparticles, such as micelles and liposomes [54,90,101]. Its spontaneous insertion into lipid membranes (Figure 8), which is controlled by hydrophobic and electrostatic interactions and driven by electrophoretic forces, is discussed in [55]. The major coat protein was used for the targeting of liposomes [81,82,87,89,90,92,93,95,96] and micelles [81,83,94,98], as well as for the encapsulation of DNA and RNA [86,102] (Figure 9). The later method mimics the natural ability of the phage coat protein to encapsulate its genomic DNA during phage reproduction [101,103]. As noted above, the proficiency of the selected landscape phage to bind and penetrate into cancer cells is translated both to their individual proteins [91] and to phage protein-targeted nanomedicines. This was proved, for example, in our experiments with paclitaxel-containing micelles targeted with phage fusion protein specific for human breast cancer cells. It was shown that nucleolin-targeted micelles bind to their cognate target cells and demonstrate a significantly higher specific cytotoxicity towards target cells in comparison with non-targeted micelle nanomedicines. Thus, cancer cell-specific phage proteins obtained by phage selection from landscape phage libraries demonstrated a high potential as a substitute antibody for polymeric micelle-based pharmaceutical preparations [92–95,97,98].

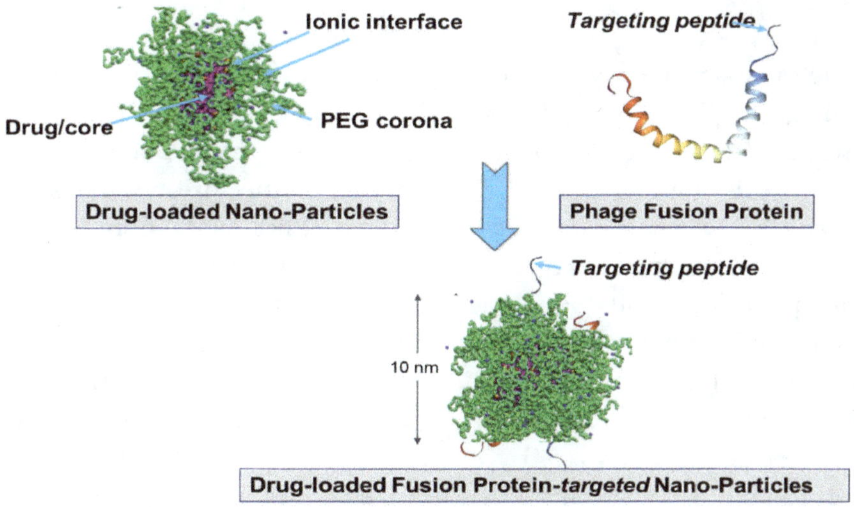

Figure 8. Preparation of landscape-phage pVIII-targeted paclitaxel-loaded PEGylated lipid micelle particles [94,98]. Equilibrated micelles contain three distinctive regions: lipid hydrophobic core, ionic interface, and poly(ethylene glycol) (PEG) corona. The fusion pVIII spans the core and displays the foreign peptide in PEG corona. The insoluble in water drug (paclitaxel) is solubilized in the core. Image of PEG-ylated sterically stabilized micelles is adopted from [104]. The model of pVIII phage protein in lipid environment is adapted from [105].

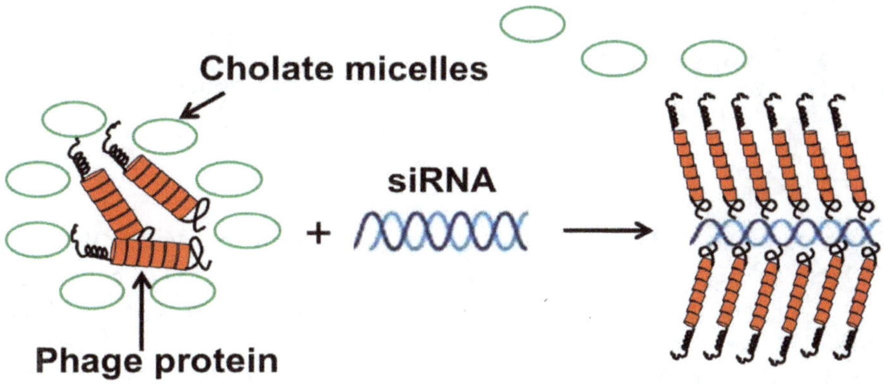

Figure 9. Preparation of nanophages from fusion phage protein and siRNA. The complex is formed during removal of protein-stabilizing cholate buffer. After the removal of cholate, the complex of RNA and protein is stabilized as a result of (a) the hydrophobic interaction of protein subunits, and (b) electrostatic interaction of positively charged C-terminus of the major coat protein and negatively charged phosphates of siRNA. Displayed N-terminus of phage protein serves as targeting ligand, which brings siRNA into the breast cancer cells. Adapted from [86].

In another scenario, fusion phage proteins were used as nanocarriers for the targeted delivery of siRNA to breast cancer cells [86]. This approach mimics the assembly of phage protein and phage DNA during different steps of the phage life cycle—and more specifically, the encapsulation of the nucleic acid into the phage capsid during secretion of the phage from the cell membrane of the host bacterium. Similarly, phage fusion proteins form a tubular vehicle for the accommodation and targeting of nucleic acids, such as siRNA, to form nanoparticles (11 nm diameter), called nanophages (Figure 9). In nanophages, the breast-cancer-targeting fusion phage protein not only protects siRNA from nucleases, but at the same time directs the delivery of a siRNA cargo to the target breast cancer cells and mediates knockdown of the targeted gene. Thus, nanophages exemplify elegant multifunctional vehicles for the targeted delivery of therapeutic nucleic acids to the site of disease.

8. Conclusions

In contrast to the most commonly used p3-type phage display vectors, which were designed to discover a desired phage-displayed peptide, the p8-type landscape-phage technology was developed with the goal to modify the properties of the whole fusion phage particle. In landscape-phage libraries, introduced in this review, a significant portion of the virion's mass (~10%) and surface (~25%) varied between distinct phage clones, comprising a billion-clone population of phage particles with unexpected emergent properties. The landscape-phage-based methodology for engineering of novel materials and devices exploits very specific mechanisms of phage biosynthesis, involving the precise self-assembly of phage particles. Landscape-phage libraries are a rich reservoir of already functionalized landscape phages and their proteins, which can be readily discovered using advanced methods of selection. Using landscape phages and phage fusion proteins for the construction of nanodevices allows for bypassing the critical and troublesome conjugation steps. No re-engineering of the landscape phage is required, as the phage particles are ready to be used as they are or can be transformed into various phage-programmed biospecific materials. The concept of a landscape phage has already contributed to various areas of biomedicine and nanotechnology, including the development of novel biosensors and monitoring devices, imaging systems, targeted nanomedicines, gene delivery platforms, universal vaccine leads, materials for bone and tissue repair, and so forth. I hope that this review will help chemists, biochemists, and bioengineers stay up-to-date with current trends in phage display and phage nanotechnology and encourage them to find new avenues in realization of their goals.

Acknowledgments: The Author acknowledge Alexey M. Eroshkin for preparation of molecular models of landscape libraries using Marvin's model [28,103], and Lee Makowski and Gregory P. Kishchenko for original images of electron microscopy and electron density models of filamentous phage.

References

1. Webster, R. Filamentous Phage Biology. In *Phage Display: A Laboratory Manual*; Barbas, C.F., Burton, D.R., Silverman, G.J., Eds.; Cold Spring Harbor Laboratory Press: Cold Spring Harbor, NY, USA, 2001; pp. 1–37.
2. Russel, M.; Model, P. Filamentous Phage. In *The Bacteriophage*, 2nd ed.; Calendar, R.L., Ed.; Oxford University Press, Inc.: Oxford, UK; New York, NY, USA, 2006; p. 746.
3. Petrenko, V.A.; Smith, G.P. (Eds.) *Phage Nanobiotechnology*; RSC Publishing: Cambridge, UK, 2011; 273p.
4. Stent, G.S. *Molecular Biology of Bacterial Viruses*; WH Freeman and Co.: San Francisco, CA, USA, 1963.
5. Summers, W.C. Bacteriophage therapy. *Annu. Rev. Microbiol.* **2001**, *55*, 437–451. [CrossRef] [PubMed]
6. Gale, E.P.; Reynolds, P.E.; Cundliffe, E.; Richmond, M.H.; Waring, M.J. *The Molecular Basis of Antibiotic Action*; Wiley-Interscience: Hoboken, NJ, USA, 1972.
7. Cairns, J.; Stent, G.S.; Watson, J.D. (Eds.) *Phage and the Origins of Molecular Biology, The Centennial Edition*; Cold Spring Harbor Laboratory Press: Cold Spring Harbor, NY, USA, 2007.
8. Petrenko, V. Evolution of phage display: From bioactive peptides to bioselective nanomaterials. *Expert Opin. Drug Deliv.* **2008**, *5*, 825–836. [CrossRef] [PubMed]
9. Glick, B.R.; Pasternak, J.J. *Molecular Biotechnology: Principles and Applications of Recombinant DNA*, 2nd ed.; ASM Press: Washington, DC, USA, 1998.
10. Rodriguez, R.L.; David, T.; Denhardt, E. *Vectors: A Survey of Molecular Cloning Vectors and Their Uses*; Butterworth-Heinemann: London, UK, 1987; 592p.
11. Petrenko, V.A.; Smith, G.P. Chapter 2: Vectors and modes of display. In *Phage Display in Biotechnology and Drug Discovery*; Gesyer, C.R., Sidhu, S.S., Eds.; CRC Press, Taylor & Francis Group: Boca Raton, FL, USA, 2005; pp. 43–74.
12. Smith, G.P.; Petrenko, V.A. Phage display. *Chem. Rev.* **1997**, *97*, 391–410. [CrossRef] [PubMed]
13. Fellouse, F.A.; Pal, G. Methods for the Construction of Phage-Displayed Libraries. In *Phage Display in Biotechnology and Drug Discovery*; Sidhu, S.S., Geyer, C.R., Eds.; CRC Press: Boca Raton, FL, USA, 2015; pp. 75–96.

14. Henry, K.A.; Arbabi-Ghahroudi, M.; Scott, J.K. Beyond phage display: Non-traditional applications of the filamentous bacteriophage as a vaccine carrier, therapeutic biologic, and bioconjugation scaffold. *Front. Microbiol.* **2015**, *6*, 755. [CrossRef] [PubMed]
15. Barbas, C.F., III; Burton, D.R.; Scott, J.K.; Silverman, G.J. *Phage Display: A. Laboratory Manual*; Cold Spring Laboratory Press: Cold Spring Harbor, NY, USA, 2001.
16. Sidhu, S.S.; Geyer, C.R. (Eds.) *Phage Display in Biotechnology and Drug Discovery*, 2nd ed.; CRC Press: Boca Raton, FL, USA, 2015; 567p.
17. Petrenko, V.A.; Smith, G.P.; Gong, X.; Quinn, T. A library of organic landscapes on filamentous phage. *Protein Eng.* **1996**, *9*, 797–801. [CrossRef] [PubMed]
18. Petrenko, V.A.; Smith, G.P. Vectors and modes of display. In *Phage Display in Biotechnology and Drug Discovery*, 2nd ed.; Sidhu, S.S., Geyer, C.R., Eds.; CRC Press, Taylor & Francis Group: Boca Raton, FL, USA, 2015; p. 567.
19. Kuzmicheva, G.A.; Jayanna, P.K.; Eroshkin, A.M.; Grishina, M.A.; Pereyaslavskaya, E.S.; Potemkin, V.A.; Petrenko, V.A. Mutations in fd phage major coat protein modulate affinity of the displayed peptide. *Protein Eng. Des. Sel.* **2009**, *22*, 631–639. [CrossRef] [PubMed]
20. Petrenko, V.A. Landscape Phage as a Molecular Recognition Interface for Detection Devices. *Microelectron. J.* **2008**, *39*, 202–207. [CrossRef] [PubMed]
21. Li, S.; Lakshmanan, R.S.; Petrenko, V.A.; Chin, B.A. Phage-Based Pathogen Biosensors. In *Phage Nanobiotechnology*; Petrenko, V.A., Smith, G.P., Eds.; RSC Publishing: Cambridge, UK, 2011; p. 273.
22. Ivanenkov, V.; Felici, F.; Menon, A.G. Uptake and intracellular fate of phage display vectors in mammalian cells. *Biochim. Biophys. Acta* **1999**, *1448*, 450–462. [CrossRef]
23. Romanov, V.I.; Durand, D.B.; Petrenko, V.A. Phage display selection of peptides that affect prostate carcinoma cells attachment and invasion. *Prostate* **2001**, *47*, 239–251. [CrossRef] [PubMed]
24. Huie, M.A.; Cheung, M.C.; Muench, M.O.; Becerril, B.; Kan, Y.W.; Marks, J.D. Antibodies to human fetal erythroid cells from a nonimmune phage antibody library. *Proc. Natl. Acad. Sci. USA* **2001**, *98*, 2682–2687. [CrossRef] [PubMed]
25. Legendre, D.; Fastrez, J. Construction and exploitation in model experiments of functional selection of a landscape library expressed from a phagemid. *Gene* **2002**, *290*, 203–215. [CrossRef]
26. Pires, D.P.; Cleto, S.; Sillankorva, S.; Azeredo, J.; Lu, T.K. Genetically Engineered Phages: A Review of Advances over the Last Decade. *Microbiol. Mol. Biol. Rev.* **2016**, *80*, 523–543. [CrossRef] [PubMed]
27. Petrenko, V.A.; Smith, G.P.; Mazooji, M.M.; Quinn, T. Alpha-helically constrained phage display library. *Protein Eng.* **2002**, *15*, 943–950. [CrossRef] [PubMed]
28. Marvin, D.A.; Hale, R.D.; Nave, C.; Citterich, M.H. Molecular-Models and Structural Comparisons of Native and Mutant Class-I Filamentous Bacteriophages Ff (Fd, F1, M13), If1 and Ike. *J. Mol. Biol.* **1994**, *235*, 260–286. [CrossRef]
29. Parmley, S.F.; Smith, G.P. Antibody-selectable filamentous fd phage vectors: Affinity purification of target genes. *Gene* **1988**, *73*, 305–318. [CrossRef]
30. Zacher, A.N.I.; Stock, C.A.; Golden, J.W.I.; Smith, G.P. A new filamentous phage cloning vector: Fd-tet. *Gene* **1980**, *9*, 127–140. [CrossRef]
31. Scott, J.K.; Smith, G.P. Searching for peptide ligands with an epitope library. *Science* **1990**, *249*, 386–390. [CrossRef] [PubMed]
32. Cwirla, S.E.; Peters, E.A.; Barrett, R.W.; Dower, W.J. Peptides on phage: A vast library of peptides for identifying ligands. *Proc. Natl. Acad. Sci. USA* **1990**, *87*, 6378–6382. [CrossRef] [PubMed]
33. McCafferty, J.; Griffiths, A.D.; Winter, G.; Chiswell, D.J. Phage antibodies: Filamentous phage displaying antibody variable domains. *Nature* **1990**, *348*, 552–554. [CrossRef] [PubMed]
34. Clackson, T.; Hoogenboom, H.R.; Griffiths, A.D.; Winter, G. Making antibody fragments using phage display libraries. *Nature* **1991**, *352*, 624–628. [CrossRef] [PubMed]
35. Kuzmicheva, G.A.; Jayanna, P.K.; Sorokulova, I.B.; Petrenko, V.A. Diversity and censoring of landscape phage libraries. *Protein Eng. Des. Sel.* **2009**, *22*, 9–18. [CrossRef] [PubMed]
36. Iannolo, G.; Minenkova, O.; Petruzzelli, R.; Cesareni, G. Modifying filamentous phage capsid: Limits in the size of the major capsid protein. *J. Mol. Biol.* **1995**, *248*, 835–844. [CrossRef] [PubMed]
37. Malik, P.; Terry, T.D.; Bellintani, F.; Perham, R.N. Factors limiting display of foreign peptides on the major coat protein of filamentous bacteriophage capsids and a potential role for leader peptidase. *FEBS Lett.* **1998**, *436*, 263–266. [CrossRef]

38. Greenwood, J.; Willis, A.E.; Perham, R.N. Multiple display of foreign peptides on a filamentous bacteriophage. Peptides from *Plasmodium falciparum* circumsporozoite protein as antigens. *J. Mol. Biol.* **1991**, *220*, 821–827. [CrossRef]
39. Malik, P.; Perham, R.N. New vectors for peptide display on the surface of filamentous bacteriophage. *Gene* **1996**, *171*, 49–51. [CrossRef]
40. Malik, P.; Terry, T.D.; Gowda, L.R.; Langara, A.; Petukhov, S.A.; Symmons, M.F.; Welsh, L.C.; Marvin, D.A.; Perham, R.N. Role of capsid structure and membrane protein processing in determining the size and copy number of peptides displayed on the major coat protein of filamentous bacteriophage. *J. Mol. Biol.* **1996**, *260*, 9–21. [CrossRef] [PubMed]
41. Iannolo, G.; Minenkova, O.; Gonfloni, S.; Castagnoli, L.; Cesareni, G. Construction, exploitation and evolution of a new peptide library displayed at high density by fusion to the major coat protein of filamentous phage. *Biol. Chem.* **1997**, *378*, 517–521. [CrossRef] [PubMed]
42. Ilyichev, A.A.; Minenkova, O.O.; Tatkov, S.I.; Karpyshev, N.N.; Eroshkin, A.M.; Petrenko, V.A.; Sandakhchiev, L.S. Construction of M13 viable bacteriophage with the insert of foreign peptides into the major coat protein. *Dokl. Biochem.* **1989**, *307*, 196–198.
43. Minenkova, O.O.; Ilyichev, A.A.; Kishchenko, G.P.; Petrenko, V.A. Design of specific immunogens using filamentous phage as the carrier. *Gene* **1993**, *128*, 85–88. [CrossRef]
44. Kishchenko, G.P.; Minenkova, O.O.; Ilichev, A.A.; Gruzdev, A.D.; Petrenko, V.A. Study of the structure of phage-M13 virions containing chimeric B-protein molecules. *Mol. Biol.-Engl. Trans.* **1991**, *25*, 1171–1176.
45. Felici, F.; Castagnoli, L.; Musacchio, A.; Jappelli, R.; Cesareni, G. Selection of antibody ligands from a large library of oligopeptides expressed on a multivalent exposition vector. *J. Mol. Biol.* **1991**, *222*, 301–310. [CrossRef]
46. Markland, W.; Roberts, B.L.; Saxena, M.J.; Guterman, S.K.; Ladner, R.C. Design, construction and function of a multicopy display vector using fusions to the major coat protein of bacteriophage M13. *Gene* **1991**, *109*, 13–19. [CrossRef]
47. Il'ichev, A.A.; Minenkova, O.O.; Tat'kov, S.I.; Karpyshev, N.N.; Eroshkin, A.M.; Petrenko, V.A.; Sandakhchiev, L.S. Poluchenie zhiznesposobnogo varianta faga M13 so vstroennym chuzherodnym peptidom v osnovnoĭ belok obolochki. *Dokl. Akad. Nauk. SSSR* **1989**, *307*, 481–483. (In Russian) [PubMed]
48. Kishchenko, G.; Batliwala, H.; Makowski, L. Structure of a foreign peptide displayed on the surface of bacteriophage M13. *J. Mol. Biol.* **1994**, *241*, 208–213. [CrossRef] [PubMed]
49. Petrenko, V.A.; Smith, G.P. Phages from landscape libraries as substitute antibodies. *Protein Eng.* **2000**, *13*, 589–592. [CrossRef] [PubMed]
50. Kishchenko, G.P.; Minenkova, O.O.; Il'ichev, A.I.; Gruzdev, A.D.; Petrenko, V.A. Izuchenie struktury virionov faga M13, soderzhashchikh molekuly khimernykh B-belkov. *Mol. Biol.* **1991**, *25*, 1497–1503. (In Russian)
51. Marvin, J.S.; Lowman, H.B. Chapter 15: Antibody Humanization and Affinity Maturation Using Phage Display. In *Phage Display in Biotechnology and Drug Discovery*, 2nd ed.; Geyer, C.R., Sidhu, S.S., Eds.; CRC Press, Taylor & Francis Group: Boca Raton, FL, USA, 2005; pp. 347–371.
52. Petrenko, V.A.; Sorokulova, I.B. Detection of biological threats. A challenge for directed molecular evolution. *J. Microbiol. Methods* **2004**, *58*, 147–168. [CrossRef] [PubMed]
53. Gross, A.L.; Gillespie, J.W.; Petrenko, V.A. Promiscuous tumor targeting phage proteins. *Protein Eng. Des. Sel.* **2016**, *29*, 93–103. [CrossRef] [PubMed]
54. Petrenko, V.A.; Jayanna, P.K. Phage-mediated Drug Delivery. In *Phage Nanobiotechnology*; Petrenko, V.A., Smith, G.P., Eds.; RSC Publishing: Cambridge, UK, 2011; pp. 55–82.
55. Petrenko, V.A.; Jayanna, P.K. Phage protein-targeted cancer nanomedicines. *FEBS Lett.* **2014**, *588*, 341–349. [CrossRef] [PubMed]
56. Petrenko, V.A.; Gillespie, J.W. Paradigm shift in bacteriophage-mediated delivery of anticancer drugs: From targeted 'magic bullets' to self-navigated 'magic missiles'. *Expert Opin. Drug Deliv.* **2016**, 1–12. [CrossRef] [PubMed]
57. Petrenko, V.A.; Vodyanoy, V.J. Phage display for detection of biological threat agents. *J. Microbiol. Methods* **2003**, *53*, 253–262. [CrossRef]
58. Petrenko, V.A.; Brigati, J.R. Phage as Biospecific Probes. In *Immunoassay and Other Bioanalytical Techniques*; Van Emon, J.M., Ed.; CRC Press, Taylor & Francis Group: Boca Raton, FL, USA, 2007.

59. Brigati, J.R.; Samoylova, T.I.; Jayanna, P.K.; Petrenko, V.A. Phage display for generating peptide reagents. *Curr. Protoc. Protein Sci.* **2008**. [CrossRef]
60. Brigati, J.; Williams, D.D.; Sorokulova, I.B.; Nanduri, V.; Chen, I.H.; Turnbough, C.L., Jr.; Petrenko, V.A. Diagnostic probes for *Bacillus anthracis* spores selected from a landscape phage library. *Clin. Chem.* **2004**, *50*, 1899–1906. [CrossRef] [PubMed]
61. Sorokulova, I.B.; Olsen, E.V.; Chen, I.H.; Fiebor, B.; Barbaree, J.M.; Vodyanoy, V.J.; Chin, B.A.; Petrenko, V.A. Landscape phage probes for *Salmonella typhimurium*. *J. Microbiol. Methods* **2005**, *63*, 55–72. [CrossRef] [PubMed]
62. Yin, L.; Luo, Y.; Liang, B.; Wang, F.; Du, M.; Petrenko, V.A.; Qiu, H.J.; Liu, A. Specific ligands for classical swine fever virus screened from landscape phage display library. *Antivir. Res.* **2014**, *109*, 68–71. [CrossRef] [PubMed]
63. Liu, P.; Han, L.; Wang, F.; Li, X.Q.; Petrenko, V.A.; Liu, A.H. Sensitive colorimetric immunoassay of *Vibrio parahaemolyticus* based on specific nonapeptide probe screening from a phage display library conjugated with MnO_2 nanosheets with peroxidase-like activity. *Nanoscale* **2018**, *10*, 2825–2833. [CrossRef] [PubMed]
64. Liu, P.; Han, L.; Wang, F.; Petrenko, V.A.; Liu, A.H. Gold nanoprobe functionalized with specific fusion protein selection from phage display and its application in rapid, selective and sensitive colorimetric biosensing of *Staphylococcus aureus*. *Biosens. Bioelectron.* **2016**, *82*, 195–203. [CrossRef] [PubMed]
65. Nanduri, V.; Sorokulova, I.B.; Samoylov, A.M.; Simonian, A.L.; Petrenko, V.A.; Vodyanoy, V. Phage as a molecular recognition element in biosensors immobilized by physical adsorption. *Biosens. Bioelectron.* **2007**, *22*, 986–992. [CrossRef] [PubMed]
66. Olsen, E.V.; Sorokulova, I.B.; Petrenko, V.A.; Chen, I.H.; Barbaree, J.M.; Vodyanoy, V.J. Affinity-selected filamentous bacteriophage as a probe for acoustic wave biodetectors of *Salmonella typhimurium*. *Biosens. Bioelectron.* **2006**, *21*, 1434–1442. [CrossRef] [PubMed]
67. Nanduri, V.; Balasubramanian, S.; Sista, S.; Vodyanoy, V.J.; Simonian, A.L. Highly sensitive phage-based biosensor for the detection of β-galactosidase. *Anal. Chim. Acta* **2007**, *589*, 166–172. [CrossRef] [PubMed]
68. Knez, K.; Noppe, W.; Geukens, N.; Janssen, K.P.; Spasic, D.; Heyligen, J.; Vriens, K.; Thevissen, K.; Cammue, B.P.; Petrenko, V.; et al. Affinity comparison of p3 and p8 peptide displaying bacteriophages using surface plasmon resonance. *Anal. Chem.* **2013**, *85*, 10075–10082. [CrossRef] [PubMed]
69. Han, L.; Liu, P.; Petrenko, V.A.; Liu, A.H. A Label-Free Electrochemical Impedance Cytosensor Based on Specific Peptide-Fused Phage Selected from Landscape Phage Library. *Sci. Rep.* **2016**, *6*, 10. [CrossRef] [PubMed]
70. Wang, F.; Liu, P.; Sun, L.; Li, C.; Petrenko, V.A.; Liu, A. Bio-mimetic nanostructure self-assembled from Au@Ag heterogeneous nanorods and phage fusion proteins for targeted tumor optical detection and photothermal therapy. *Sci. Rep.* **2014**, *4*, 6808. [CrossRef] [PubMed]
71. Ogata, A.F.; Edgar, J.M.; Majumdar, S.; Briggs, J.S.; Patterson, S.V.; Tan, M.X.; Kudlacek, S.T.; Schneider, C.A.; Weiss, G.A.; Penner, R.M. Virus-Enabled Biosensor for Human Serum Albumin. *Anal. Chem.* **2017**, *89*, 1373–1381. [CrossRef] [PubMed]
72. Wan, J.; Johnson, M.; Guntupalli, R.; Petrenko, V.A.; Chin, B.A. Detection of *Bacillus anthracis* spores in liquid using phage-based magnetoelastic micro-resonators. *Sens. Actuators B Chem.* **2007**, *127*, 559–566. [CrossRef]
73. Wan, J.; Shu, H.; Huang, S.; Fiebor, B.; Chen, I.-H.; Petrenko, V.A.; Chin, B.A. Phage-based Magnetoelastic wireless biosensors for detecting *Bacillus anthracis* spores. *IEEE Sens. J.* **2007**, *7*, 470–477. [CrossRef]
74. Lakshmanan, R.S.; Guntupalli, R.; Hu, J.; Kim, D.J.; Petrenko, V.A.; Barbaree, J.M.; Chin, B.A. Phage immobilized magnetoelastic sensor for the detection of *Salmonella typhimurium*. *J. Microbiol. Methods* **2007**, *71*, 55–60. [CrossRef] [PubMed]
75. Fu, L.; Li, S.; Zhang, K.; Chen, I.H.; Petrenko, V.A.; Cheng, Z. Magnetostrictive Microcantilever as an Advanced Transducer for Biosensors. *Sensors* **2007**, *7*, 2929–2941. [CrossRef] [PubMed]
76. Samoylov, A.; Cochran, A.; Schemera, B.; Kutzler, M.; Donovan, C.; Petrenko, V.; Bartol, F.; Samoylova, T. Humoral immune responses against gonadotropin releasing hormone elicited by immunization with phage-peptide constructs obtained via phage display. *J. Biotechnol.* **2015**, *216*, 20–28. [CrossRef] [PubMed]
77. Kouzmitcheva, G.A.; Petrenko, V.A.; Smith, G.P. Identifying diagnostic peptides for Lyme disease through epitope discovery. *Clin. Diagn. Lab. Immunol.* **2001**, *8*, 150–160. [CrossRef] [PubMed]

78. Lang, Q.; Wang, F.; Yin, L.; Liu, M.; Petrenko, V.A.; Liu, A. Specific probe selection from landscape phage display library and its application in enzyme-linked immunosorbent assay of free prostate-specific antigen. *Anal. Chem.* **2014**, *86*, 2767–2774. [CrossRef] [PubMed]
79. Han, L.; Xia, H.; Yin, L.; Petrenko, V.A.; Liu, A. Selected landscape phage probe as selective recognition interface for sensitive total prostate-specific antigen immunosensor. *Biosens. Bioelectron.* **2018**, *106*, 1–6. [CrossRef] [PubMed]
80. Qi, H.; Wang, F.; Petrenko, V.A.; Liu, A. Peptide microarray with ligands at high density based on symmetrical carrier landscape phage for detection of cellulase. *Anal. Chem.* **2014**, *86*, 5844–5850. [CrossRef] [PubMed]
81. Sanchez-Purra, M.; Ramos, V.; Petrenko, V.A.; Torchilin, V.P.; Borros, S. Double-targeted polymersomes and liposomes for multiple barrier crossing. *Int. J. Pharm.* **2016**, *511*, 946–956. [CrossRef] [PubMed]
82. Wang, T.; Hartner, W.C.; Gillespie, J.W.; Praveen, K.P.; Yang, S.; Mei, L.A.; Petrenko, V.A.; Torchilin, V.P. Enhanced tumor delivery and antitumor activity in vivo of liposomal doxorubicin modified with MCF-7-specific phage fusion protein. *Nanomedicine* **2014**, *10*, 421–430. [CrossRef] [PubMed]
83. Wang, T.; Yang, S.; Mei, L.A.; Parmar, C.K.; Gillespie, J.W.; Praveen, K.P.; Petrenko, V.A.; Torchilin, V.P. Paclitaxel-loaded PEG-PE-based micellar nanopreparations targeted with tumor-specific landscape phage fusion protein enhance apoptosis and efficiently reduce tumors. *Mol. Cancer Ther.* **2014**, *13*, 2864–2875. [CrossRef] [PubMed]
84. Gillespie, J.W.; Wei, L.; Petrenko, V.A. Selection of Lung Cancer-Specific Landscape Phage for Targeted Drug Delivery. *Comb. Chem. High Throughput Screen.* **2016**, *19*, 412–422. [CrossRef] [PubMed]
85. Gillespie, J.W.; Gross, A.L.; Puzyrev, A.T.; Bedi, D.; Petrenko, V.A. Combinatorial synthesis and screening of cancer cell-specific nanomedicines targeted via phage fusion proteins. *Front. Microbiol.* **2015**, *6*, 628. [CrossRef] [PubMed]
86. Bedi, D.; Gillespie, J.W.; Petrenko, V.A., Jr.; Ebner, A.; Leitner, M.; Hinterdorfer, P.; Petrenko, V.A. Targeted delivery of siRNA into breast cancer cells via phage fusion proteins. *Mol. Pharm.* **2013**, *10*, 551–559. [CrossRef] [PubMed]
87. Bedi, D.; Musacchio, T.; Fagbohun, O.A.; Gillespie, J.W.; Deinnocentes, P.; Bird, R.C.; Bookbinder, L.; Torchilin, V.P.; Petrenko, V.A. Delivery of siRNA into breast cancer cells via phage fusion protein-targeted liposomes. *Nanomedicine* **2011**, *7*, 315–323. [CrossRef] [PubMed]
88. Jayanna, P.K.; Bedi, D.; Deinnocentes, P.; Bird, R.C.; Petrenko, V.A. Landscape phage ligands for PC3 prostate carcinoma cells. *Protein Eng. Des. Sel.* **2010**, *23*, 423–430. [CrossRef] [PubMed]
89. Jayanna, P.K.; Bedi, D.; Gillespie, J.W.; DeInnocentes, P.; Wang, T.; Torchilin, V.P.; Bird, R.C.; Petrenko, V.A. Landscape phage fusion protein-mediated targeting of nanomedicines enhances their prostate tumor cell association and cytotoxic efficiency. *Nanomedicine* **2010**, *6*, 538–546. [CrossRef] [PubMed]
90. Jayanna, P.K.; Torchilin, V.P.; Petrenko, V.A. Liposomes targeted by fusion phage proteins. *Nanomedicine* **2009**, *5*, 83–89. [CrossRef] [PubMed]
91. Fagbohun, O.A.; Bedi, D.; Grabchenko, N.I.; Deinnocentes, P.A.; Bird, R.C.; Petrenko, V.A. Landscape phages and their fusion proteins targeted to breast cancer cells. *Protein Eng. Des. Sel.* **2012**, *25*, 271–283. [CrossRef] [PubMed]
92. Wang, T.; D'Souza, G.G.; Bedi, D.; Fagbohun, O.A.; Potturi, L.P.; Papahadjopoulos-Sternberg, B.; Petrenko, V.A.; Torchilin, V.P. Enhanced binding and killing of target tumor cells by drug-loaded liposomes modified with tumor-specific phage fusion coat protein. *Nanomedicine* **2010**, *5*, 563–574. [CrossRef] [PubMed]
93. Wang, T.; Kulkarni, N.; Bedi, D.; D'Souza, G.G.; Papahadjopoulos-Sternberg, B.; Petrenko, V.A.; Torchilin, V.P. In vitro optimization of liposomal nanocarriers prepared from breast tumor cell specific phage fusion protein. *J. Drug Target.* **2011**, *19*, 597–605. [CrossRef] [PubMed]
94. Wang, T.; Petrenko, V.A.; Torchilin, V.P. Paclitaxel-loaded polymeric micelles modified with MCF-7 cell-specific phage protein: Enhanced binding to target cancer cells and increased cytotoxicity. *Mol. Pharm.* **2010**, *7*, 1007–1014. [CrossRef] [PubMed]
95. Wang, T.; Yang, S.; Petrenko, V.A.; Torchilin, V.P. Cytoplasmic delivery of liposomes into MCF-7 breast cancer cells mediated by cell-specific phage fusion coat protein. *Mol. Pharm.* **2010**, *7*, 1149–1158. [CrossRef] [PubMed]
96. Wang, T.; Kulkarni, N.; D'Souza, G.G.; Petrenko, V.A.; Torchilin, V.P. On the mechanism of targeting of phage fusion protein-modified nanocarriers: Only the binding peptide sequence matters. *Mol. Pharm.* **2011**, *8*, 1720–1728. [CrossRef] [PubMed]

97. Wang, J.; Dong, B.; Chen, B.; Jiang, Z.; Song, H. Selective photothermal therapy for breast cancer with targeting peptide modified gold nanorods. *Dalton Trans.* **2012**, *41*, 11134–11144. [CrossRef] [PubMed]
98. Wang, T.; Petrenko, V.A.; Torchilin, V.P. Optimization of Landscape Phage Fusion Protein-Modified Polymeric PEG-PE Micelles for Improved Breast Cancer Cell Targeting. *J. Nanomed. Nanotechnol.* **2012**, 008. [CrossRef]
99. Houten, N.E.; Scott, J.K. Chapter 6: Phage Libraries for Developing Antibody-Targeted Diagnostics and Vaccines. In *Phage Display in Biotechnology and Drug Discovery*, 2nd ed.; Geyer, C.R., Sidhu, S.S., Eds.; CRC Press, Taylor & Francis Group: Boca Raton, FL, USA; 2005; pp. 123–179.
100. Smith, A.; Manoli, H.; Jaw, S.; Frutoz, K.; Epstein, A.L.; Khawli, L.A.; Theil, F.P. Unraveling the Effect of Immunogenicity on the PK/PD, Efficacy, and Safety of Therapeutic Proteins. *J. Immunol. Res.* **2016**, *2016*, 2342187. [CrossRef] [PubMed]
101. Opella, S.J. The Roles of Structure, Dynamics and Assembly in the Display of Peptides on Filamentous Bacteriophage. In *Phage Nanobiotechnology*; Petrenko, V.A., Smith, G.P., Eds.; RSC Publishing: Cambridge, UK, 2011; pp. 12–32.
102. Mount, J.D.; Samoylova, T.I.; Morrison, N.E.; Cox, N.R.; Baker, H.J.; Petrenko, V.A. Cell targeted phagemid rescued by preselected landscape phage. *Gene* **2004**, *341*, 59–65. [CrossRef] [PubMed]
103. Marvin, D.A.; Symmons, M.F.; Straus, S.K. Structure and assembly of filamentous bacteriophages. *Prog. Biophys. Mol. Biol.* **2014**, *114*, 80–122. [CrossRef] [PubMed]
104. Vukovic, L.; Khatib, F.A.; Drake, S.P.; Madriaga, A.; Brandenburg, K.S.; Kral, P.; Onyuksel, H. Structure and dynamics of highly PEG-ylated sterically stabilized micelles in aqueous media. *J. Am. Chem. Soc.* **2011**, *133*, 13481–13488. [CrossRef] [PubMed]
105. Papavoine, C.H.; Christiaans, B.E.; Folmer, R.H.; Konings, R.N.; Hilbers, C.W. Solution structure of the M13 major coat protein in detergent micelles: A basis for a model of phage assembly involving specific residues. *J. Mol. Biol.* **1998**, *282*, 401–419. [CrossRef] [PubMed]

7

Ultrasensitive and Fast Diagnostics of Viable *Listeria* Cells by CBD Magnetic Separation Combined with A511::*luxAB* Detection

Jan W. Kretzer, Mathias Schmelcher and Martin J. Loessner *

Institute of Food, Nutrition and Health, ETH Zurich, Schmelzbergstrasse 7, 8092 Zurich, Switzerland; Jan.Kretzer@reg-ob.bayern.de (J.W.K.); mathias.schmelcher@hest.ethz.ch (M.S.)
* Correspondence: martin.loessner@ethz.ch

Abstract: The genus *Listeria* includes foodborne pathogens that cause life-threatening infections in those at risk, and sensitive and specific methods for detection of these bacteria are needed. Based on their unrivaled host specificity and ability to discriminate viable cells, bacteriophages represent an ideal toolbox for the development of such methods. Here, the authors describe an ultrasensitive diagnostic protocol for *Listeria* by combining two phage-based strategies: (1) specific capture and concentration of target cells by magnetic separation, harnessing cell wall-binding domains from *Listeria* phage endolysins (CBD-MS); and (2) highly sensitive detection using an adaptation of the A511::*luxAB* bioluminescent reporter phage assay in a microwell plate format. The combined assay enabled direct detection of approximately 100 bacteria per ml of pure culture with genus-level specificity in less than 6 h. For contaminated foods, the procedure included a 16 h selective enrichment step, followed by CBD-MS separation and A511::*luxAB* detection. It was able to consistently detect extremely low numbers (0.1 to 1.0 cfu/g) of viable *Listeria* cells, in a total assay time of less than 22 h. These results demonstrate the superiority of this phage-based assay to standard culture-based diagnostic protocols for the detection of viable bacteria, with respect to both sensitivity and speed.

Keywords: bacteriophage; diagnostics; *Listeria monocytogenes*; endolysin; magnetic separation; reporter phage

---

1. Introduction

The opportunistic human pathogen *Listeria monocytogenes* is exclusively transmitted via contaminated food, and poses a serious threat to immunocompromised individuals, children, and the elderly. Listeriosis is a disease which can cause severe symptoms such as meningitis, septicemia, and, in pregnant women, spontaneous abortion [1]. Although the incidence of listeriosis is relatively low, it is still of major concern due to reported average mortality rates of up to 30% [1–3]. As *L. monocytogenes* can multiply at refrigeration temperatures, even low contamination levels are considered a potential risk. Since foods that harbor *Listeria* generally feature a highly abundant and diverse background flora, highly sensitive diagnostic protocols are required. Standard culture-based methods for the detection of *Listeria* in foods such as the International Organization for Standardization (ISO) standard plating method (ISO 11290-1:2017) are still considered as the gold standard; however, they require a minimum of 96 h to obtain preliminary results. This emphasizes the need for alternative rapid screening and diagnostic protocols. A number of culture-independent detection methods for bacterial pathogens have been described, including polymerase chain reaction (PCR)-based protocols, mass spectrometry (MS), and immunological techniques. However, each of these methods has important drawbacks, such as the necessity for lengthy pre-enrichment procedures, the inability to distinguish between live and dead cells (PCR), the requirement for expensive devices (MS), or a lack in sensitivity and/or specificity (immunological assays) [4–6]. Over the past two decades, bacteriophages have emerged as ideal tools for bacterial diagnostics, particularly due to their extraordinary specificity for their target cells, their robustness and inexpensive production, and their ability to selectively detect viable cells [7]. A unique approach for the rapid detection of viable *Listeria* is the luciferase reporter bacteriophage A511::*luxAB* [8]. This derivative of the broad-host-range virulent *Listeria* phage A511 was constructed by inserting a fused *luxAB* gene from *Vibrio harveyi* downstream of the strongly expressed major capsid protein gene of the phage [8]. Upon infection of *Listeria* cells present in a given sample, a measurable bioluminescence signal is produced within approximately 60–150 min. A511::*luxAB* was then used for the detection of *L. monocytogenes* from artificially contaminated food and environmental samples [9]. With a total assay time of 24 to 48 h, the method was significantly faster than the standard plating protocols, while featuring similar sensitivity. However, the original assay format was somewhat impractical due to the use of single large plastic tubes with all manual handling, and could not be automated. In addition, the lower detection limit of approximately 1×10^3 cfu/mL did not meet the stringent requirements in food safety testing.

The aims of this study were to adapt the reporter phage assay to a 96-well microwell plate format to enable high throughput screening of samples; to enhance the luminescent light signal by the addition of sodium azide (NaN_3); and to combine the assay with upstream bead-based magnetic separation utilizing cell wall-binding domains from endolysins specific for *Listeria* spp. (CBD-MS) [6] in an effort to improve detection sensitivity and reduce the overall assay time. The functionalized magnetic beads allow rapid and efficient capturing and magnetic separation of *Listeria* cells from contaminated food samples, with genus-level specificity and high target cell affinity.

2. Materials and Methods

2.1. Bacteria and Culture Conditions

All bacteria used in this study were taken from the Weihenstephan *Listeria* collection (WSLC). *L. monocytogenes* strains EGDe (serovar 1/2 a), WSLC 1001 (ATCC 19112) (sv 1/2c), WSLC 1685 (ScottA) (sv 4b), WSLC 2012 (ATCC 33091) (sv 6b), and *Listeria ivanovii* WSLC 3009 (sv 5) were grown in half-concentrated brain heart infusion medium (BHI $\frac{1}{2}$) (Oxoid, Hampshire, UK), at 30 °C. Cultures were incubated for 16 h, diluted fivefold in fresh medium, incubated for another 2 h, and subsequently diluted in phosphate buffered saline (PBS) supplemented with Tween 20 (Sigma-Aldrich, Buchs, Switzerland) (PBST; 50 mM Na_2HPO_4; 120 mM NaCl; 0.1% Tween 20; pH 8.0) to the desired concentration of cells.

2.2. Production of CBD Fusion Proteins and Coating of Magnetic Beads

N-terminally 6xHis-tagged fusion proteins consisting of the green fluorescent protein (GFP) and cell wall-binding domains (CBDs) from bacteriophage endolysins Ply118 and Ply500 (HGFP_CBD118 and HGFP_CBD500) were expressed in *Escherichia coli* and purified as previously described [6,10,11]. No glycerol was added to the purified protein preparations. Concentrations of purified proteins were adjusted to 2.5 mg/mL by the addition of PBS (50 mM Na_2HPO_4, 120 mM NaCl, pH 8.0). Protein preparations were stored at −20 °C until use. Paramagnetic polystyrene beads (Dynabeads M-270 Epoxy; Dynal, Oslo, Norway) were activated and coated with the purified CBDs as described previously [6].

2.3. Propagation of A511::*luxAB*

Bacteriophage A511::*luxAB* was propagated as described previously [9]. Phage concentration was adjusted to 1.0×10^9 pfu/mL by the addition of SM buffer (SMB; 94 mM NaCl, 8 mM $MgSO_4 \times 7H_2O$, 100 mM Tris, pH 7.4), and phage stocks were stored at 4 °C until use.

2.4. Downsizing the A511::*luxAB* Reporter Phage Assay from Single Tube to Microwell Format

For the adaptation of the A511::*luxAB* luciferase reporter phage assay [8,9] to microplate format, the optimum conditions concerning the bacteria/phage ratio, incubation time, and incubation temperature were determined. Experiments were carried out using pure cultures of *L. monocytogenes* ScottA, and the preparation of bacterial cells was as described above, with a slight modification: dilutions of subcultures were prepared using BHI ($\frac{1}{2}$) medium instead of PBST to ensure metabolic fitness of bacterial target cells during phage infection. Dilutions of subcultures with bacterial concentrations ranging from 3.0×10^3 to 1.0×10^6 cfu/mL were incubated with varying concentrations of phage particles ranging from 1.0×10^6 to 3.0×10^8 pfu/mL. The incubation temperature was varied in a range from 20 °C to 30 °C, whereas the incubation time was varied from 120 to 180 min. For each experiment, 200 μL of bacterial suspensions were mixed with 20 μL of phage suspensions in individual wells of a 96-well microwell plate (Black/White Isoplates (black frames with white well inserts), Perkin Elmer, Schwerzenbach, Switzerland). Measurements were performed using a multilabel microplate reader equipped with dual injectors (Viktor$_3$, Perkin Elmer). Following preliminary experiments, the consecutive steps of the optimized measurement protocol were as follows: injection of 12 μL of a sodium azide (NaN_3; Sigma-Aldrich) solution (1% in H_2O, resulting in a final concentration of 7.6 mM), immediately followed by the injection of 12 μL of a nonanal (nonyl aldehyde, Sigma-Aldrich) solution (0.25% *v*/*v*, in 70% ethanol) (Loessner et al., 1996a). Measurement of light emission was performed using 10 s intervals for integration and calculation of relative light units (RLUs). The average value of RLUs per second was displayed as cps (counts per second).

2.5. Effect of NaN_3

To determine the effect of NaN_3 on the signal intensity in the A511::*luxAB* reporter phage assay, experiments with *L. monocytogenes* WSLC ScottA subcultures were carried out essentially as described above. Cell concentrations in samples ranged from 5.0×10^4 to 5.0×10^6 cfu/mL. The treatment of samples and subsequent measurements were performed at optimum conditions as determined before, with and without the use of NaN_3 in the assay. Injection of 12 μL of a NaN_3 solution (stock concentration: 1%, corresponding to 153.8 mM; final assay concentration: 7.6 mM) prior to the injection of 12 μL nonanal solution was then incorporated into the optimized protocol (see above).

2.6. Evaluation of the Combined CBD-MS/A511::*luxAB* Assay

One ml aliquots of *Listeria* cultures of the different strains tested (*L. monocytogenes* WSLC 1001, *Listeria innocua* WSLC 2012, *L. monocytogenes* ScottA, *L. monocytogenes* EGDe, and *L. ivanovii* WSLC 3009) containing approximately 1.0×10^2–1.0×10^5 cfu/mL, were mixed with 20 μL of CBD118-coated beads and 20 μL of CBD500-coated beads (2×10^7 beads in total). The magnetic separation procedure was performed as previously described [6]. In brief, samples were incubated in an overhead rotator (NeoLab, Heidelberg, Germany) at 10 rpm for 40 min, and subsequently magnetic beads were separated using a MPC magnetic tube holder (Dynal, Oslo, Norway). Following careful removal of the supernatant, beads were washed with 1 mL of PBST, separated again, and finally resuspended in 200 μL of BHI ($\frac{1}{2}$). The bead suspension with captured bacterial cells was incubated in a horizontal shaker (Kühner, Birsfelden, Switzerland) at 37 °C and 180 rpm for 180 min and subsequently mixed with 20 μL of A511::*luxAB* to yield a final assay concentration of 1.0×10^8 pfu/mL, followed by incubation at 24 °C for 140 min. Since preliminary testing indicated that the presence of beads could interfere with light emission, beads were magnetically separated as described above, and supernatants were collected and transferred to a microwell plate. To determine background luminescence, negative control samples containing either bacteriophages (1.0×10^8 pfu/mL) but no *Listeria* cells, or *Listeria* cells (1.0×10^6 cfu/mL) but no phage, were included in the experiments. All experiments were carried out in triplicate. The exact numbers of cells in each experiment used were determined by triplicate plating of 100 μL aliquots of appropriate dilutions of *Listeria* subcultures on BHI agar, and subsequent incubation at 37 °C for 16 h.

2.7. Testing Artificially Contaminated Foods

A set of nine different food items, including sliced iceberg lettuce, chocolate milk, mozzarella cheese, Swiss "Vacherin" red smear soft cheese, Swiss "Tomme" white mold soft cheese, ready-to-eat shrimp, minced meat, smoked salmon, and smoked turkey breast, was used in these experiments. All food samples were purchased at local retailers. First, each sample was aseptically divided into portions of 25 g each in a laminar flow hood. While one portion of each food was tested for the presence of *Listeria* (i.e., natural contamination) by standard procedures, all others were packed into sterile polypropylene plastic bags and immediately frozen at −80 °C. Samples that were free of *Listeria* were used for artificial contamination. Samples were thawed and inoculated with 1 mL each of diluted subcultures of either *L. monocytogenes* EGDe (serovar 1/2 a) or *L. monocytogenes* ScottA (serovar 4b) to obtain initial contamination levels ranging from 0.1 to 100 cfu/g food. In the case of solid foods, the inoculum was distributed on the surface of the sample by carefully massaging the bag. Exact initial contamination levels of food samples were determined by triplicate plating of 100 μL aliquots of appropriate dilutions of subcultures used for contamination. To determine the background, blank samples were prepared by adding 1 mL of PBST containing no *Listeria* cells to one sample of each food item.

All samples were then stored at 4 °C for 22–24 h, in order to simulate more realistic storage and sampling conditions. The authors did not observe any growth of the spiked *Listeria* cells during this period, which confirms other studies that have reported no significant increase in *L. monocytogenes* viable counts during initial storage at 4 °C for 24 h [12,13]. Subsequently, the 25 g food samples were homogenized with 50 mL of citrate buffer (for dairy products; 58 mM tri-sodium citrate dihydrate in H_2O_{dd}, pH 7.5) or PBS (other products) using a stomacher laboratory blender (Seward, West Sussex, UK). Samples were subjected to 16 h of selective enrichment in 175 mL tryptic soy broth (TSB, pH 7.5; Biolife, Milan, Italy) containing 12 mg/L acriflavine, 52 mg/L nalidixic acid and 64 mg/L cycloheximide (all from Sigma-Aldrich) (IDF 143A:1995).

Detection of *Listeria* spp. in enrichment cultures was performed essentially as described above. However, 100 μL of 10× PBST (tenfold concentrated) was added to each 1 mL sample taken from the enrichment cultures to adjust the pH values. Individual samples of each food item, contaminated with different numbers of cells of either *L. monocytogenes* EGDe or ScottA, were prepared. Analysis of each sample was carried out in duplicate.

2.8. Statistical Analysis

One-way analysis of variance (ANOVA) with a post hoc Dunnett's multiple comparisons test was applied for comparison of log-transformed cps values with the respective control values in the combined CBD-MS/A511::*luxAB* assay. Two-way ANOVA followed by Sidak's multiple comparisons test was used for comparing mean normalized cps values from reporter phage assays with and without the use of NaN_3.

3. Results

3.1. Adaption of the Luciferase Reporter Phage Assay to Microwell Plates

In order to allow higher throughput processing of samples, the A511::*luxAB* reporter phage assay, developed for use in a single tube luminometer [8,9], was adapted to a 96-well microplate format for processing in a semi-automated multi-label microplate reader equipped with programmable injectors. The different types of microwell plates initially tested yielded very different signal-to-noise ratios; the best results were obtained using black frame plates with white well inserts. Next, the impact of sample volume, bacteria-phage ratio, incubation temperature, and incubation time on light signal emission was determined. The results obtained with the reduced sample volumes (200 µL) in the microplate format only slightly differed from those measured in standard single tube-based measurements using 1 mL samples [8,9]. An incubation time of 140 min yielded the highest signal intensity, which confirmed that phage infection and luciferase-catalyzed light emission are largely independent on the test volume and luminescence detector. An increase of incubation temperature from 20 °C to 24 °C yielded a slightly higher signal intensity, without affecting the stability of the heat-sensitive LuxAB fusion protein synthesized in phage-infected target cells [6,8]. The variation of the phage concentration in the assay ratio revealed that 1.0×10^8 pfu/mL is sufficient for the infection of *Listeria* target cells at variable concentrations up to 1.0×10^6 cfu/mL. With the addition of more phage, no increase in signal intensity could be achieved, whereas lower concentrations resulted in a decrease in signal intensity (Figure 1).

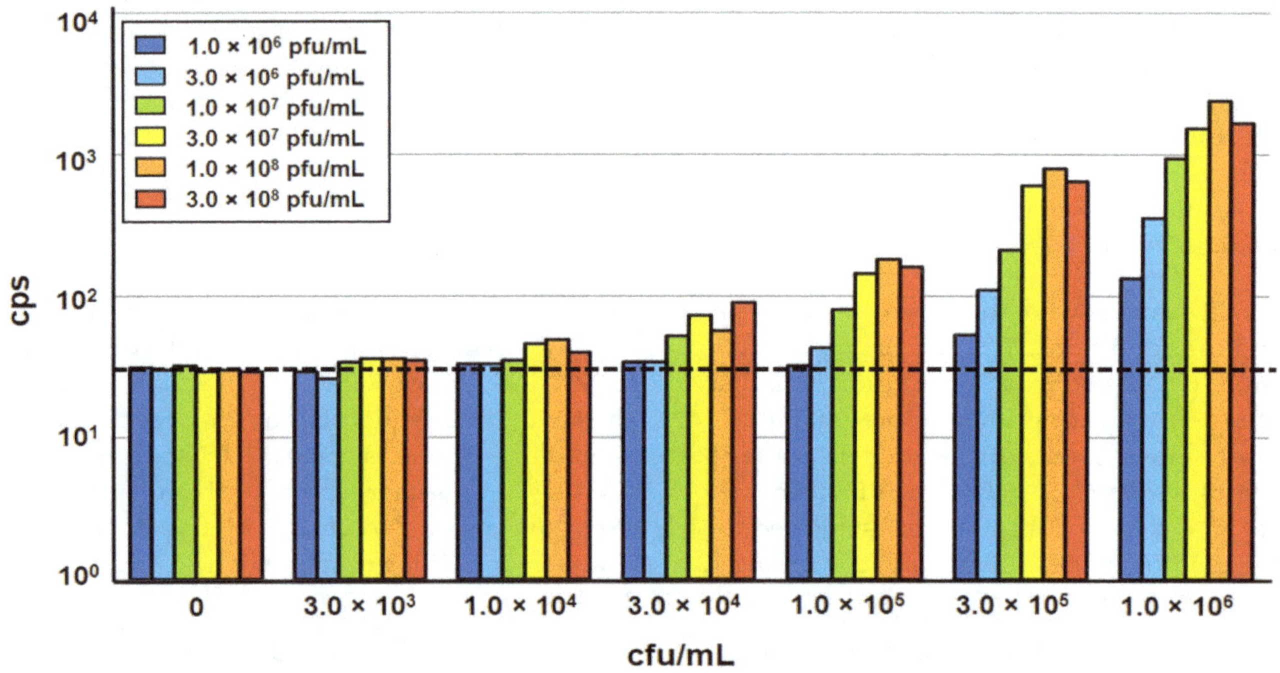

Figure 1. Evaluation of optimum bacteria:phage ratios in a 96-well microplate-based A511::*luxAB* assay. Increasing concentrations of *Listeria monocytogenes* ScottA (3.0×10^3 to 1.0×10^6 cfu/mL) were incubated with luciferase reporter phage A511::*luxAB* (1.0×10^6 to 3.0×10^8 pfu/mL). Bacteria were exposed to phage in 200 µL of half-concentrated brain heart infusion medium (BHI $\frac{1}{2}$) in 96-well plates, and incubated at 24 °C for 140 min. Bioluminescence measurements were performed over a total period of 10 s, and are reported as counts per second (cps). The dotted horizontal line indicates the level of background light emission (noise) of the assay, determined in the negative control sample (without phage).

3.2. Sodium Azide Enhances the Bioluminescence Signal

The azide anion (N_3^-) from NaN_3 inhibits the enzyme cytochrome oxidase in the respiratory chain [14,15]. This inhibition leads to a shift in the equilibrium of the reaction FMN $\rightarrow$ $FMNH_2$ to its reduced form $FMNH_2$. Here, $FMNH_2$ is a co-factor required for the luciferase-driven oxidation of the aldehyde substrate in the A511::*luxAB* assay [8], and a limiting factor of the reaction. Therefore, an increase in $FMNH_2$ concentration should result in increased light emission. To test this hypothesis, the effect of NaN_3 on light emission from the reporter phage assay was evaluated. Following several rounds of optimization, the authors found that the injection of 7.6 mM NaN_3 immediately prior to the injection of the nonanal luciferase aldehyde substrate significantly ($p < 0.0001$) increased the bioluminescence signal at all cell concentrations tested (Figure 2).

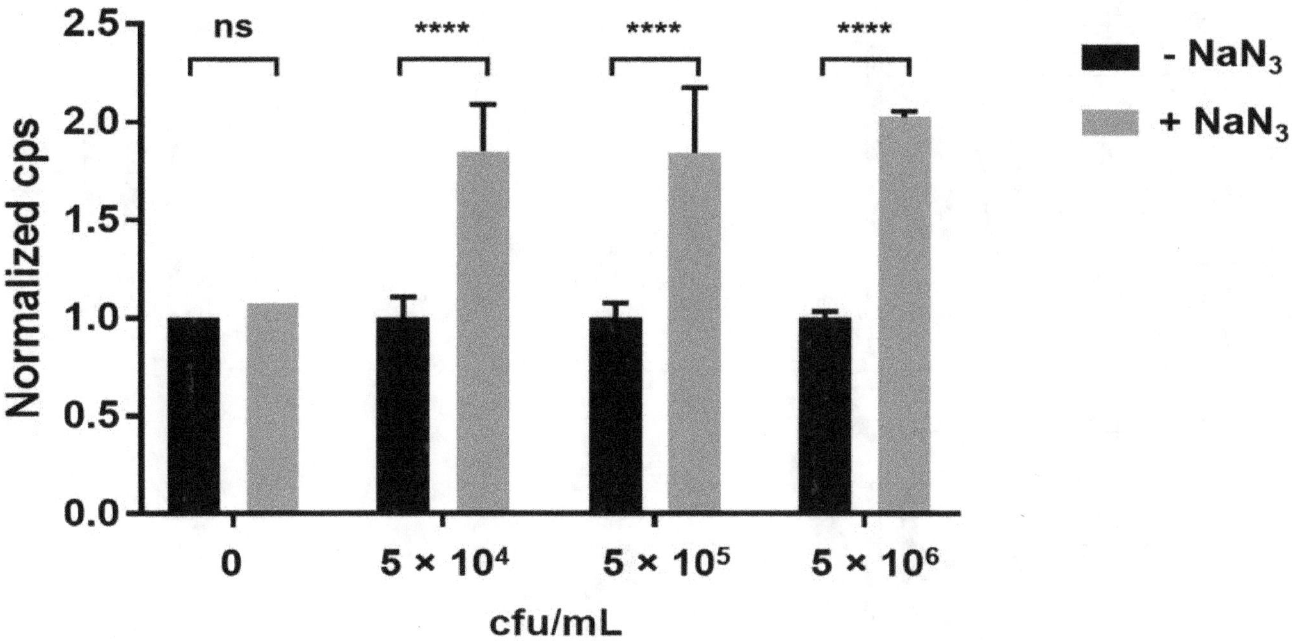

Figure 2. Effect of sodium azide (NaN_3) addition on bioluminescent signal intensity. Different concentrations of *Listeria monocytogenes* ScottA cells ranging from 5.0×10^4 to 5.0×10^6 cfu/mL were infected with 1.0×10^8 pfu/mL of A511::*luxAB*. Bacteria were incubated with phages in 200 μL of BHI $\frac{1}{2}$ in microwell plates at 24 °C for 140 min. Measurements were performed in parallel with identical samples with and without the addition of 7.6 mM NaN_3 as signal enhancer. For each bacterial concentration, bioluminescence signals have been normalized to the control sample (without NaN_3). Asterisks indicate significant differences between values obtained with and without NaN_3. ****, $p < 0.0001$; ns, non-significant.

3.3. Combining CBD-MS Listeria Cell Separation with A511::luxAB Infection Yields Superior Sensitivity

To further increase sensitivity and shorten the assay time, the authors combined the previously described CBD-MS method [6] with the microwell A511::*luxAB* assay. For this, paramagnetic microbeads coated with CBD118 and CBD500 from two different *Listeria* phages [11] were used to capture and separate *Listeria* cells of different strains, from dilute suspensions. The combined binding range of the two CBDs used here covered strains from all *Listeria* species and serovars [11].

Immediately following separation, the bacterial cells immobilized on the beads were resuspended in fresh media in the microplate wells, and allowed to recover for 3 h at 30 °C. Infection by the reporter phage and incubation at 24 °C were followed by bioluminescence measurement.

The combined assay was able to reliably detect cell concentrations $\leq 3 \times 10^2$ cfu/mL. The minimum bacterial concentrations yielding cps values significantly higher than the control were 3.0×10^2 cfu/mL for *L. monocytogenes* ScottA, 1.0×10^2 cfu/mL for *L. monocytogenes* EGDe, 2.0×10^2 cfu/mL for *L. monocytogenes* WSLC 1001, 1.6×10^2 cfu/mL for *L. ivanovii* WSLC 3009, and 0.9×10^2 cfu/mL for *L. innocua* WSLC 2012, within a total time of 6 h or less (Figure 3). Generally, an increase in bacterial concentrations resulted in an increased luminescent signal intensity measured in the reporter phage assay. Negative control blank samples containing either no *Listeria* cells (Figure 3), or no bacteriophages yielded background signals (noise) ranging from 24 to 30 cps (mean = 28.1, SD = 1.83).

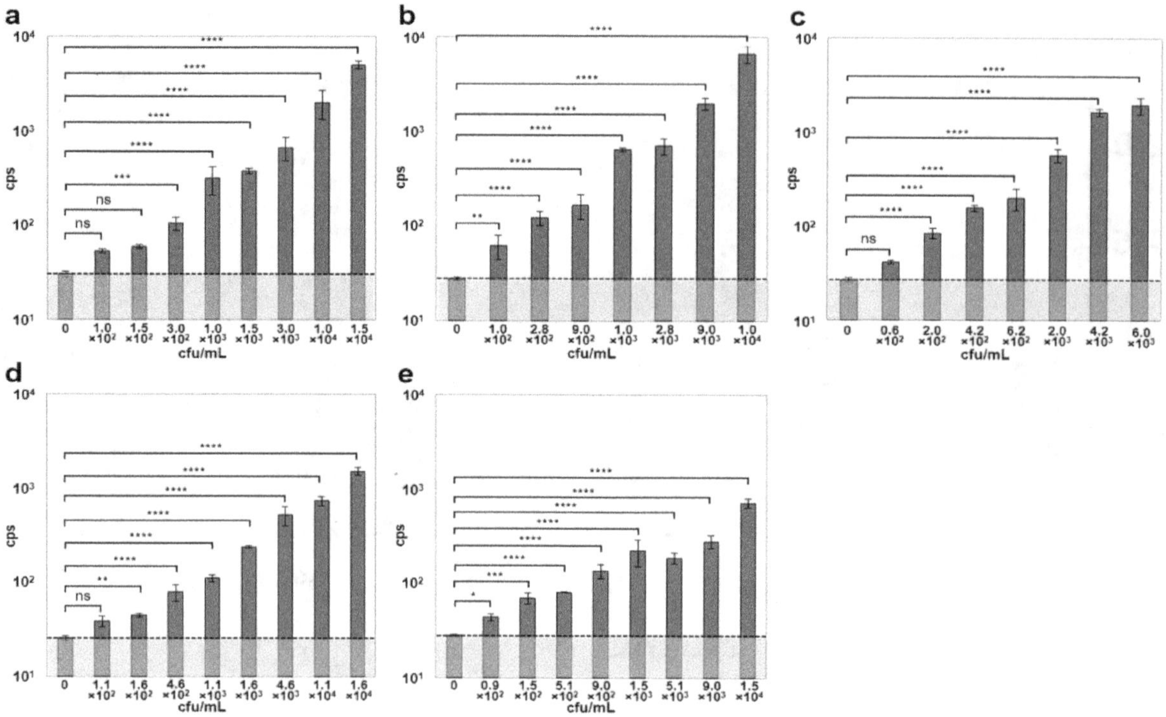

Figure 3. Detection of *Listeria* spp. strains employing the combined cell wall-binding domain-based magnetic separation (CBD-MS)/A511::*luxAB* assay. Dilutions of pure cultures of five different *Listeria* strains representing different serovars were used for evaluating the combined CBD-MS/A511::*luxAB* assay. (**a**) *L. monocytogenes* ScottA (serovar [SV] 4b); (**b**) *L. monocytogenes* EGDe (SV 1/2a); (**c**) *L. monocytogenes* WSLC 1001 (SV 1/2c); (**d**) *L. ivanovii* WSLC 3009 (SV 5); (**e**) *L. innocua* WSLC 2012 (SV 6b). The dotted horizontal bar indicates the level of background light emission (noise) of the assay, determined in the negative control sample (no bacteria). Asterisks indicate significant differences between values obtained from samples with bacteria and the respective control values. ****, $p < 0.0001$; ***, $p < 0.001$; **, $p < 0.01$; *, $p < 0.05$; ns, non-significant.

3.4. Detection of Listeria in Artificially Contaminated Food

Finally, the authors evaluated the combined assay using nine different food items spiked with *Listeria* EGDe (serovar 1/2a) or ScottA (serovar 4b), at viable cell concentrations ranging from 0.1 to 100 cfu/g. The exact initial numbers added to the food samples was determined by serial plating on selective agar plates and yielded levels between 0.1 and 0.15 cfu/g for the lowest, and between 100 and 150 cfu/g for the highest contamination for both strains. The tested foods included chocolate milk, mozzarella cheese, sliced iceberg lettuce, ready-to-eat shrimp, minced meat, smoked salmon, smoked turkey breast, Vacherin-type red smear soft cheese, and Tomme-type white mold soft cheese. Following inoculation, the food samples were stored at 4 °C for 24 h to simulate more realistic conditions and give the bacteria time to accommodate to the environmental conditions. Samples of 25 g were then taken, and following a 16 h selective enrichment step, subjected to the combined CBD-MS/A511::*luxAB* assay, as depicted in Figure 4

The detection limits in these trials ranged from 0.1 cfu/g (for both strains in smoked salmon) to 100 cfu/g (for EGDe in "Tomme" soft cheese), and was between 0.1 and 1.0 cfu/g for most of the tested food items, except for the two soft cheeses, which yielded higher detection limits (Figure 5). The total assay time was 22 h. The data indicate that the signal intensity obtained from a specific sample nicely correlates with the number of target cells present, and also confirm that the different food items used in the experiments provide different growth conditions for the bacteria.

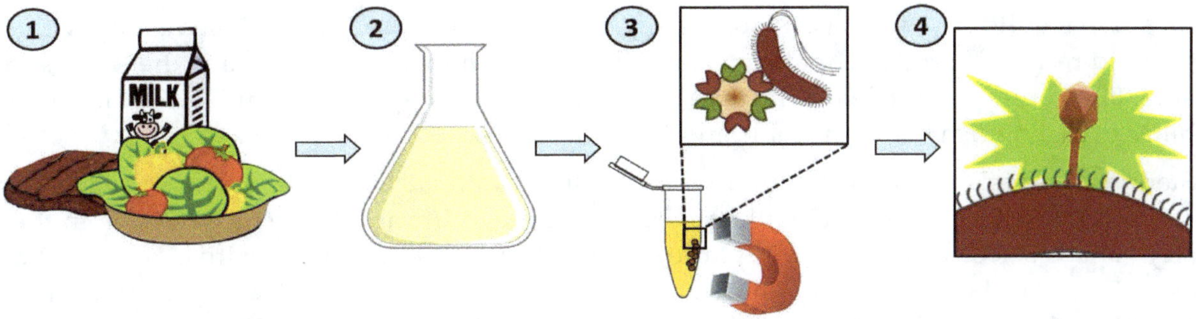

Figure 4. Schematic representation of the workflow of the combined CBD-MS/A511::*luxAB* assay. (**1**) food sample; (**2**) selective enrichment; (**3**) CBD-based magnetic separation; (**4**) A511::*luxAB* phage infection and generation of bioluminescent signal.

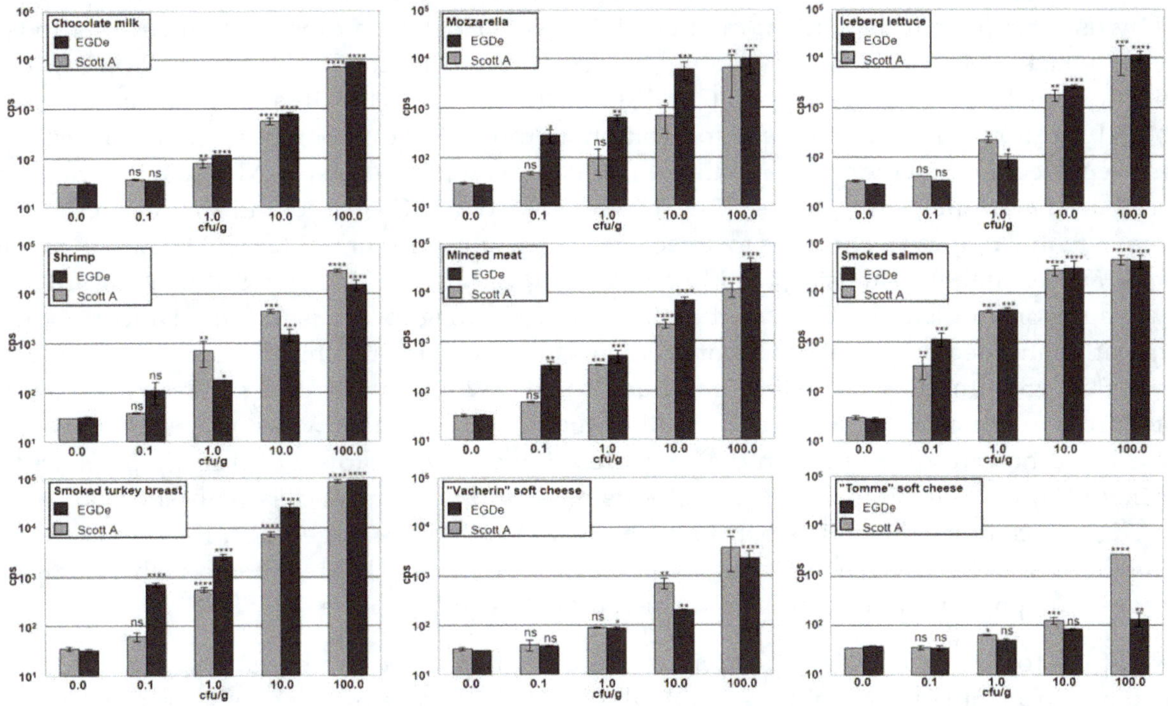

Figure 5. Detection of different concentrations of cells of *L. monocytogenes* ScottA and *L. monocytogenes* EGDe, in artificially contaminated food samples with the CBD-MS/A511::*luxAB* assay. Eight samples for each food item were spiked with bacteria of either ScottA or EGDe strains, at contamination levels of 0.1, 1.0, 10.0, or 100.0 cfu/g. Additionally, one blank sample for each food item was prepared by adding sterile buffer without bacteria. All samples were tested for the presence of *Listeria* spp. in parallel. Error bars represent SD from two replicates. Asterisks indicate significant differences between values obtained from samples contaminated with bacteria and the respective blank samples. ****, $p < 0.0001$; ***, $p < 0.001$; **, $p < 0.01$; *, $p < 0.05$; ns, non-significant.

4. Discussion

The detection of pathogenic bacteria from potentially contaminated foods represents a major challenge for food producers, particularly when regulations allow only low bacterial concentrations or even require the complete absence of the pathogen (zero tolerance). The Draft Guidance for Industry on the "Control of *Listeria monocytogenes* in Ready-To-Eat Foods" by the U.S. Food and Drug Administration (FDA, 2017) recommends the implementation of listeriocidal process steps that ensure that *L. monocytogenes* is non-detectable in the final product. The current European Union legislation (Commission Regulation (EC) No 2073/2005) requires the absence of *Listeria monocytogenes* in 25 g of any food product that can support its growth until the end of shelf life. Therefore, any diagnostic protocols for *Listeria* must be highly sensitive and specific in order to prevent false-negative and false-positive results, respectively. Additionally, testing should ideally produce results within one day, to avoid recalls of products that have already reached the market by the time the test results are available. Further important factors include ease of use to allow application in the food industry and the ability to distinguish between viable and inactivated cells.

Standard selective plating methods like the International Dairy Federation (IDF) Standard 143A:1995 or the ISO 11290-1:2017 still represent the state of the art of *Listeria* detection in the food industry. Those selective enrichment/plating methods are specific and sensitive, and they can be applied in any standard laboratory. However, with any of these standard procedures, preliminary results will not be available within less than 96 h. Furthermore, insufficient recovery of stressed cells during food processing and slow or no growth on agars containing selective agents represent a possible disadvantage of these methods [16].

Over the past decades, many efforts have been made to improve the detection of *L. monocytogenes* in food by using culture-independent methods [17]. Nevertheless, almost all of these methods have certain drawbacks. Most currently available molecular detection methods are based on nucleic acid amplification, including conventional PCR [18], multiplex PCR [19,20], and quantitative real-time PCR [21,22]. Cost-effective alternative amplification methods that do not require thermal cycling have also been reported, and include loop-mediated isothermal amplification (LAMP) [23] and nucleic acid sequence-based amplification (NASBA) [24]. The popularity of PCR-based and similar methods likely lies in the potential of these techniques to detect very low concentrations of target DNA with maximum specificity, at reasonable cost, using readily available apparatus, with the possibility for automation. However, a major problem of PCR-based methods in food diagnostics is that the DNA polymerases are frequently inhibited by common food ingredients. To overcome this hurdle, some researchers have combined PCR with immunomagnetic separation (IMS) [25–28]. Using specific IMS, target cells may be separated from inhibitory substances and simultaneously concentrated, which increases the sensitivity of the assay. Although some of the studies reported detection of as low as 1.0 cfu/g from artificially contaminated foods [26], all PCR-based methods have another crucial disadvantage that cannot be eliminated: they cannot differentiate between inactivated and viable target cells. Even though primary contaminations of many raw food ingredients are usually eliminated by treatments such as temperature or high-pressure pasteurization, samples would still produce a positive signal.

Examples for other PCR-independent molecular diagnostic methods developed for *Listeria* include mass spectrometry techniques (MALDI-TOF) [29,30], assays that use biosensors measuring surface plasmon resonance [31,32], or methods based on flow cytometry [33,34]. Although these techniques may be of particular scientific interest, and some of them provide low detection limits, their application in the food industry has been limited due to the requirement for sophisticated and expensive equipment, and specifically trained operators.

Other approaches combined standard plating methods with antibody-based IMS or IMS-like procedures to improve sensitivity and reduce assay time [35–38]. Although a reduction of the required time down to 48 h could be achieved in these studies, and detection limits ranged from 1.0 cfu/25 g [38] to 1.0 cfu/g [39], the approaches have other disadvantages. A major problem of IMS is the lack of proper specificity of the antibodies used, and undesired cross-reactivity with other organisms than the target organism is quite frequent [36,37,39].

The authors addressed this problem by replacing antibodies with cell wall-binding domains (CBDs), which are recombinant high-affinity domains derived from *Listeria* phage endolysins. They work extremely well for the functionalization of paramagnetic beads, and for the subsequent magnetic capture and separation of *L. monocytogenes* from artificially and naturally contaminated food samples [6]. The CBDs generally feature high affinity and extreme specificity for their target cells, with no cross-reaction with other bacteria. This CBD-MS assay was shown to be superior compared to the standard plating protocols in both time requirement (48 h) and sensitivity, with relatively low detection limits for most of the food items tested. However, the assay still relied on surface plating following magnetic separation, which accounted for half of the total assay time. To further shorten the time required to obtain results, the authors employed in this study A511::*luxAB* reporter phage detection of the captured bacteria, including scaling to microplate format, and further optimizations such as optimized incubation temperature, phage concentration, and use of a light emission enhancer. Altogether, the combined CBD-MS/A511::*luxAB* assay consisted of four steps: (i) selective separation of *Listeria* cells from an enrichment culture or other sample by CBD-MS; (ii) washing and concentration of *Listeria* cells immobilized on beads, followed by enrichment in non-selective medium in order to improve the sensitivity of the assay and resuscitate injured cells; (iii) infection with A511::*luxAB*; and (iv) removal of magnetic beads and measurement of light emission in a semiautomatic plate reader, following injection of signal enhancer and luciferase substrate.

Using pure bacterial cultures, as few as 90 cfu/mL could be detected in as little as 6 h by the assay. When contaminated foods are examined, the assay includes a selective pre-enrichment step. This selective step is necessary because of the high number of organisms in the background flora of many food products, which would otherwise overgrow the *Listeria* cells. However, the selective pre-enrichment is significantly shorter than in the conventional culture-based protocols, and the authors' method achieved a detection limit of 0.1 to 1.0 cfu/g in less than 24 h for most food items. This renders the assay significantly faster than the initial CBD-MS [6] and A511::*luxAB* reporter phage methods [9], with improved sensitivity.

Besides sensitivity and speed, the CBD-MS/A511::*luxAB* assay has another important advantage compared to magnetic-separation methods that use plating on selective agar as an end-point detection. Due to food processing, microorganisms present in food are often subjected to sublethal stresses. As these cells might be non-cultivable in selective media, they may remain undetected by the plating method. In this study, the incubation of *Listeria* immobilized and separated on CBD-coated beads serves as a resuscitation step and certainly improves detection of physiologically injured cells.

An important advantage of this bacteriophage-based assay is its extreme specificity, essentially eliminating undesired cross-reactivity with other bacterial genera by including three independent *Listeria*-specific selection steps: (i) selective enrichment for *Listeria* cells; (ii) CBD-MS using CBDs that have been demonstrated to be specific for *Listeria* [11]; and (iii) infection by a reporter phage based on the strictly genus-specific A511 phage [40]. It should be noted, however, that the assay as described here will not only signal the presence of *L. monocytogenes,* but it will also indicate other *Listeria* species sensitive to A511 infection. Yet, the authors believe that information on the presence of *Listeria* species is of great value, whereas testing for *L. monocytogenes* alone may not provide the full picture, that is, information on insufficient processing or potential contamination sources and routes is missing. Yet, if such specificity is desirable, a combination of the CBD-based enrichment and *L. monocytogenes*-specific PCR can also be used, as reported earlier [41].

A limitation of the assay is that A511::*luxAB* infects approximately 95% of all relevant *Listeria monocytogenes* strains of serovars 1/2 and 4 [40,42].Therefore, up to 5% of potential target strains may not be detected by this approach. However, it should be noted that available evidence suggests that the resistance of *Listeria* to phages such as A511 and others is due to loss of cell-wall associated carbohydrate decorations in their wall teichoic acids [43], and that these strains reveal a strongly attenuated virulence [44].

It is therefore concluded that, due to their unrivaled specificity for their hosts, genetically engineered reporter phages and phage-derived affinity proteins such as the CBDs represent very versatile and powerful tools for the detection of target host cells, including pathogenic bacteria. In this study, the authors combined two such phage-based techniques into one detection protocol, thereby harnessing the extraordinary specificity of bacteriophage in different ways. In terms of both sensitivity and time requirement, the assay seems superior to previous methods, including culture-based protocols.

Author Contributions: J.W.K. and M.J.L. conceived and designed the experiments; J.W.K. performed the experiments; J.W.K. and M.S. analyzed the data; M.J.L. contributed reagents/materials/analysis tools; J.W.K., M.J.L., and M.S. wrote the paper.

References

1. Vazquez-Boland, J.A.; Kuhn, M.; Berche, P.; Chakraborty, T.; Dominguez-Bernal, G.; Goebel, W.; Gonzalez-Zorn, B.; Wehland, J.; Kreft, J. *Listeria* pathogenesis and molecular virulence determinants. *Clin. Microbiol. Rev.* **2001**, *14*, 584–640. [CrossRef] [PubMed]
2. Bula, C.J.; Bille, J.; Glauser, M.P. An epidemic of food-borne listeriosis in western Switzerland: Description of 57 cases involving adults. *Clin. Infect. Dis.* **1995**, *20*, 66–72. [CrossRef] [PubMed]
3. Lomonaco, S.; Nucera, D.; Filipello, V. The evolution and epidemiology of *Listeria monocytogenes* in Europe and the United States. *Infect. Genet. Evol.* **2015**, *35*, 172–183. [CrossRef] [PubMed]
4. Rees, C.E.; Dodd, C.E. Phage for rapid detection and control of bacterial pathogens in food. *Adv. Appl. Microbiol.* **2006**, *59*, 159–186. [PubMed]
5. Schmelcher, M.; Loessner, M. Bacteriophage: Powerful Tools for the Detection of Bacterial Pathogens. In *Principles of Bacterial Detection: Biosensors, Recognition Receptors and Microsystems*; Zourob, M., Elwary, S., Turner, A., Eds.; Springer: New York, NY, USA, 2008; Volume 2, pp. 731–754.
6. Kretzer, J.W.; Lehmann, R.; Schmelcher, M.; Banz, M.; Kim, K.P.; Korn, C.; Loessner, M.J. Use of high-affinity cell wall-binding domains of bacteriophage endolysins for immobilization and separation of bacterial cells. *Appl. Environ. Microbiol.* **2007**, *73*, 1992–2000. [CrossRef] [PubMed]
7. Schmelcher, M.; Loessner, M.J. Application of bacteriophages for detection of foodborne pathogens. *Bacteriophage* **2014**, *4*, e28137. [CrossRef] [PubMed]
8. Loessner, M.J.; Rees, C.E.; Stewart, G.S.; Scherer, S. Construction of luciferase reporter bacteriophage A511::*luxAB* for rapid and sensitive detection of viable *Listeria* cells. *Appl. Environ. Microbiol.* **1996**, *62*, 1133–1140. [PubMed]
9. Loessner, M.J.; Rudolf, M.; Scherer, S. Evaluation of luciferase reporter bacteriophage A511::*luxAB* for detection of *Listeria monocytogenes* in contaminated foods. *Appl. Environ. Microbiol.* **1997**, *63*, 2961–2965. [PubMed]
10. Loessner, M.J.; Schneider, A.; Scherer, S. Modified *Listeria* bacteriophage lysin genes (*ply*) allow efficient overexpression and one-step purification of biochemically active fusion proteins. *Appl. Environ. Microbiol.* **1996**, *62*, 3057–3060. [PubMed]
11. Loessner, M.J.; Kramer, K.; Ebel, F.; Scherer, S. C-terminal domains of *Listeria monocytogenes* bacteriophage murein hydrolases determine specific recognition and high-affinity binding to bacterial cell wall carbohydrates. *Mol. Microbiol.* **2002**, *44*, 335–349. [CrossRef] [PubMed]
12. Papageorgiou, D.K.; Marth, E.H. Behavior of *Listeria monocytogenes* at 4 and 22 °C in Whey and Skim Milk Containing 6 or 12% Sodium-Chloride. *J. Food Prot.* **1989**, *52*, 625–630. [CrossRef]
13. Zhang, L.; Moosekian, S.R.; Todd, E.C.; Ryser, E.T. Growth of *Listeria monocytogenes* in different retail delicatessen meats during simulated home storage. *J. Food Prot.* **2012**, *75*, 896–905. [CrossRef] [PubMed]
14. Bowler, M.W.; Montgomery, M.G.; Leslie, A.G.; Walker, J.E. How azide inhibits ATP hydrolysis by the F-ATPases. *Proc. Natl. Acad. Sci. USA* **2006**, *103*, 8646–8649. [CrossRef] [PubMed]
15. Stannard, J.N.; Horecker, B.L. The in vitro inhibition of cytochrome oxidase by azide and cyanide. *J. Biol. Chem.* **1948**, *172*, 599–608. [PubMed]
16. Fratamico, P.M.; Bayles, D.O. Molecular approaches for detection, identification, and analysis of food-borne pathogens. In *Foodborne Pathogens: Microbiology and Molecular Biology*; Fratamico, P.M., Bhunia, A.K., Smith, J.L., Eds.; Caister Academic Press: Norfolk, VA, USA, 2005; pp. 1–13.
17. Law, J.W.; Ab Mutalib, N.S.; Chan, K.G.; Lee, L.H. An insight into the isolation, enumeration, and molecular detection of *Listeria monocytogenes* in food. *Front. Microbiol.* **2015**, *6*, 1227. [CrossRef] [PubMed]
18. Khan, J.A.; Rathore, R.S.; Khan, S.; Ahmad, I. In vitro detection of pathogenic *Listeria monocytogenes* from food sources by conventional, molecular and cell culture method. *Braz. J. Microbiol.* **2013**, *44*, 751–758. [CrossRef] [PubMed]
19. Liu, H.Q.; Lu, L.Q.; Pan, Y.J.; Sun, X.H.; Hwang, C.A.; Zhao, Y.; Wu, V.C.H. Rapid detection and differentiation of *Listeria monocytogenes* and *Listeria* species in deli meats by a new multiplex PCR method. *Food Control* **2015**, *52*, 78–84. [CrossRef]
20. Rawool, D.B.; Malik, S.V.S.; Shakuntala, I.; Sahare, A.M.; Barbuddhe, S.B. Detection of multiple virulence-associated genes in *Listeria monocytogenes*, isolated from bovine mastitis cases. *Int. J. Food Microbiol.* **2007**, *113*, 201–207. [CrossRef] [PubMed]

21. de Oliveira, M.A.; Ribeiro, E.G.A.; Bergamini, A.M.M.; De Martinis, E.C.P. Quantification of *Listeria monocytogenes* in minimally processed leafy vegetables using a combined method based on enrichment and 16S rRNA real-time PCR. *Food Microbiol.* **2010**, *27*, 19–23. [CrossRef] [PubMed]
22. Gattuso, A.; Gianfranceschi, M.V.; Sonnessa, M.; Delibato, E.; Marchesan, M.; Hernandez, M.; De Medici, D.; Rodriguez-Lazaro, D. Optimization of a Real Time PCR based method for the detection of *Listeria monocytogenes* in pork meat. *Int. J. Food Microbiol.* **2014**, *184*, 106–108. [CrossRef] [PubMed]
23. Cho, A.R.; Dong, H.J.; Seo, K.H.; Cho, S. Development of a loop-mediated isothermal amplification assay for detecting *Listeria monocytogenes prfA* in milk. *Food Sci. Biotechnol.* **2014**, *23*, 467–474. [CrossRef]
24. Uyttendaele, M.; Schukkink, R.; van Gemen, B.; Debevere, J. Development of NASBA, a nucleic acid amplification system, for identification of *Listeria monocytogenes* and comparison to ELISA and a modified FDA method. *Int. J. Food Microbiol.* **1995**, *27*, 77–89. [CrossRef]
25. Fluit, A.C.; Torensma, R.; Visser, M.J.; Aarsman, C.J.; Poppelier, M.J.; Keller, B.H.; Klapwijk, P.; Verhoef, J. Detection of *Listeria monocytogenes* in cheese with the magnetic immuno-polymerase chain reaction assay. *Appl. Environ. Microbiol.* **1993**, *59*, 1289–1293. [PubMed]
26. Hudson, J.A.; Lake, R.J.; Savill, M.G.; Scholes, P.; McCormick, R.E. Rapid detection of *Listeria monocytogenes* in ham samples using immunomagnetic separation followed by polymerase chain reaction. *J. Appl. Microbiol.* **2001**, *90*, 614–621. [CrossRef] [PubMed]
27. Luo, D.; Huang, X.; Mao, Y.; Chen, C.; Li, F.; Xu, H.; Xiong, Y. Two-step large-volume magnetic separation combined with PCR assay for sensitive detection of *Listeria monocytogenes* in pasteurized milk. *J. Dairy Sci.* **2017**, *100*, 7883–7890. [CrossRef] [PubMed]
28. Yang, Y.; Xu, F.; Xu, H.; Aguilar, Z.P.; Niu, R.; Yuan, Y.; Sun, J.; You, X.; Lai, W.; Xiong, Y.; et al. Magnetic nano-beads based separation combined with propidium monoazide treatment and multiplex PCR assay for simultaneous detection of viable *Salmonella* Typhimurium, *Escherichia coli* O157:H7 and *Listeria monocytogenes* in food products. *Food Microbiol.* **2013**, *34*, 418–424. [CrossRef] [PubMed]
29. Jadhav, S.; Sevior, D.; Bhave, M.; Palombo, E.A. Detection of *Listeria monocytogenes* from selective enrichment broth using MALDI-TOF Mass Spectrometry. *J. Proteomics* **2014**, *97*, 100–106. [CrossRef] [PubMed]
30. Jadhav, S.; Gulati, V.; Fox, E.M.; Karpe, A.; Beale, D.J.; Sevior, D.; Bhave, M.; Palombo, E.A. Rapid identification and source-tracking of *Listeria monocytogenes* using MALDI-TOF mass spectrometry. *Int. J. Food Microbiol.* **2015**, *202*, 1–9. [CrossRef] [PubMed]
31. Morlay, A.; Roux, A.; Templier, V.; Piat, F.; Roupioz, Y. Label-Free Immuno-Sensors for the Fast Detection of *Listeria* in Food. *Methods Mol. Biol.* **2017**, *1600*, 49–59. [PubMed]
32. Zhang, X.; Tsuji, S.; Kitaoka, H.; Kobayashi, H.; Tamai, M.; Honjoh, K.I.; Miyamoto, T. Simultaneous Detection of *Escherichia coli* O157:H7, *Salmonella enteritidis*, and *Listeria monocytogenes* at a Very Low Level Using Simultaneous Enrichment Broth and Multichannel SPR Biosensor. *J. Food. Sci.* **2017**, *82*, 2357–2363. [CrossRef] [PubMed]
33. Hibi, K.; Abe, A.; Ohashi, E.; Mitsubayashi, K.; Ushio, H.; Hayashi, T.; Ren, H.; Endo, H. Combination of immunomagnetic separation with flow cytometry for detection of *Listeria monocytogenes*. *Anal. Chim. Acta* **2006**, *573–574*, 158–163. [CrossRef] [PubMed]
34. Kim, J.S.; Anderson, G.P.; Erickson, J.S.; Golden, J.P.; Nasir, M.; Ligler, F.S. Multiplexed Detection of Bacteria and Toxins Using a Microflow Cytometer. *Anal. Chem.* **2009**, *81*, 5426–5432. [CrossRef] [PubMed]
35. Bauwens, L.; Vercammen, F.; Hertsens, A. Detection of pathogenic *Listeria* spp. in zoo animal faeces: Use of immunomagnetic separation and a chromogenic isolation medium. *Vet. Microbiol.* **2003**, *91*, 115–123. [CrossRef]
36. Skjerve, E.; Rorvik, L.M.; Olsvik, O. Detection of *Listeria monocytogenes* in foods by immunomagnetic separation. *Appl. Environ. Microbiol.* **1990**, *56*, 3478–3481. [PubMed]
37. Uyttendaele, M.; Van Hoorde, I.; Debevere, J. The use of immuno-magnetic separation (IMS) as a tool in a sample preparation method for direct detection of *L. monocytogenes* in cheese. *Int. J. Food Microbiol.* **2000**, *54*, 205–212. [CrossRef]
38. Wadud, S.; Leon-Velarde, C.G.; Larson, N.; Odumeru, J.A. Evaluation of immunomagnetic separation in combination with ALOA *Listeria* chromogenic agar for the isolation and identification of *Listeria monocytogenes* in ready-to-eat foods. *J. Microbiol. Methods* **2010**, *81*, 153–159. [CrossRef] [PubMed]
39. Kaclikova, E.; Kuchta, T.V.; Kay, H.; Gray, D. Separation of *Listeria* from cheese and enrichment media using antibody-coated microbeads and centrifugation. *J. Microbiol. Methods* **2001**, *46*, 63–67. [CrossRef]

40. Loessner, M.J.; Busse, M. Bacteriophage typing of *Listeria* species. *Appl. Environ. Microbiol.* **1990**, *56*, 1912–1918. [PubMed]
41. Walcher, G.; Stessl, B.; Wagner, M.; Eichenseher, F.; Loessner, M.J.; Hein, I. Evaluation of paramagnetic beads coated with recombinant *Listeria* phage endolysin-derived cell-wall-binding domain proteins for separation of *Listeria monocytogenes* from raw milk in combination with culture-based and real-time polymerase chain reaction-based quantification. *Foodborne Pathog. Dis.* **2010**, *7*, 1019–1024. [PubMed]
42. Loessner, M.J. Improved procedure for bacteriophage typing of *Listeria* strains and evaluation of new phages. *Appl. Environ. Microbiol.* **1991**, *57*, 882–884. [PubMed]
43. Eugster, M.R.; Morax, L.S.; Huls, V.J.; Huwiler, S.G.; Leclercq, A.; Lecuit, M.; Loessner, M.J. Bacteriophage predation promotes serovar diversification in *Listeria monocytogenes*. *Mol. Microbiol.* **2015**, *97*, 33–46. [CrossRef] [PubMed]
44. Carvalho, F.; Atilano, M.L.; Pombinho, R.; Covas, G.; Gallo, R.L.; Filipe, S.R.; Sousa, S.; Cabanes, D. L-Rhamnosylation of *Listeria monocytogenes* Wall Teichoic Acids Promotes Resistance to Antimicrobial Peptides by Delaying Interaction with the Membrane. *PLoS Pathog.* **2015**, *11*, e1004919. [CrossRef] [PubMed]

8

Strategies to Encapsulate the *Staphylococcus aureus* Bacteriophage phiIPLA-RODI

Eva González-Menéndez [1], Lucía Fernández [1], Diana Gutiérrez [1], Daniel Pando [2], Beatriz Martínez [1], Ana Rodríguez [1] and Pilar García [1,*]

[1] Instituto de Productos Lácteos de Asturias (IPLA-CSIC), Paseo Río Linares s/n, 33300 Villaviciosa, Spain; eva.gm@ipla.csic.es (E.G.-M.); lucia.fernandez@ipla.csic.es (L.F.); dianagufer@ipla.csic.es (D.G.); bmf1@ipla.csic.es (B.M.); anarguez@ipla.csic.es (A.R.)

[2] Nanovex Biotechnologies S.L., Parque Tecnológico de Asturias, CEEI, 33428 Llanera, Spain; pando@nanovexbiotech.com

* Correspondence: pgarcia@ipla.csic.es

Abstract: The antimicrobial properties of bacteriophages make them suitable food biopreservatives. However, such applications require the development of strategies that ensure stability of the phage particles during food processing. In this study, we assess the protective effect of encapsulation of the *Staphylococcus aureus* bacteriophage phiIPLA-RODI in three kinds of nanovesicles (niosomes, liposomes, and transfersomes). All these systems allowed the successful encapsulation of phage phiIPLA-RODI with an efficiency ranged between 62% and 98%, regardless of the concentration of components (like phospholipids and surfactants) used for vesicle formation. Only niosomes containing 30 mg/mL of surfactants exhibited a slightly lower percentage of encapsulation. Regarding particle size distribution, the values determined for niosomes, liposomes, and transfersomes were 0.82 ± 0.09 µm, 1.66 ± 0.21 µm, and 0.55 ± 0.06 µm, respectively. Importantly, bacteriophage infectivity was maintained during storage for 6 months at 4 °C for all three types of nanovesicles, with the exception of liposomes containing a low concentration of components. In addition, we observed that niosomes partially protected the phage particles from low pH. Thus, while free phiIPLA-RODI was not detectable after 60 min of incubation at pH 4.5, titer of phage encapsulated in niosomes decreased only 2 log units. Overall, our results show that encapsulation represents an appropriate procedure to improve stability and, consequently, antimicrobial efficacy of phages for application in the food processing industry.

Keywords: bacteriophages; encapsulation; niosomes; transfersomes; liposomes; *Staphylococcus aureus*

---

1. Introduction

Bacteriophages, viruses that infect and kill bacteria, are ubiquitous in the environment. Indeed, they are considered to be the most abundant organisms in the biosphere, playing an important role in biogeochemical cycles and in the development of microbial communities [1]. Over the last few years, there has been a notable interest in exploiting the antimicrobial properties of phages for the control of bacterial pathogens (phage therapy), as a strategy to curtail the relentless increase in antibiotic-resistant bacteria [2–4]. Moreover, phages are harmless to humans, animals, and plants, making them a safe alternative to conventional antimicrobials [5].

Another potential application of bacteriophages is as biocontrol agents along the food chain [6]. The main advantages of bacteriophages in a food context are related to their specificity and safety, and several studies have already confirmed their effectiveness. For example, bacteriophages infecting the most important foodborne pathogens have been successfully used to reduce the microbial load in livestock, thereby decreasing the risk of transmission of zoonotic bacteria. Also, phages can be directly applied to foods as preservatives to prevent the development of undesirable bacteria [7–9]. Additionally, bacterial viruses have been proposed as promising disinfectants to reduce the risk of food contamination caused by biofilms in food processing industries [10]. Nonetheless, widespread use of bacteriophages in agriculture, farming or food industrial settings will not be possible until they receive full approval by the regulatory authorities. In that sense, it is worth mentioning the recent approval of several phage-based products by the U. S. Food and Drug Administration to be used as food preservatives. Thus, two companies, Intralytix, Inc. (http://www.intralytix.com) and Micreos (http://www.micreos.com) currently market an array of phage-based products against some of the most important foodborne pathogenic bacteria. Some examples include *Listeria monocytogenes* (ListShield™ and PhageGuard Listex), *E. coli* O157:H7 (EcoShield™), *Shigella* (ShigaShield™) and *Salmonella* (SalmoFresh™ and PhageGuard S).

Depending on the specific use, phages can be delivered by oral administration to animals or by direct spraying on food or industrial surfaces [7,11,12]. In many cases, bacteriophages will encounter harsh physicochemical conditions following the application of phage-based products. For instance, the presence of bile salts and low pH in the gastrointestinal tract of farm animals or the UV light and low pH on the surface of some fruits may lead to inactivation of the viral particles. This would unavoidably result in a loss of infectivity against their target bacterium and, ultimately, failure of the disinfection procedure.

A feasible possibility to overcome this problem is the encapsulation of phages to protect them from environmental challenges so that they can reach their target microbes. (Micro- or nano-) encapsulation is a technology that allows packaging solid, liquid, or gaseous materials in miniature capsules or vesicles that can release their contents at controlled rates, sometimes triggered by specific environmental cues (pH, temperature, etc.) [13]. Indeed, this technique is often used in the food industry for the delivery of bioactive compounds or probiotics [14]. Thus, nanoencapsulation of different compounds like vitamins, minerals, and proteins confers several advantages since the molecules inside the nanovesicles are protected from hazardous environmental conditions and display increased stability as well as greater intestinal and epidermal absorption [15–18]. Indeed, there are several examples in literature where microencapsulation increased the stability of probiotic bacteria in dairy products [19], or the viability of these bacteria during their passage through the gastrointestinal tract [20,21]. Similarly, nanoencapsulation of an antimicrobial peptide (nisin) or lysozyme allowed overcoming stability issues and prevented their interaction with food components when used as additives for dairy products to control the growth of pathogenic bacteria [22,23]. Thus far, there are some studies regarding the encapsulation of bacteriophages in emulsions as well as in micro and nanovesicles prepared using different techniques based on chemical (polymerization) or physical (drying or extrusion) processes. However, there is only limited information about the encapsulation of bacteriophages in liposomes and none concerning the use of other types of nanovesicles (reviewed by [24,25]).

Staphylococcus aureus is an important pathogenic bacterium responsible for serious infections and food-borne diseases in humans [26,27]. The European Food Safety Authority (EFSA) reported a total of 434 food-borne outbreaks caused by staphylococcal toxins in 2015 [28]; meanwhile, 360 outbreak-associated illnesses and 27 hospitalizations were caused by *S. aureus* toxins in the US in 2016 [29]. Furthermore, the increase in methicillin-resistant (MRSA) and vancomycin-resistant strains (VRSA) has fostered research on the development of new weapons to fight against this pathogen [30]. Among these weapons, bacteriophages and phage-derived proteins are an interesting alternative to treat *S. aureus* infections [31,32] and prevent food contamination [33,34].

Our previous work had already demonstrated the ability of bacteriophages to reduce *S. aureus* contamination in dairy products including milk, curd, and cheese [35–37]. Moreover, there is evidence that they can also be used as a helper when combined with hydrostatic high pressure (HHP) to improve the safety of dairy products [38]. More recently, we have isolated and characterized the polyvalent *Myoviridae* phage phiIPLA-RODI, which is able to infect a broad range of staphylococcal species including *S. aureus* strains from food industry origin [39]. Its high infectivity against both planktonic cultures and biofilms makes it a good candidate for the biocontrol of *S. aureus* in food-related settings. However, the poor stability of phiIPLA-RODI under certain environmental conditions, might compromise its use during processing of fermented products.

The aim of this study was to evaluate the stability of phage phiIPLA-RODI under different encapsulation conditions compatible with the delivery of phage to foods. We also assessed the viability of the encapsulated phage during storage as well as its resistance under extreme conditions.

2. Materials and Methods

2.1. Bacterial Strains

S. aureus IPLA1 was used as the host strain of phage phiIPLA-RODI [39]. Isolated colonies of this strain were obtained by streaking out a frozen stock onto Baird–Parker (BP) agar plates. Subsequent bacterial cultures were routinely grown in tryptic soy broth (TSB; Scharlau, Barcelona, Spain) with shaking or on plates containing TSB supplemented with 2% (*w*/*v*) bacteriological agar (TSA). The growth temperature for all experiments was 37 °C.

2.2. Phage Propagation and Enumeration

Bacteriophage phiIPLA-RODI was routinely propagated as previously described [39]. Briefly, early exponential cultures of *S. aureus* IPLA1 (OD_{600} = 0.1) were infected with the phage at a multiplicity of infection (MOI) of 1. The infected cultures were then incubated for 3 h at 37 °C with vigorous shaking. Phage lysates were obtained by centrifugation of these cultures and subsequent filtration (0.45 µm cellulose acetate Sartolab® RF Vacuum Filters). Afterwards, the phage lysate was partially purified by adding NaCl (0.5 M, final concentration) and PEG 8000 (10%, final concentration). These samples were incubated at 4 °C for 24 h, and then centrifuged at 10,000 rpm at 4 °C for 30 min. The pellet containing the phage particles was suspended in SM buffer (20 mM Tris HCl, 10 mM $MgSO_4$, 10 mM $Ca(NO_3)_2$ and 0.1 M NaCl, pH 7.5).

Phage titer was calculated by the double-layer plaque assay. Briefly, 0.1 mL from a 1:10 dilution of a *S. aureus* IPLA1 stationary culture (10^8 CFU/mL) were mixed with several dilutions of individual phage suspensions in 3 mL of molten TSB supplemented with 0.7% agar and the mixture was poured onto TSA plates and incubated at 37 °C for 18 h. The phage titer was expressed as PFU/mL. All assays were performed using biological triplicates and values were presented as the mean ± standard deviation.

2.3. Encapsulation Processes

Liposomes are bilayer vesicles prepared with phospholipids and containing cholesterol as a stabilizing agent. Their size ranges depending on the type of vesicles formed are the following: multilamellar vesicles (size > 200 nm), large unilamellar vesicles (100–400 nm), small unilamellar vesicles (<100 nm).

Niosomes and transfersomes are similar to liposomes in terms of structure and physical properties. They can be unilamellar or multilamellar vesicles prepared by adding surfactants and other amphiphilic molecules, alternative to phospholipids, such as non-ionic surfactants [40].

Liposomes and niosomes were prepared using Pronanosome Lipo-N™ and Pronanosome Nio-N™ (Nanovex Biotechnologies, Llanera, Spain), respectively. Briefly, different amounts of Pronanosomes were employed to reach different final concentrations (30, 50 and 70 mg/mL). The products were hydrated with a solution of 0.05 M HEPES buffer pH 7.5 (Alfa Aesar, Karlsruhe, Germany) and then the bacteriophage suspension (3.2×10^8 PFU/mL) was added at a final concentration of 4% (1.3×10^7 PFU/mL). Then, the samples were mixed in two steps: a first step of manual shaking for 3 min followed by a homogenization step using a homogenizer (Heidolph, Schwabach, Germany) at different speeds and times (Table 1).

Table 1. Composition, characteristics (homogenization and duration, size, zeta potential) and viability (measured as log reduction in phage titer after treatment comparing with the initial titer) of vesicles containing phage phiIPLA-RODI.

Vesicles	Composition		Homogenization (rpm)/Duration (min)	Z-Average (μm)	ζ-Potential (mV)	Viability Loss (Log Units)
	Components	Concentration *				
Niosome	Pronanosome Nio-N™	30 mg/mL, 50 mg/mL, or 70 mg/mL	8000/5	0.83 ± 0.11 0.85 ± 0.12 0.80 ± 0.07	−34.3 ± 1.0 −33.4 ± 0.2 −35.6 ± 1.3	0.5 ± 0.1 1.0 ± 0.2 1.1 ± 0.2
Liposome	Pronanosome Lipo-N™	30 mg/mL, 50 mg/mL, or 70 mg/mL	5000/5	1.51 ± 0.17 1.60 ± 0.17 1.89 ± 0.03	−14.1 ± 1.0 −14.1 ± 0.5 −13.5 ± 0.1	1.3 ± 0.1 1.2 ± 0.2 1.2 ± 0.2
Transfersome	Phospholipon 90G and Span 60 (1:1)	30 mg/mL, 50 mg/mL, or 70 mg/mL	8000/5	0.51 ± 0.07 0.55 ± 0.03 0.58 ± 0.06	−30.3 ± 1.2 −30.8 ± 2.0 −28.6 ± 0.2	1.4 ± 0.3 1.2 ± 0.0 1.3 ± 0.1

Notes: Each value represents the mean ± standard deviation of three samples. * In the case of transfersomes, the concentration refers to the total of the two components.

Transfersomes were prepared by the thin film hydration (TFH) or dry film method [41] with minor modifications. Different amounts (30, 50 and 70 mg/mL) of Phospholipon 90G (P90; Lipoid, Ludwigshafen, Germany) and Span 60 (Alfa Aesar, Karlsruhe, Germany) (1:1) were dissolved in chloroform. Chloroform was subsequently removed by incubation at 40 °C, under reduced pressure in a rotary evaporator (Heidolph, Schwabach, Germany) until the formation of a dry film, which was subsequently hydrated by a solution of 0.05 M HEPES buffer pH 7.5. Then, bacteriophages were added at a final concentration of 4% (1.3×10^7 PFU/mL). The nanovesicles were subsequently generated after being submitted to 3 min of manual shaking and homogenized for 5 min at 8000 rpm (Table 1).

2.4. Characterization of Nanovesicles

Mean (Z-Average) particle size in the different nanovesicles (niosomes, liposomes and transfersomes) was determined by using a Zetasizer Nano ZS (Malvern Instruments Ltd., Worcestershire, UK), via Dynamic Light Scattering (DSL).

The ζ-potential is strongly linked to vesicle stability, with high absolute values of ζ-potential indicating electrostatic repulsion between vesicles. To determine the ζ-potential the M3-PALS (Phase Analysis Light Scattering) technique was used [15]. Three independent samples were taken from each formulation at room temperature and values were expressed as mean ± standard deviation.

2.5. Viability and Encapsulation Efficiency

The number of bacteriophages that did not survive the formulation process for niosomes, liposomes or transfersomes, was expressed as log reduction in viability:

$$\text{Viability} = \log_{10}\left(\frac{\text{Total phage titer after treatment}\left(\frac{\text{PFU}}{\text{mL}}\right)}{\text{Initial titer}\left(\frac{\text{PFU}}{\text{mL}}\right)}\right) \quad (1)$$

Phage viability after the formation of niosomes, liposomes and transfersomes was obtained by determining the total phage titer (encapsulated + non encapsulated). To determine the titer of non-encapsulated phages or free phages (F), aliquots containing 1.5 mL of niosomes, liposomes and transfersomes were centrifuged at 13,200 rpm and 4 °C for 60 min and the supernatant was titrated as described above. To determine the titer of encapsulated phages, the pellet containing the nanovesicles was washed twice with PBS buffer (137 mM NaCl, 2.7 mM KCl, 10 mM 138 Na_2HPO_4 and 2 mM KH_2PO_4; pH 7.4) following supernatant removal, and centrifuged again under the same conditions. Again, the supernatant was removed and the pellet was kept for further processing. Then, pellets were treated with 30 μL of chloroform and vortexed for 5 s to disrupt the vesicles. Finally, SM buffer was added up to a final volume of 1.5 mL. This mixture was centrifuged for 15 min at 10,000 rpm and 4 °C. Serial dilutions of the supernatant were then plated by using the double layer technique for phage titration.

The efficiency of encapsulation (EE) was calculated as the percentage of phages encapsulated inside nanovesicles compared to the total phage titer:

$$EE = \left(\frac{\text{Encapsulated phage (E) (PFU/mL)}}{\text{Total phage (PFU/mL)}}\right) \times 100 \quad (2)$$

where, total phage = encapsulated phage (E) + non encapsulated or free phage (F).

2.6. Stability of Encapsulated Phages During Storage

The stability of phages entrapped into nanovesicles (niosomes, liposomes, and transfersomes) was determined by storage of samples at 4 °C for 6 months. Every two months the viability of free and encapsulated phages was determined. A philPLA-RODI suspension (1.49×10^8 PFU/mL) in SM was used as control. Three independent samples were taken from each formulation and from the control phage suspension. The titer was presented as the mean of these values ± standard deviation.

2.7. Stability of Encapsulated Phages Under Extreme Conditions

The stability of phages encapsulated in niosomes, liposomes, and transfersomes at a pH of 4.5 was determined by preparing a 1:2 dilution of the nanovesicles in Britton–Robinson pH universal buffer (40 mM H_3PO_4, 40 mM CH_3COOH, 40 mM H_3BO_3, adjusted to pH 4.5) and then incubating the samples at room temperature for 60 min. Similarly, nanovesicles were incubated in the following conditions: (1) dilution (1/10) in NaCl 4.5 M for 60 min at room temperature, and (2) at 60 °C for 90 min. For control purposes, a phage suspension in SM buffer was subjected to the same treatments. After treatment, samples were centrifuged and the supernatant was titrated to determine the number of non-encapsulated phages. Viability of encapsulated phages was quantified after disruption of the nanovesicles as indicated above. The initial titer of samples before treatment was taken as a control. All assays were performed by triplicate.

2.8. Statistical Analysis

Statistical analyses for encapsulation efficiency and phage stability were performed using the statistical package IBM SPSS Statistics for Windows Version 23 (IBM Corp., Armonk, NY, USA). The differences in data related to phage encapsulation efficiency and its stability were subjected to one-way analysis of variance (ANOVA). The Student–Newman–Keuls (SNK) test was used for comparison of means of the stability of philPLA-RODI in the different nanovesicles at a level of significance $p < 0.05$. Three biological replicates were used in all the assays.

3. Results

3.1. Bacteriophage philPLA-RODI Can Be Successfully Entrapped in Different Nanovesicles

With the aim of seeking an optimal formulation for the application of phage philPLA-RODI, we examined the efficacy of different encapsulation techniques. In a first step, we determined the stability of the phage during different encapsulation processes and assessed the properties of the resulting nanovesicles. To do that, a phage suspension (3.2×10^8 PFU/mL) was encapsulated into niosomes, liposomes, and transfersomes. Then, the size of the prepared vesicles was measured using Dynamic Light Scattering. The nanovesicles formed had an average size in the range of 0.5 μm to 2 μm (0.82 ± 0.09 μm for niosomes, 1.66 ± 0.21 μm for liposomes and 0.55 ± 0.06 μm for transfersomes). Also, the net charge of the nanovesicles surface was estimated by determination of the ζ-potential, with liposomes showing the lowest absolute values (Table 1). Additionally, the viability of phages after the encapsulation process was determined. The results obtained in phage titration assays indicated that philPLA-RODI phage particles could withstand all the encapsulation methods tested, as all nanovesicles contained infective phages. However, the titer of phages released after the disruption of nanovesicles was variable, with a reduction ranging from 0.5 to 1.4 log units compared to the initial phage titer (Table 1).

3.2. The Efficiency of phiIPLA-RODI Encapsulation Is Not Greatly Influenced by Component Concentration

Once established that phage phiIPLA-RODI could be successfully encapsulated in nanovesicles, we examined whether differences in the composition of the vesicles may affect the encapsulation efficiency. In order to do that, the three types of nanovesicles (niosomes, liposomes and transfersomes) were prepared by using different concentrations (30, 50 and 70 mg/mL) of their respective components. The results of this experiment indicated that the percentage of encapsulation was largely independent of the concentration of components or the nanovesicle type ($p > 0.05$) (Table 2). The only exception to this was that, for niosomes, the highest encapsulation efficiencies (99% and 94%) were obtained by using 30 and 50 mg/mL of Pronanosome Nio-N™, while only a 62% was observed when niosomes were made with 70 mg/mL. Although not statistically significant, it is worth noting that liposomes formed using 70 mg/mL of Pronanosome Lipo-N™ and transfersomes containing 30 mg/mL of components (Phospholipon 90G and Span 60) showed variable results between different replicates (Table 2). This suggests that encapsulation of phiIPLA-RODI using those concentrations would not consistently result in high encapsulation efficiencies.

Table 2. Encapsulation efficiency for bacteriophage phiIPLA-RODI in niosomes, liposomes and transfersomes prepared using different concentrations of components.

Concentration of Components (mg/mL)	Encapsulation Efficiency (% PFU/mL)		
	Niosomes	Liposomes	Transfersomes
30	99.8.0 ± 0.03	98.6 ± 0.47	76.9 ± 21.49
50	94.5 ± 3.29	95.2 ± 4.30	95.6 ± 4.96
70	62.3 ± 14.35 *,#	85.5 ± 9.04	96.6 ± 2.89

Notes: Each value represents the mean ± standard deviation of three samples. The asterisk (*) indicates a significantly different efficiency of encapsulation as a function of the concentration and the pound (#) as a function of the type of nanovesicle ($p < 0.05$; ANOVA).

3.3. Stability of Encapsulated Phage Particles during Storage at Low Temperature

One of the main challenges associated to the development of phage-based products is to ensure their stability under storage conditions. With this in mind, we assessed whether encapsulated and non-encapsulated phiIPLA-RODI particles remained viable during storage at 4 °C throughout a 6-month period. As a control, we tested the stability of a phage suspension in SM buffer (1.49×10^8 PFU/mL).

A high stability was observed for phages encapsulated in all three types of nanovesicles regardless of the nature and concentration of their components, with decreases in phage titer below 2 log units (Figure 1). The only exception was observed for phages encapsulated in liposomes formed with 30 mg/mL of Pronanosome Lipo-N™. In this case, there was a 3-log reduction in encapsulated phage titer after four months, and a similar decrease in non-encapsulated phage titer after only two months. In general, the stability of free phages was lower than that of encapsulated phages in all the formulations tested. Indeed, in the case of transfersomes, the differences in phage titer after six months between the encapsulated and non-encapsulated fractions ranged between 2 and 4 log units depending on the component concentration (Figure 1). In contrast, the differences observed for niosomes and liposomes were always below 1 log unit (Figure 1). Surprisingly, the titer of control phage suspension was reduced only by 1 log unit (7.14 ± 0.25 $\log_{10}$ PFU/mL) after 6 months of storage, which means that, only for phages inside niosomes (50 and 70 mg/mL) and liposomes (50 mg/mL), the stability of phage was improved ($p < 0.05$).

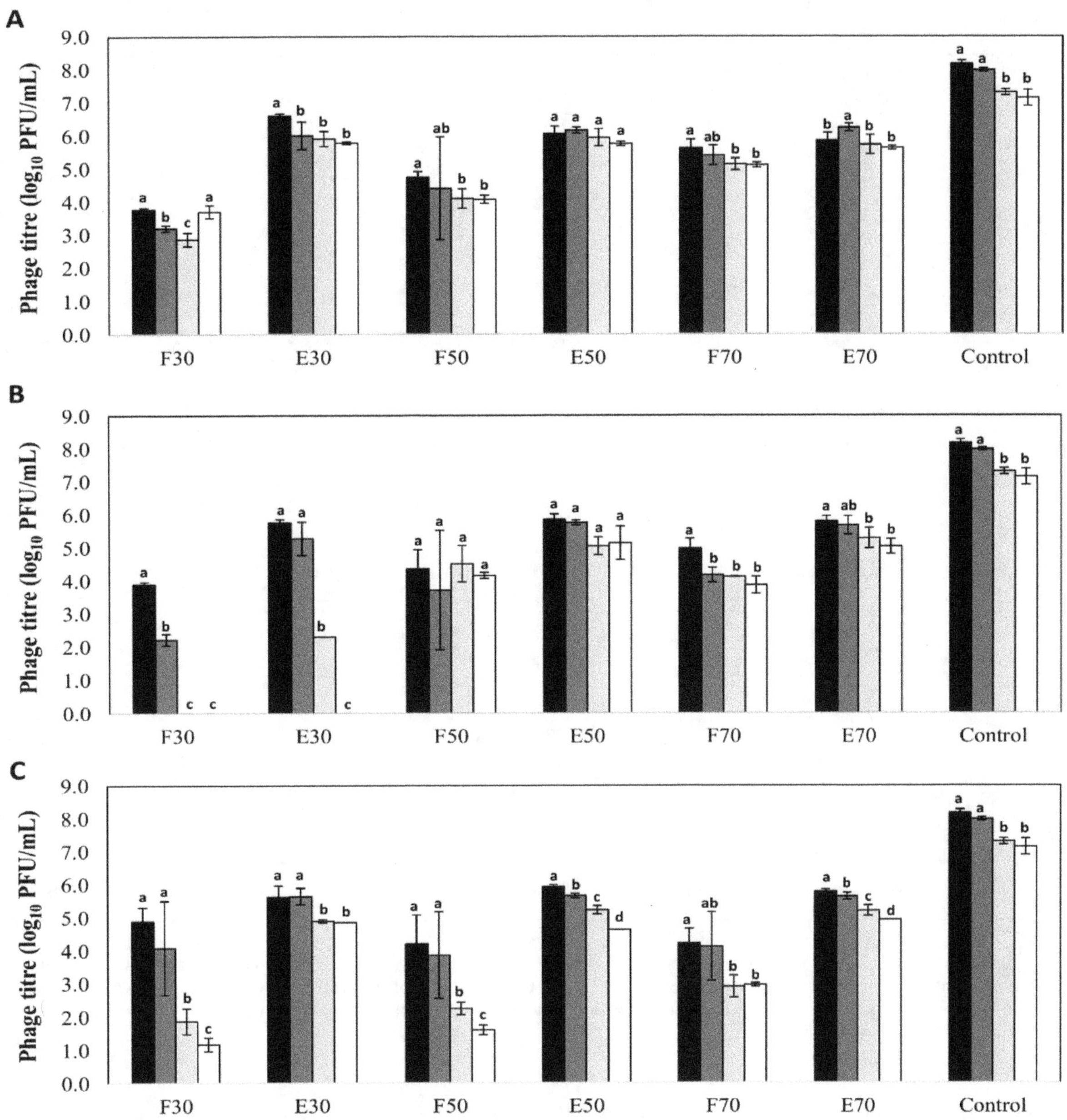

Figure 1. Stability (log_{10} PFU/mL) of bacteriophage philPLA-RODI encapsulated in different types of nanovesicles and in SM buffer (control), after storage at 4 °C: (**A**) niosomes, (**B**) liposomes and (**C**) transfersomes. Phage titer was determined after encapsulation (black bars) and also after storage for 2 months (dark grey bars), 4 months (light grey bars) and 6 months (white bars). (F): non encapsulated or free phage; (E): encapsulated phage. Numbers (30, 50 or 70) indicates the concentration of components expressed in mg/mL. Bars represent mean ± standard deviation of three biological replicates. Different letters indicate differences in stability ($p < 0.05$; ANOVA and SNK post-hoc comparison).

3.4. Niosomes Protect Phages from Low pH and High Temperature

A major advantage of nanovesicles for the application of bacteriophages in the food industry would be the protection of the viral particles from adverse physicochemical conditions commonly found in food-processing settings. Here, we examined if encapsulation of the philPLA-RODI in niosomes, liposomes or transfersomes had a protective effect against low pH. The results obtained indicated that niosomes effectively protect phages at a pH of 4.5. Thus, while non-encapsulated phages and control phage suspension were completely inactivated under these conditions, phages entrapped inside niosomes retained part of their infectivity, exhibiting only a reduction of ~2 log units (Table 3).

Table 3. Stability (log_{10} PFU/mL) of bacteriophage philPLA-RODI to different treatments.

Nanovesicles	Component (mg/mL)	Phage philPLA-RODI	Initial Titer log_{10} (PFU/mL)	pH 4.5 60 min	Tª 60 °C 90 min	NaCl 4.5 M 60 min
Niosomes	30	F	3.73 ± 0.20	–		4.74 ± 0.08 *
		E	5.78 ± 0.03	3.78 ± 0.1 *	3.54 ± 0.50 *	4.52 ± 0.13 *
		T	5.78 ± 0.03	3.78 ± 0.1 *	3.54 ± 0.50 *	4.95 ± 0.02 *
	50	F	4.09 ± 0.12	–	N/A	5.32 ± 0.18 *
		E	5.76 ± 0.07	4.00 ± 0.19 *	N/A	4.50 ± 0.10 *
		T	5.77 ± 0.06	4.00 ± 0.19 *	N/A	5.38 ± 0.15 *
	70	F	5.11 ± 0.07	–	N/A	5.79 ± 0.17 *
		E	5.63 ± 0.06	3.85 ± 0.16 *	N/A	4.44 ± 0.11 *
		T	5.74 ± 0.04	3.85 ± 0.16 *	N/A	5.81 ± 0.16
Liposomes	30	F	3.53 ± 0.15	–	–	5.03 ± 0.32 *
		E	5.80 ± 0.85	–	–	4.85 ± 0.22
		T	5.80 ± 0.84	–	–	5.25 ± 0.25
	50	F	4.16 ± 0.09	–	N/A	5.05 ± 0.12 *
		E	5.14 ± 0.51	3.46 ± 0.28 *	–	4.03 ± 0.37 *
		T	5.18 ± 0.22	3.46 ± 0.28 *	N/A	5.19 ± 0.14
	70	F	3.87 ± 0.26	–	N/A	4.71 ± 0.68
		E	5.03 ± 0.22	2.30 ± 0.24 *	N/A	4.48 ± 0.54
		T	5.06 ± 0.18	2.30 ± 0.24 *	N/A	4.91 ± 0.62
Transfersomes	30	F	–	–	–	4.93 ± 0.53 *
		E	4.88 ± 0.21	–	–	4.43 ± 0.34
		T	4.88 ± 0.21	–	–	5.01 ± 0.48
	50	F	–	–	N/A	5.01 ± 0.43 *
		E	4.63 ± 0.15	–	N/A	4.18 ± 0.22 *
		T	4.63 ± 0.15	–	N/A	4.75 ± 0.52
	70	F	2.95 ± 0.02	–	N/A	5.05 ± 0.19 *
		E	4.93 ± 0.05	–	N/A	4.60 ± 0.05 *
		T	4.94 ± 0.10	–	N/A	5.18 ± 0.17
Control phage in SM buffer	N/A	N/A	8.52 ± 0.10	–	4.75 ± 0.22 *	8.08 ± 0.48

Note: Each value represents the mean ± standard deviation of three samples. (F): free or non-encapsulated phage; (E): encapsulated phage; (T): Total phage = F + E. Initial titer: titer of phage before treatment process. (–) Below bacteriophage threshold (10^2 PFU/mL). The asterisk (*) indicates a statistically significant difference with respect to the initial titer. N/A: not applicable.

Overall, liposomes turned out to be less stable at low pH values, as only those vesicles containing 50 and 70 mg/mL of Pronanosome Lipo-N™ were able to protect the encapsulated phages, showing a reduction of nearly 2–3 log units with respect to the initial phage titer. No protective effect from low pH was achieved in the case of transfersomes, as no active phage particles could be detected after the treatment (Table 3).

Regarding temperature stability, phage encapsulation in liposomes and transfersomes did not offer protection in treatments at 60 °C for 90 min. Indeed, a total loss of viability was observed for both encapsulated and non-encapsulated phages. Similarly, an equivalent reduction in phage titer (about 4 log units) was recorded for the control phage suspension. In contrast, niosomes exerted better protection of the encapsulated phages since lower reduction in phage titer (2–3 log units) was detected (Table 3).

Although high NaCl concentrations had not any appreciable effect on phage stability, we tested the impact on nanovesicles of 4.5 M NaCl for 60 min, since this salt concentration could cause vesicles destabilization. Indeed, we observed a slight reduction (~1 log unit) in encapsulated phage titer and an increase in non-encapsulated phages, regardless of the type of nanovesicle and the surfactant concentration used (Table 3). These results suggest a slight destabilization of the nanovesicles in the presence of a high salt concentration that results in the partial release of encapsulated phages.

4. Discussion

The increasing number of studies confirming the success of phages as antimicrobials along with the current trend in the consumption of healthy, chemical-free foods has boosted the interest in bacteriophages as natural biopreservatives [42] and disinfectants [10]. However, for phage-based products to be successful, it is necessary to design proper formulations that meet certain stability requirements that standard phage suspensions do not currently have. Microencapsulation techniques including emulsification, extrusion, spray-drying, electrospun nanofibers and whey protein films have been proposed as feasible alternatives to solve this problem [24]. In addition, the use of food-compatible nanomaterials together with an approved preparation procedure, including food-grade solvents and detergents, is also an essential requirement [14]. In the present study, we attempted to find the most effective techniques for the encapsulation of phage particles to be used in food industry applications.

In a first step, we confirmed that phage phiIPLA-RODI remained stable after different encapsulation processes including niosomes, liposomes and transfersomes.

Phages encapsulated into niosomes (50 and 70 mg/mL of Pronanosome Nio-N™) and liposomes (50 mg/mL of Pronanosome Lipo-N™) retained stability for longer periods of time than those kept in SM suspension. All studies published so far report the use of liposomes as vehicles to encapsulate phages [8,43], but there are no reports about the use of niosomes and transfersomes. To determine the optimal conditions that result in the highest encapsulation efficiency for phage phiIPLA-RODI, we prepared nanovesicles using different concentrations of surfactants and phospholipids. A previous study reported that total lipid concentration was an important factor for the loading efficiency of liposomes [44]. However, we did not usually observe this trend when encapsulating phiIPLA-RODI in different types of nanovesicles. Indeed, a reduction in component concentration only improved encapsulation of phiIPLA-RODI into niosomes, but did not significantly modify the encapsulation efficiency in liposomes or transfersomes. Additionally, we could not establish any correlation between encapsulation efficiency and the average particle size associated to each vesicle type. For instance, the average size of transfersomes (~0.5 μm) was three times smaller than that of liposomes (~1.6 μm), but similar phage encapsulation efficiencies were obtained with both types of nanovesicles. In contrast, Leung et al. [45], found that encapsulation efficiency increased with vesicle size in liposomes obtained by using microfluidics, although this trend was only observed for small vesicles (up to 0.5 μm).

The use of encapsulated phages for therapeutic [25] or food applications [46] requires optimization of the formulation process. For example, it is paramount to control the release of phages from nanovesicles and maximize the stability of the encapsulated phage particles. Release of phages can be controlled by external parameters such as temperature or pH, provided they do not negatively affect the phage, or by spontaneous rupture in contact with components of food matrix [25]. Overall, we found that the stability during storage of the encapsulated phages was always higher than that of free or non-encapsulated phages in the same suspension, but the stability was only higher than that of the control phage suspension in SM buffer in some cases. A high stability of encapsulated phages is in accordance (consonance) with previous findings where nanoencapsulation in liposomes increased phage stability during storage conditions as well as during their delivery to animals [8,43]. Moreover, the use of phages inside liposomes facilitates their entry into macrophages and also protects phages from neutralizing antibodies [47]. More specifically, we found that encapsulation of phage phiIPLA-RODI into niosomes and liposomes led to a slightly higher stability during storage at 4 °C than encapsulation into transfersomes. Thus, viability of phages encapsulated in transfersomes (70 mg/mL of Phospholipon 90G and Span 60) after six months of storage was lower than that obtained for niosomes. It does not appear that the low stability of phages in transfersomes is related with the ζ-potential of these nanovesicles, as it was quite similar to that of niosomes. It is also worth noting the deleterious effect for phage phiIPLA-RODI of storage in liposomes containing 30 mg/mL of Pronanosome Lipo-N™. In fact, even non-encapsulated phages were inactivated in these suspensions. It does not seem, however, that this result is due to a low stability of phiIPLA-RODI at low temperatures since this phage is highly stable under refrigeration conditions. We can speculate that components from liposomes might bind to the phage proteins necessary for the recognition of and/or binding to the cell receptor, thereby inhibiting their interaction with host bacteria. An increase in the concentration of components might help to stabilize liposomes, as can be deduced from the higher protection of phages by liposomes containing 50 and 70 mg/mL of Pronanosome Lipo-N^{TM}. Therefore, the low stability of liposomes containing 30 mg/mL would result in the release of these components to the suspension. These components may interact with phage proteins and potentially inactivate the phage particles.

Despite the well-established antimicrobial activity of bacteriophages in some food applications, their use might be limited under certain processing conditions that inactivate the phage particles. For instance, the fermentation of milk by lactic acid bacteria results in a pH ≤ 4.5, which leads to the inactivation of many bacteriophages including phiIPLA-RODI [39]. For this reason, we investigated here if encapsulation was an effective method for the protection of phage phiIPLA-RODI from low pH values. Our results confirmed the protective effect of niosomes under acidic conditions. However, we observed that liposomes only offered partial protection to the phage particles, with a decrease in phage viability of up to 3 log units. Regarding transfersomes, our results indicated that they had a negligible contribution to phage stability maintenance. Moreover, our results show that transfersomes themselves were not stable at low pH values, which may be related to their specific content in Phospholipon 90G and Span 60. In fact, it is a common practice to modulate liposome stability at different pH values by changing their composition as a means to control the release of the vesicle content [48]. This has great relevance for the delivery of drugs, enzymes, and other therapeutics into vesicles that will only release their content at the pH of the target organ [48]. In an agro-food context, protection of phages against extremely acidic conditions was an approach used by Colom [8] to deliver *Salmonella* phages to the chicken stomach and to increase their residence time in the intestinal tract. These authors found that phages encapsulated into liposomes were stable after exposure to a pH of 2.8 and, although the phage titer was reduced by 3–5 log units, protection against *Salmonella* by the encapsulated phages persisted for at least 1 week.

In relation to the higher stability to temperature of niosomes, their composition with non-ionic surfactants might explain this result because they are more stable than phospholipids [49]. Indeed, phage phiIPLA-RODI inside niosomes turned out to be less sensitive to temperature than in a SM buffer suspension, as a reduction of 2–3 log-units was observed compared to 4 log units reduction in the control suspension.

Furthermore, we explore the stability of nanovesicles under external conditions that, without having an effect on phage viability, might destabilize their structure and, therefore, expose phages to further undesirable conditions. For example, high salt concentrations are frequently used in the processing of some foods such as cheese, but do not seem to affect the viability of phiIPLA-RODI. Indeed, all nanovesicles turned out to be quite stable in the presence of NaCl. However, in some cases, we observed a slight increase in non-encapsulated phages and a reduction in encapsulated phages, which suggests that some nanovesicles opened spontaneously, releasing the encapsulated phage particles. The spontaneous release of compounds when nanovesicles are applied to food might be a feasible approach for phage delivery. It has been previously used to incorporate cheese-ripening enzymes to milk in order to accelerate cheese proteolysis [50,51]. The rate of release could be affected by some factors such as cheese fat content, being stimulated by increasing the fat content (0% to 20%) [52].

It is worth noting that nanovesicles are considered to have a promising future in the food industry. Indeed, several commercialized liposome products have already been approved for different applications including food supplements and food preservatives as well as products for pathogen and pesticide detection [53].

Overall, our results show that nanovesicles are also suitable candidates for the production of phiIPLA-RODI-based formulations that will help to maintain phage stability during food processing conditions or in the gastrointestinal tract. More specifically, our results thus far seem to indicate that niosomes are the most interesting nanovesicles for the encapsulation of phiIPLA-RODI. Indeed, in addition to offering the greatest protection to the phage particles, they are also less costly than liposomes [40]. Future work will validate the effectiveness of encapsulated phages in a food matrix as long as the niosomes synthesis procedures are ready for scaling up, in order to make these formulations a real alternative to the food industry.

Author Contributions: E.G.-M., L.F., D.G., D.P., B.M., A.R. and P.G. conceived and designed the experiments; and E.G.-M. performed the experiments; E.G.-M. and P.G. analyzed the data; E.G.-M., L.F., D.G., D.P., B.M., A.R. and P.G. wrote the paper.

Acknowledgments: PG, BM, and AR are members of the FWO Vlaanderen funded "Phagebiotics" research community (WO.016.14) and the bacteriophage network FAGOMA. We thank R. Calvo for technical assistance.

References

1. Clokie, M.R.; Millard, A.D.; Letarov, A.V.; Heaphy, S. Phages in nature. *Bacteriophage* **2011**, *1*, 31–45. [CrossRef] [PubMed]
2. Wright, A.; Hawkins, C.H.; Anggård, E.E.; Harper, D.R. A controlled clinical trial of a therapeutic bacteriophage preparation in chronic otitis due to antibiotic-resistant *Pseudomonas aeruginosa*; a preliminary report of efficacy. *Clin. Otolaryngol.* **2009**, *34*, 349–357. [CrossRef] [PubMed]
3. Saez, A.C.; Zhang, J.; Rostagno, M.H.; Ebner, P.D. Direct feeding of microencapsulated bacteriophages to reduce *Salmonella* colonization in pigs. *Foodborne Pathog. Dis.* **2011**, *8*, 1269–1274. [CrossRef] [PubMed]
4. Borie, C.; Albala, I.; Sánchez, P.; Sánchez, M.L.; Ramírez, S.; Navarro, C.; Morales, M.A.; Retamales, A.J.; Robeson, J. Bacteriophage treatment reduces *Salmonella* colonization of infected chickens. *Avian Dis.* **2008**, *52*, 64–67. [CrossRef] [PubMed]
5. Kutter, E.; de Vos, D.; Gvasalia, G.; Alavidze, Z.; Gogokhia, L.; Kuhl, S.; Abedon, S.T. Phage therapy in clinical practice: Treatment of human infections. *Curr. Pharm. Biotechnol.* **2010**, *11*, 69–86. [CrossRef] [PubMed]
6. García, P.; Martínez, B.; Obeso, J.M.; Rodríguez, A. Bacteriophages and their application in food safety. *Lett. Appl. Microbiol.* **2008**, *47*, 479–485. [CrossRef] [PubMed]
7. Bueno, E.; García, P.; Martínez, B.; Rodríguez, A. Phage inactivation of *Staphylococcus aureus* in fresh and hard-type cheeses. *Int. J. Food Microbiol.* **2012**, *158*, 23–27. [CrossRef] [PubMed]
8. Colom, J.; Cano-Sarabia, M.; Otero, J.; Cortés, P.; Maspoch, D.; Llagostera, M. Liposome-encapsulated bacteriophages for enhanced oral phage therapy against *Salmonella* spp. *Appl. Environ. Microbiol.* **2015**, *81*, 4841–4849. [CrossRef] [PubMed]
9. Sheng, H.; Knecht, H.J.; Kudva, I.T.; Hovde, C.J. Application of bacteriophages to control intestinal *Escherichia coli* O157:H7 levels in ruminants. *Appl. Environ. Microbiol.* **2006**, *72*, 5359–5366. [CrossRef] [PubMed]
10. Gutiérrez, D.; Rodríguez-Rubio, L.; Martínez, B.; Rodríguez, A.; García, P. Bacteriophages as Weapons against Bacterial Biofilms in the Food Industry. *Front. Microbiol.* **2016**, *7*, 825. [CrossRef] [PubMed]
11. Hammerl, J.A.; Jäckel, C.; Alter, T.; Janzcyk, P.; Stingl, K.; Knüver, M.T.; Hertwig, S. Reduction of *Campylobacter jejuni* in broiler chicken by successive application of group II and group III phages. *PLoS ONE* **2014**, *9*, e114785. [CrossRef] [PubMed]
12. Soni, K.A.; Nannapaneni, R. Removal of *Listeria monocytogenes* biofilms with bacteriophage P100. *J. Food Protect.* **2010**, *73*, 1519–31524. [CrossRef]
13. Anal, A.K.; Singh, H. Recent advances in microencapsulation of probiotics for industrial applications and targeted delivery. *Trends Food Sci. Technol.* **2007**, *18*, 240–251. [CrossRef]
14. Dias, M.I.; Ferreira, I.C.; Barreiro, M.F. Microencapsulation of bioactives for food applications. *Food Funct.* **2015**, *6*, 1035–1052. [CrossRef] [PubMed]
15. Pando, D.; Beltrán, M.; Gerone, I.; Matos, M.; Pazos, C. Resveratrol entrapped niosomes as yoghurt additive. *Food Chem.* **2015**, *170*, 281–287. [CrossRef] [PubMed]
16. Davis, J.L.; Paris, H.L.; Beals, J.W.; Binns, S.E.; Giordano, G.R.; Scalzo, R.L.; Bell, C. Liposomal-encapsulated ascorbic acid: Influence on vitamin C bioavailability and capacity to protect against ischemia–reperfusion injury. *Nutr. Metab. Insights* **2016**, *9*, 25–30. [CrossRef] [PubMed]
17. Gutiérrez, G.; Matos, M.; Barrero, P.; Pando, D.; Iglesias, O.; Pazos, C. Iron-entrapped niosomes and their potential application for yogurt fortification. *LWT Food Sci. Technol.* **2016**, *74*, 550–556. [CrossRef]

18. Parmentier, J.; Thewes, B.; Gropp, F.; Fricker, G. Oral peptide delivery by tetraether lipid liposomes. *Int. J. Pharm.* **2011**, *415*, 150–157. [CrossRef] [PubMed]
19. Schoina, V.; Terpou, A.; Angelika-Ioanna, G.; Koutinas, A.; Kanellaki, M.; Bosnea, L. Use of Pistacia terebinthus resin as immobilization support for *Lactobacillus casei* cells and application in selected dairy products. *J. Food Sci. Technol.* **2015**, *52*, 5700–5708. [CrossRef] [PubMed]
20. D'Orazio, G.; Di Gennaro, P.; Boccarusso, M.; Presti, I.; Bizzaro, G.; Giardina, S.; Michelotti, A.; Labra, M.; La Ferla, B. Microencapsulation of new probiotic formulations for gastrointestinal delivery: In vitro study to assess viability and biological properties. *Appl. Microbiol. Biotechnol.* **2015**, *99*, 9779–9789. [CrossRef] [PubMed]
21. Rodklongtan, A.; La-Ongkham, O.; Nitisinprasert, S.; Chitprasert, P. Enhancement of *Lactobacillus reuteri* KUB-AC5 survival in broiler gastrointestinal tract by microencapsulation with alginate-chitosan semi-interpenetrating polymer networks. *J. Appl. Microbiol.* **2014**, *117*, 227–238. [CrossRef] [PubMed]
22. Malheiros, P.S.; Sant'Anna, V.; Utpott, M.; Brandelli, A. Antilisterial activity and stability of nanovesicle-encapsulated antimicrobial peptide P34 in milk. *Food Control* **2012**, *23*, 42–47. [CrossRef]
23. Were, L.M.; Bruce, B.; Davidson, P.M.; Weiss, J. Encapsulation of nisin and lysozyme in liposomes enhances efficacy against Listeria monocytogenes. *J. Food Prot.* **2004**, *67*, 922–927. [CrossRef] [PubMed]
24. Choińska-Pulit, A.; Mituła, P.; Śliwka, P.; Łaba, W.; Kurzępa-Skaradzińska, A. Bacteriophage encapsulation: Trends and potential applications. *Trends Food Sci. Technol.* **2015**, *45*, 212–221. [CrossRef]
25. Malik, D.J.; Sokolov, I.J.; Vinner, G.K.; Mancuso, F.; Cinquerrui, S.; Vladisavljevic, G.T.; Clokie, M.R.J.; Garton, N.J.; Stapley, A.G.F.; Kirpichnikova, A. Formulation, stabilisation and encapsulation of bacteriophage for phage therapy. *Adv. Colloid Interface Sci.* **2017**, *249*, 100–133. [CrossRef] [PubMed]
26. Lowy, F.D. *Staphylococcus aureus* infections. *N. Engl. J. Med.* **1998**, *339*, 520–532. [CrossRef] [PubMed]
27. Le Loir, Y.; Baron, F.; Gautier, M. *Staphylococcus aureus* and food poisoning. *Genet. Mol. Res.* **2003**, *2*, 63–76. [PubMed]
28. European Food Safety Authority (EFSA); European Centre for Disease Prevention and Control (ECDC). The European Union summary report on trends and sources of zoonoses, zoonotic agents and food-borne outbreaks in 2015. *EFSA J.* **2016**, *14*, 4634. [CrossRef]
29. Centers for Disease Control and Prevention (CDC). *National Outbreak Reporting System (NORS)*; Centers for Disease Control and Prevention (CDC): Atlanta, GA, USA, 2017.
30. Kurosu, M.; Siricilla, S.; Mitachi, K. Advances in MRSA drug discovery: Where are we and where do we need to be? *Expert Opin. Drug Discov.* **2013**, *8*, 1095–1116. [CrossRef] [PubMed]
31. Rhoads, D.D.; Wolcott, R.D.; Kuskowski, M.A.; Wolcott, B.M.; Ward, L.S.; Sulakvelidze, A. Bacteriophage therapy of venous leg ulcers in humans: Results of a phase I safety trial. *J. Wound Care* **2009**, *18*, 237–243. [CrossRef] [PubMed]
32. Fischetti, V.A. Using phage lytic enzymes to control pathogenic bacteria. *BMC Oral Health* **2006**, *6*, S16. [CrossRef] [PubMed]
33. El Haddad, L.; Roy, J.P.; Khalil, G.E.; St-Gelais, D.; Champagne, C.P.; Labrie, S.; Moineau, S. Efficacy of two *Staphylococcus aureus* phage cocktails in cheese production. *Int. J. Food Microbiol.* **2016**, *217*, 7–13. [CrossRef] [PubMed]
34. Guo, T.; Xin, Y.; Zhang, C.; Ouyang, X.; Kong, J. The potential of the endolysin Lysdb from *Lactobacillus delbrueckii* phage for combating *Staphylococcus aureus* during cheese manufacture from raw milk. *Appl. Microbiol. Biotechnol.* **2016**, *100*, 3545–3554. [CrossRef] [PubMed]
35. García, P.; Madera, C.; Martínez, B.; Rodríguez, A. Biocontrol of *Staphylococcus aureus* in curd manufacturing processes using bacteriophages. *Int. Dairy J.* **2007**, *17*, 1232–1239. [CrossRef]
36. Obeso, J.M.; García, P.; Martínez, B.; Arroyo-López, F.N.; Garrido-Fernández, A.; Rodríguez, A. Use of logistic regression for prediction of the fate of *Staphylococcus aureus* in pasteurized milk in the presence of two lytic phages. *Appl. Environ. Microbiol.* **2010**, *76*, 6038–6046. [CrossRef] [PubMed]
37. García, P.; Madera, C.; Martinez, B.; Rodríguez, A.; Suarez, J.E. Prevalence of bacteriophages infecting *Staphylococcus aureus* in dairy samples and their potential as biocontrol agents. *J. Dairy Sci.* **2009**, *92*, 3019–3026. [CrossRef] [PubMed]
38. Tabla, R.; Martinez, B.; Rebollo, J.E.; Gonzalez, J.; Ramirez, M.R.; Roa, I.; Rodriguez, A.; Garcia, P. Bacteriophage performance against *Staphylococcus aureus* in milk is improved by high hydrostatic pressure treatments. *Int. J. Food Microbiol.* **2012**, *156*, 209–213. [CrossRef] [PubMed]

39. Gutiérrez, D.; Vandenheuvel, D.; Martínez, B.; Rodríguez, A.; Lavigne, R.; García, P. Two Phages, philPLA-RODI and philPLA-C1C, lyse mono- and dual-species staphylococcal biofilms. *Appl. Environ. Microbiol.* **2015**, *81*, 3336–3348. [CrossRef] [PubMed]
40. Marianecci, C.; Di Marzio, L.; Rinaldi, F.; Celia, C.; Paolino, D.; Alhaique, F.; Esposito, S.; Carafa, M. Niosomes from 80s to present: The state of the art. *Adv. Colloid Interface Sci.* **2014**, *205*, 187–206. [CrossRef] [PubMed]
41. Bangham, A.D.; Standish, M.M.; Watkins, J.C. Diffusion of univalent ions across the lamellae of swollen phospholipids. *J. Mol. Biol.* **1965**, *13*, 238–252. [CrossRef]
42. Coffey, B.; Mills, S.; Coffey, A.; McAuliffe, O.; Ross, R.P. Phage and their lysins as biocontrol agents for food safety applications. *Annu. Rev. Food Sci. Technol.* **2010**, *1*, 449–468. [CrossRef] [PubMed]
43. Singla, S.; Harjai, K.; Raza, K.; Wadhwa, S.; Katare, O.P.; Chhibber, S. Phospholipid vesicles encapsulated bacteriophage: A novel approach to enhance phage biodistribution. *J. Virol. Methods* **2016**, *236*, 68–76. [CrossRef] [PubMed]
44. Nallamothu, R.; Wood, G.C.; Kiani, M.F.; Moore, B.M.; Horton, F.P.; Thoma, L.A. A targeted liposome delivery system for combretastatin A4: Formulation optimization through drug loading and in vitro release studies. *PDA J. Pharm. Sci. Technol.* **2006**, *60*, 144–155. [PubMed]
45. Leung, S.S.; Morales, S.; Britton, W.; Kutter, E.; Chan, H.K. Microfluidic-assisted bacteriophage encapsulation into liposomes. *Int. J. Pharm.* **2018**, *545*, 176–182. [CrossRef] [PubMed]
46. Ahmadi, H.; Wang, Q.; Lim, L.T.; Balamurugan, S. Encapsulation of Listeria Phage A511 by Alginate to Improve Its Thermal Stability. *Methods Mol. Biol.* **2018**, *1681*, 89–95. [CrossRef] [PubMed]
47. Singla, S.; Harjai, K.; Katare, O.P.; Chhibber, S. Encapsulation of bacteriophage in liposome accentuates its entry in to macrophage and shields it from neutralizing antibodies. *PLoS ONE* **2016**, *11*, e0153777. [CrossRef] [PubMed]
48. Drummond, D.C.; Zignani, M.; Leroux, J.C. Current status of pH-sensitive liposomes in drug delivery. *Prog. Lipid Res.* **2000**, *39*, 409–460. [CrossRef]
49. Kumar, G.P.; Rajeshwarrao, P. Nonionic surfactant vesicular systems for effective drug delivery—An overview. *Acta Pharm. Sin. B* **2001**, *1*, 208–219. [CrossRef]
50. Picon, A.; Gaya, P.; Medina, M.; Nunez, M. Proteinases encapsulated in stimulated release liposomes for cheese ripening. *Biotechnol. Lett.* **1997**, *19*, 345–348. [CrossRef]
51. Kheadr, E.E.; Vuillemard, J.C.; El Deeb, S.A. Accelerated Cheddar cheese ripening with encapsulated proteinases. *Int. J. Food Sci. Technol.* **2000**, *35*, 483–495. [CrossRef]
52. Laloy, E.; Vuillemard, J.C.; Dufour, P.; Simard, R. Release of enzymes from liposomes during cheese ripening. *J. Control. Release* **1998**, *54*, 213–222. [CrossRef]
53. Shukla, S.; Haldorai, Y.; Hwang, S.K.; Bajpai, V.K.; Huh, Y.S.; Han, Y.K. Current Demands for Food-Approved Liposome Nanoparticles in Food and Safety Sector. *Front. Microbiol.* **2017**, *8*, 2398. [CrossRef] [PubMed]

9

Exploiting a Phage-Bacterium Interaction System as a Molecular Switch to Decipher Macromolecular Interactions in the Living Cell

Éva Viola Surányi [1,2,*,†], Rita Hírmondó [2,*,†], Kinga Nyíri [1,2], Szilvia Tarjányi [2], Bianka Kőhegyi [1,2], Judit Tóth [2] and Beáta G. Vértessy [1,2,*]

1 Department of Applied Biotechnology and Food Sciences, Budapest University of Technology and Economics, H-1111 Budapest, Hungary; nyiri.kinga@ttk.mta.hu (K.N.); bianka.kohegyi@gmail.com (B.K.)

2 Institute of Enzymology, RCNS, Hungarian Academy of Sciences, H-1111 Budapest, Hungary; kicsiszilvi.tarjanyi@gmail.com (S.T.); toth.judit@ttk.mta.hu (J.T.)

* Correspondence: eva.suranyi@mail.bme.hu (É.V.S.); hirmondo.rita@ttk.mta.hu (R.H.); vertessy@mail.bme.hu or vertessy.beata@ttk.mta.hu (B.G.V.)

† These authors contributed equally to this work as first authors.

Abstract: Pathogenicity islands of *Staphylococcus aureus* are under the strong control of helper phages, where regulation is communicated at the gene expression level via a family of specific repressor proteins. The repressor proteins are crucial to phage-host interactions and, based on their protein characteristics, may also be exploited as versatile molecular tools. The Stl repressor from this protein family has been recently investigated and although the binding site of Stl on DNA was recently discovered, there is a lack of knowledge on the specific protein segments involved in this interaction. Here, we develop a generally applicable system to reveal the mechanism of the interaction between Stl and its cognate DNA within the cellular environment. Our unbiased approach combines random mutagenesis with high-throughput analysis based on the *lac* operon to create a well-characterized gene expression system. Our results clearly indicate that, in addition to a previously implicated helix-turn-helix segment, other protein moieties also play decisive roles in the DNA binding capability of Stl. Structural model-based investigations provided a detailed understanding of Stl:DNA complex formation. The robustness and reliability of our novel test system were confirmed by several mutated Stl constructs, as well as by demonstrating the interaction between Stl and dUTPase from the Staphylococcal φ11 phage. Our system may be applied to high-throughput studies of protein:DNA and protein:protein interactions.

Keywords: gene expression regulation; molecular probe; macromolecular interactions; phage-host interaction

---

1. Introduction

Staphylococcus aureus (*S. aureus*) pathogenicity islands (SaPIs) operate in the fine-tuned control mechanism between phages and the host bacterial cell. SaPIs play an important role in spreading virulence factors [1]. These mobile genetic elements usually encode toxins, antibiotic resistance genes, and antigens responsible for the pathogenicity of the given strain. Pathogenicity islands are considered to be similar to phages since these mobile elements can also move via horizontal transfer from one strain to the other, hijacking the phage capsid. Under normal circumstances, the pathogenicity island is under the repression of a transcription regulatory factor encoded by the island. The master repressor of a specific SaPI, namely $SaPI_{bov1}$, is the Stl protein. Stl binding to a specific DNA segment within SaPI [2,3] prevents the excision and replication of the mobile genetic elements. SaPIs can be mobilized by infecting the bacteria with certain staphylococcal bacteriophages (e.g., Φ11 and 80α phages) or by the induction of endogenous prophages [4]. Following bacteriophage infection, the excision and replication of SaPIs is induced by the formation of a repressor:derepressor complex constituting the master repressor and another phage-related protein. As a result, the repression is terminated and SaPI becomes mobilized, i.e., the SaPI DNA is replicated and packaged into parallel maturing phage particles in the bacterium [5]. In recent years, more and more of these repressor:derepressor interactions have been discovered. Firstly, dUTPases from Φ11 and 80α phages were identified as specific derepressors of Stl function. It has also been shown that Φ11 dUTPase has a stronger effect on the mobilization of $SaPI_{bov1}$ as compared to 80α dUTPase [4]. Interestingly, in addition to the Φ11 and 80α phage dUTPases, both belong to the trimeric dUTPase enzyme family, structurally very different dimeric dUTPases (encoded by different phages) are shown to be also able to derepress the same SaPI island [6,7].

The $SaPI_{bov1}$ master repressor, Stl, has separate domains for DNA and protein binding [8]. The DNA binding function of Stl is performed through a helix-turn-helix (HTH) motif [8,9], as in case of the λcI repressor protein of the lambda phage [10,11]. A similar domain architecture was identified in other representatives of the family of Stl-like repressors encoded in different SaPIs as well [8]. These repressor proteins are highly diverse considering their amino acid sequence and higher order structure; however, there is considerable sequence conservation within their HTH segments. Using in silico prediction tools, the HTH motif could be found in eight out of the 12 investigated master repressors within the diverse *S. aureus* pathogenicity islands. In these repressors, the HTH motif was invariably located at the *N*-terminal part, similar to Stl. The *C*-terminal domain of the $SaPI_{bov1}$ Stl repressor was proposed to contribute to the binding of protein interaction partner(s) [8]. The specific DNA segment recognized by the HTH motif of Stl was also recently characterized in detail [3] (cf. Figure 1 DNA sequence indicated in green). It was shown that Stl binds to a palindromic sequence, constituted by two 6-mer repeat sequences interrupted with a five-nucleotide long non-specific segment. Such palindromes can be found at the *stl-str* and the *str-xis* intergenic regions of the *S. aureus* genome, which is the regulation site of Stl controlling the Str and Xis expression [2,3].

The identification of the cognate DNA binding segments for the Stl repressor offers a basis for the design of a molecular switch. Switchable systems based on e.g., bacterial operons, transcription factors, or repressor proteins have been bioengineered for decades and are used for basic research as well as for various biotechnological applications. One of the first widely used systems for gene expression control in bacteria is the lactose repressor system (the *lac* operon) described by Jacob and Monod in 1961 [12]. Another extensively used switch-system for the regulation of recombinant protein production relies on the Tet repressor (TetR) [13,14] that drives the transcription of a family of tetracycline (tc) resistance determinants in Gram-negative bacteria. TetR can be used for the selective control of the expression of single genes in some organisms (plants and lower eukaryotes) without further modifications [14]. The several other commonly used inducible promoters include P_{BAD} [15], P_{tac} [16], P_{trc} [17], and P_{T7} [18]. Besides these, popular two-hybrid reporter systems may also apply transcription factors. In the bacterial two-hybrid system, one of the target proteins can be fused to the dimeric bacteriophage λcI repressor [19].

Our aim was to design and implement a switchable system in which the macromolecular interactions between Stl and its cognate DNA binding segments can be revealed. Despite our numerous trials, neither the flexible Stl protein on its own nor its complex with DNA could be crystallized. Hence, we needed another experimental approach to gain insights into the Stl:DNA interaction. Using the Stl-based reporter system, the molecular components responsible for Stl binding to DNA can be characterized in a high-throughput manner. We used *Mycobacterium smegmatis* (*M. smegmatis*) as the cellular host system for our studies, as Stl expression in mycobacterial cells is already implemented [20] and the large evolutionary distance between mycobacteria and *S. aureus* (Actinobacteria vs. Firmicutes, respectively) provides an opportunity to investigate the macromolecular interactions of Stl without putative perturbing effects. Finally, we extended our test system to investigate the Stl interaction with a phage protein, Φ11 dUTPase.

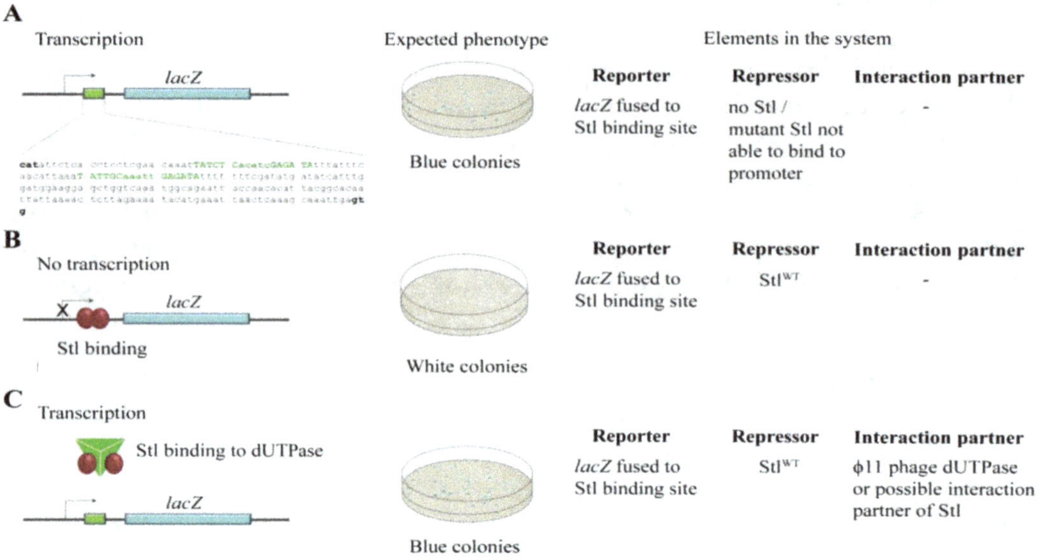

Figure 1. The design of the switch system. (**A**) In the absence of Stl-binding to its recognition sequence, *lacZ* can be expressed in the cell, leading to blue colonies in the experimental setup. The exact sequence from the $SaPI_{bov1}$ genome (13733–13933) cloned into the *lacZ* promoter is shown. Specific Stl binding sites are labeled with green, conserved sites are shown with capital letters; (**B**) Stl binding to the promoter inhibits *lacZ* expression, leading to white colonies in the experimental setup; (**C**) The phage dUTPase protein sequestrates Stl^{WT} from the promoter DNA segment, leading to blue colonies.

2. Materials and Methods

2.1. Bacterial Strains, Media, and Growth Conditions

The *E. coli* XL1-Blue and BL21 Rosetta strains were used for cloning and in vitro protein expression, respectively. The *M. smegmatis* mc^2155 strain used for further experiments was grown in Lemco liquid culture or on solid plates with the addition of 15 g L^{-1} Bacto agar as described previously [21]. Kanamycin was added at 20 μg/mL, hygromycin B at 50 μg/mL, and chloramphenicol at 34 μg/mL. X-gal (5-bromo-4-chloro-3-indolyl-β-D-galactopyranoside) was used at 40 μg/mL, and tetracycline at 20 ng/mL for the induction of protein expression in *M. smegmatis*.

2.2. Cloning and Mutagenesis

For the construction of the p2NIL-$LacZ^{Str}$-INT vector, the Stl-regulated promoter region [3] was PCR-amplified and cloned into the p2NIL vector using SalI and HindIII restriction sites. The *lacZ* gene was PCR-amplified from pGOAL17 [22] and cloned upstream of the promoter containing Stl binding sites using BglII and SalI restriction sites to generate the p2NIL-$LacZ^{Str}$ vector. The integration cassette from the pUC-Gm-Int plasmid [23] was PCR amplified and cloned into the p2NIL-$LacZ^{Str}$ plasmid using NotI restriction site to generate the p2NIL-$LacZ^{Str}$-INT construct.

The pKW08-$Stl^{C\text{-}term}$ and the pKW08-Stl^{AA} were generated by inserting the truncated or mutant Stl sequence amplified from pGEX-4T-1-Stl-CTD and pGEX-4T-1-Stl^{AA}, respectively [8], into the pKW08-Lx vector using HindIII and BamHI restriction sites.

The colony PCR product of the random mutagenesis experiment was inserted into the pKW08 vector using BamHI and HindIII restriction sites to generate the pKW08-Stl^{A236T} plasmid.

The pGex-4T-1-Stl^{A236T} plasmid was generated by site-directed mutagenesis (QuikChange method, Stratagene) [24].

The pSJ27-φdut plasmid was constructed from the pSJ27 plasmid (kind gift of Graham F. Hatfull [25]) using recombination cloning. The φ11 dut gene sequence was synthesized by GenScript.

Primers used in this study are summarized in Supplementary Table S2. Successful cloning and mutagenesis was verified in all cases by sequencing of the appropriate region of the plasmid. All of the plasmids used in this study are summarized in Supplementary Table S3.

2.3. *Construction of the M. smegmatis Reporter Strain*

p2NIL-LacZ^{Str}-INT plasmid (0.5 μg) was electroporated into electrocompetent wild-type *M. smegmatis* to generate the LacZ^{Str}-carrying *M. smegmatis* strain. For inducible Stl expression, 0.5 μg of pKW08-Stl generated in Reference [20] was electroporated into electrocompetent LacZ^{Str} carrying *M. smegmatis* strain. Protein expression was verified by Western blot as previously described [20].

2.4. *Analysis of Different Stl Mutants in the Reporter System*

pKW08-Stl, pKW08-Stl$^{C\text{-}term}$, pKW08-Stl^{AA}, pKW08-Stl^{A236T} or pKW08-Stl^{MUT} (0.5 μg) was electroporated into the electrocompetent LacZ^{Str}-carrying *M. smegmatis* strain. Then, 100 μL of the transformants was parallelly plated on kanamycin, hygromycin B, and X-gal or on kanamycin, hygromycin, X-gal, and tetracycline containing agar plates and incubated for 5 days at 37 °C. Blue colonies indicate *lacZ* transcription (no repression of the promoter), while white colonies indicate the functional repression of the promoter.

2.5. *Random Mutagenesis of Stl*

Random mutagenesis events were elicited using error-prone PCR conditions [26] comprising am extra 6 mM $MgCl_2$ and 0.5 mM $MnCl_2$, resulting in less faithful replication by the RedTaq polymerase (Sigma, St. Louis, MO, USA). The mutagenized product was cloned in pKW08-Lx using BamHI and HindIII restriction sites generating the pKW08-Stl^{MUT} library. The pKW08-Stl^{MUT} library was then electroporated into the LacZ^{Str}-carrying *M. smegmatis* strain reviewed in the previous paragraph. The inserted *stl* sequence was amplified by colony PCR from selected blue colonies and subjected to sequencing. Primers used for cloning, PCR, and sequencing are compiled in Supplementary Table S2.

2.6. *Protein Expression and Purification*

The *E. coli* strain BL21 Rosetta (DE3) was transformed with the pGex-4T1-Stl and pGex-4T-1-Stl^{A236T} vectors and propagated in 500 ml Luria-Bertani broth (LB) to an OD_{600} of 0.6. The culture was then cooled to 30 °C and induced with 0.5 mM iso-propyl-β-D-thiogalactoside for 4 h. Subsequent manipulation of the harvested cells was carried out on ice. For Stl purification, cell pellets were solubilized in 30 mL of lysis buffer (50 mM HEPES (pH 7.5), 200 mM NaCl) supplemented with 2 mM dithiothreitol (DTT), 2 μg/mL RNase and DNase, and ethylenediaminetetraacetic acid (EDTA) free protease inhibitor (cOmplete ULTRA Tablets, Mini). Cell suspensions were sonicated (60 s, four times) and centrifuged (16,000× *g* for 30 min). Supernatant was loaded on a glutathione-agarose affinity-chromatography column. The column was washed with 10 volumes of lysis buffer. On-column cleavage was then performed by the addition of 80 cleavage unit thrombin to remove the glutathione-S-transferase tag. Following overnight cleavage, Stl was eluted from the column at >95% purity, as estimated by sodium dodecyl sulfate polyacrylamide gel electrophoresis (SDS-PAGE).

2.7. *Electrophoretic Mobility Shift Assay (EMSA)*

EMSA experiments were performed using a 229-mer dsDNA oligonucleotide of the Stl regulated promoter region based on our previous results [3] (cf. Supplementary Table S2). Complementary oligonucleotides were custom synthesized by Eurofins MWG Operon and hybridized by controlled gradual cooling after 5 min of incubation at 95 °C. The investigated proteins were mixed with 100 ng DNA in 20 µL total volume. After incubation for 5 min at room temperature, samples were loaded onto 8% polyacrylamide gel. Pre-electrophoresis was performed for 1 h at 150 V in tris-borate-EDTA (TBE) buffer, followed by sample electrophoresis for 70 min at room temperature at 100 V. Bands were detected using GelRed (Biotium, Fremont, CA, USA) and a Bio-Rad gel documentation system. The EMSA experiment was repeated three times.

2.8. Construction of the Double Switch System to Test Protein:Protein Interaction

For the construction of the Stl:φ11 dUTPase double switch system, pSJ27-φdut vector was electroporated into the Stl- and $LacZ^{Str}$-carrying *M. smegmatis* strain.

3. Results

3.1. Design and Construction of a Reporter System to Characterize Different Functions of the Stl Protein

The exact DNA sequence of the cognate Stl recognition site within the SaPI promoter region has already been determined [3]. Based on this knowledge, we constructed a switchable gene expression system using the Stl:DNA interaction.

To explore the Stl:DNA interaction in the cellular environment, the reporter system shown in Figure 1 was constructed in *M. smegmatis*. The reporter system is based on the widely used blue-white colony selection strategy, i.e., the activity of a β-galactosidase reporter gene (*lacZ*) leads to the conversion of the substrate X-gal to a blue-colored product providing a straightforward readout. We fused the *lacZ* gene to the specific Stl binding sequence [3,27] so that the Stl:DNA interaction inhibits *lacZ* transcription (white colonies). In the absence of Stl or when Stl:DNA binding is compromised, i.e., when either a mutant Stl or a phage protein able to sequestrate Stl is present, *lacZ* is expressed, leading to blue colonies (Figure 1).

The reporter plasmid also contained a L5 integration cassette [23] enabling it to integrate into the mycobacterial genome to the specific attP site (Figure 2). As a result of stable integration, transformed mycobacterium cells possess only one copy of the reporter system and therefore the color intensity of the β-galactosidase reaction is not dependent on the copy number of the vector. For inducible Stl expression, the cells containing the reporter construct were transformed with the plasmid pKW08-Stl (generated in Reference [20]).

3.2. Validation of the Switch System Using Wild-Type and Mutant Stl Constructs

To test the system functionality, several Stl constructs with various DNA-binding abilities were expressed. Full-length (i.e., wild-type) Stl was used as positive control. As described previously, both specific truncation and rationally designed point-mutations within the Stl protein sequence could lead to a significant reduction of the DNA-binding capability [8]. The $Stl^{C\text{-}term}$ truncated construct is practically unable to bind to DNA since it lacks the N-terminus of the protein encoding the HTH motif, the bona fide segment for DNA binding. The mutant Stl^{AA} containing mutations within the HTH motif was shown to be significantly perturbed in DNA-binding as well [8]. The results of the validation experiment are shown in Figure 3. The Stl protein is presented as a molecular model in the figure. This model was constructed based on vacuum circular dichroism (CD) experiments and model building [8]. Blue colonies formed in all cases without the induction of Stl expression. Upon the induction of Stl expression, white colonies formed with the full-length wild-type Stl as it is able to fully repress the transcription of the *lacZ* gene (colony counting revealed that 70% of the colonies were white). The expression of both the $Stl^{C\text{-}term}$ and Stl^{AA}, having diminished DNA-binding ability in vitro, leads to the formation of blue *M. smegmatis* colonies, confirming that the system is functional (colony counting data: 100% blue colonies, i.e., no white colonies were observed).

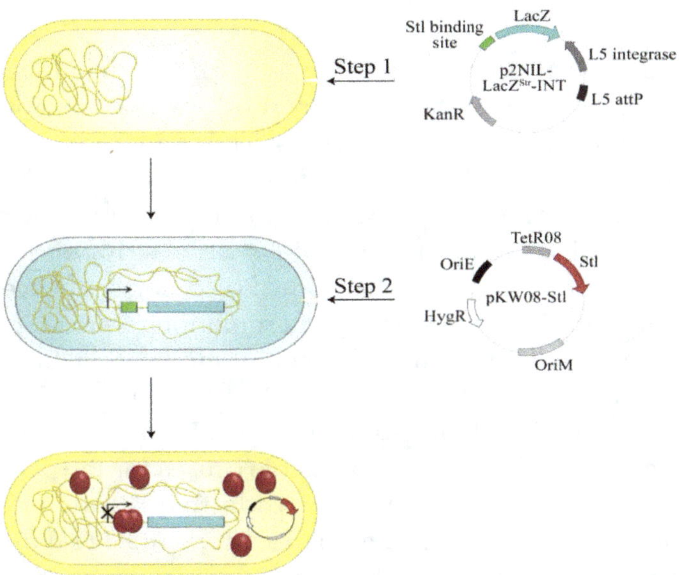

Figure 2. Workflow of the in vivo Stl:DNA binding assay: The p2NIL-LacZ^{Str}-int vector carrying the desired Stl binding site in the promoter region of *lacZ* gene was electroporated into *M. smegmatis* cells (Step 1). After electroporation, this vector integrates into the genome. This is followed by Step 2: electroporation of pKW08-Stl vector into the Stl switch system. Stl expression is inducible in a tetracycline-dependent way.

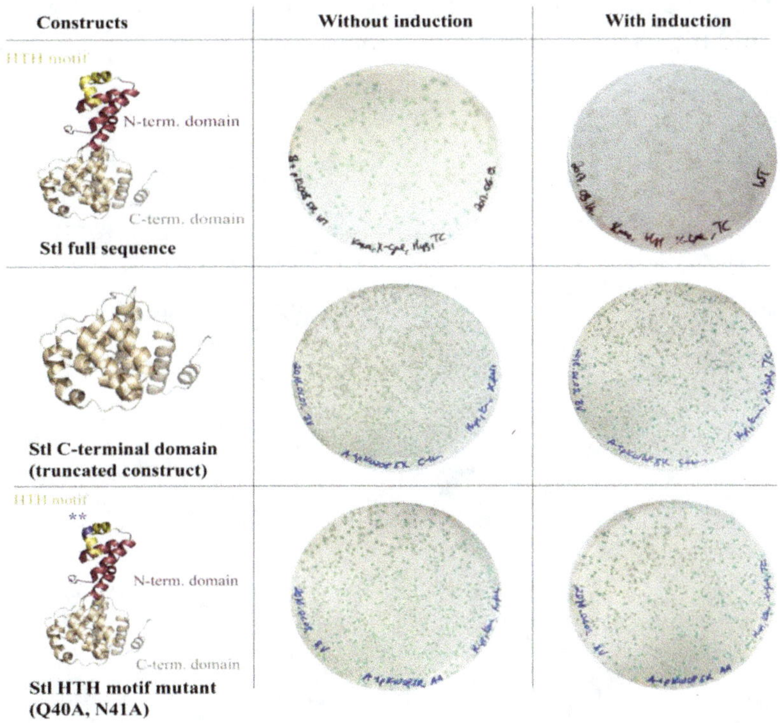

Figure 3. Validation of the switch system: To test the system functionality, several Stl constructs with various DNA-binding abilities were used. Without induction, i.e., in the absence of the Stl expression inducer tetracycline, cells form blue colonies in all cases. Upon the induction of the full-length wild-type Stl, white colonies are formed as Stl is able to fully repress the *lacZ* gene. Both the truncated Stl C-terminal domain construct and the HTH motif mutant Stl have diminished DNA-binding ability [8] and, as expected, their expression in *M. smegmatis* leads to the formation of blue colonies. Note that the difference in the number of colonies between variants represents the transformation efficiency.

3.3. Random Mutagenesis-Based Search for Stl Mutations Affecting DNA-Binding

We wished to apply this switch system as a non-biased high-throughput approach to characterize the Stl:DNA interaction with a focus on the protein residues that are involved in DNA-binding. The Stl was therefore subjected to random mutagenesis. The random mutagenesis PCR reaction mix contained excess Mg^{2+} and Mn^{2+} ions for a less faithful replication of the DNA by the polymerase [26]. The mutagenized product was cloned and electroporated into mycobacteria hosting the $LacZ^{Str}$ reporter system. Blue colonies indicating the lack of *lacZ* transcription were selected. These blue colonies represented inefficient Stl:DNA interaction. The Stl coding sequence was therefore amplified by colony PCR and sent to sequencing. A wide range of mutations were identified (cf. Supplementary Table S1 for a full list). Figure 4A focuses on the several mutations occurring as single mutation events in the Stl sequence (Stl^{E59K}, Stl^{I123T}, Stl^{V144A}, Stl^{R177H}, $Stl^{K214Stop}$, Stl^{A236T}, Stl^{K238E}). We projected these mutations upon the previously published three-dimensional (3D) structural model of Stl [8] (Figure 4B). The analysis of the 3D structural model revealed plausible causes for the perturbing effects for several of the listed mutations. For example, the E59K mutation that results in changing the negatively charged glutamate to a positively charged lysine in the vicinity of the HTH motif presumably alters the electrostatic potential and hence may perturb DNA binding. Two of the mutations, I123T and R177H, lead to steric clashes in the model suggesting a strongly perturbed protein structure. The effect of the V144A mutation is not straightforwardly explained by in silico data only. The K214Stop mutant may be too short to be functional. The K238E again represents a large change in the charge of the replaced side chain. The hydrophobic to polar change in the A236T mutant may have impact on intramolecular interactions that contribute to the flexibility of the 3D structure, although the functional effect of this mutation predicted in the structural model is minor compared to most of the other mutations. We chose to further investigate this hit both in vivo and in vitro in order to verify that the random mutagenesis approach resulted in valid hits. This is indicative of the sensitivity of our screening system if even this minor structural change represents a valid hit.

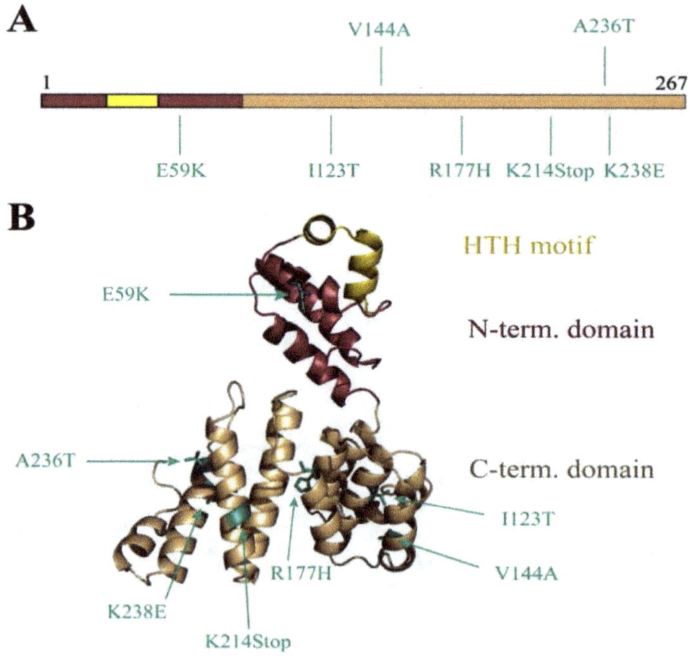

Figure 4. Probing DNA binding by random mutagenesis: Random mutagenesis PCR was carried out on the Stl coding sequence generating various single mutations. The PCR product was cloned into pKW08 vector and electroporated into the Stl switch system. The depicted mutants showed blue phenotype in the designed Stl switch system, suggesting defective DNA-binding ability. (**A**) The location of the hits in the Stl full-length protein; (**B**) Mutants mapped on the Phyre2 structural model of Stl [8].

3.4. Analysis of the Stl^{A236T} DNA Binding Properties In Vivo and In Vitro

To exclude any background effect in the previous random mutagenesis experiment, we cloned and expressed the Stl^{A236T} mutant in the Stl switch system (Figure 5A). The exclusive appearance of blue colonies clearly showed that the mutation disrupted the DNA binding ability of Stl in the cellular environment. As an in vitro test for DNA binding, we carried out EMSA using purified proteins and a 229-mer dsDNA oligonucleotide of the Stl-regulated promoter region (Figure 5B shows a representative experiment). It is clearly seen in the gel that the band corresponding to free DNA disappears in conjunction with the appearance of various DNA-protein complexes upon the increase of the protein concentration (Figure 5B). The densitometric analysis of the EMSA gel shows that the free DNA disappears at 1.5 μM Stl concentration when wild-type Stl is added, while free DNA is still present at even 4 μM concentration of the Stl^{A236T} mutant (Figure 5C). This indicates the diminished DNA binding capability of the Stl^{A236T} mutant compared to the wild-type Stl. The fact that the Stl^{A236T} does not lose its DNA binding capability completely in the EMSA indicates that its overall structure is probably not affected by the mutation. This result also suggests that a relatively minor difference in the structure and a moderate decrease in the in vitro DNA binding affinity could explain the loss of in vivo DNA binding.

Both the in vivo and in vitro validation experiments using the Stl^{A236T} mutant demonstrated that our Stl switch system is suitable for the identification of protein segments responsible for DNA binding.

A

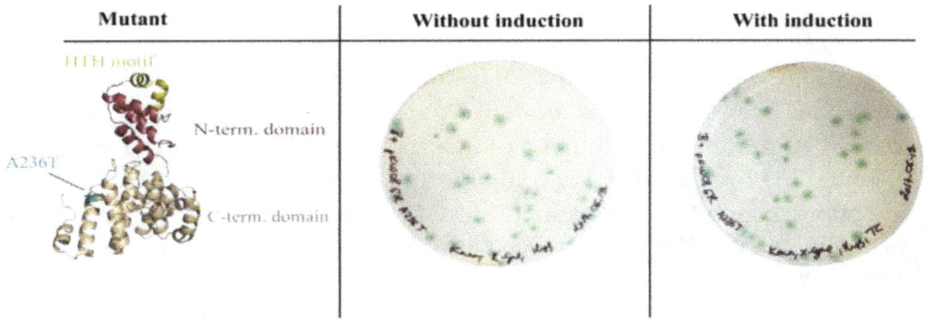

B

DNA	+	+	+	+	+	+	+	+	+
Stl^{WT} (μM)	-	0.5	1.5	3	4	-	-	-	-
Stl^{A236T} (μM)	-	-	-	-	-	0.5	1.5	3	4

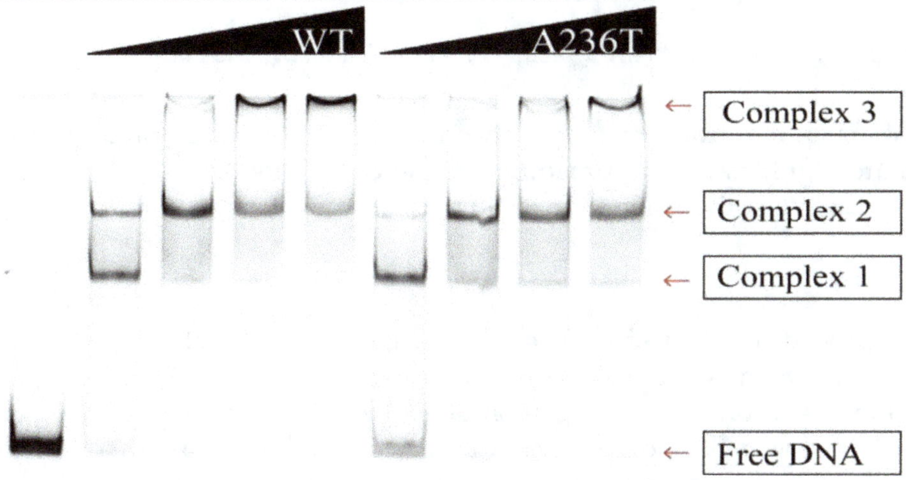

Figure 5. *Cont.*

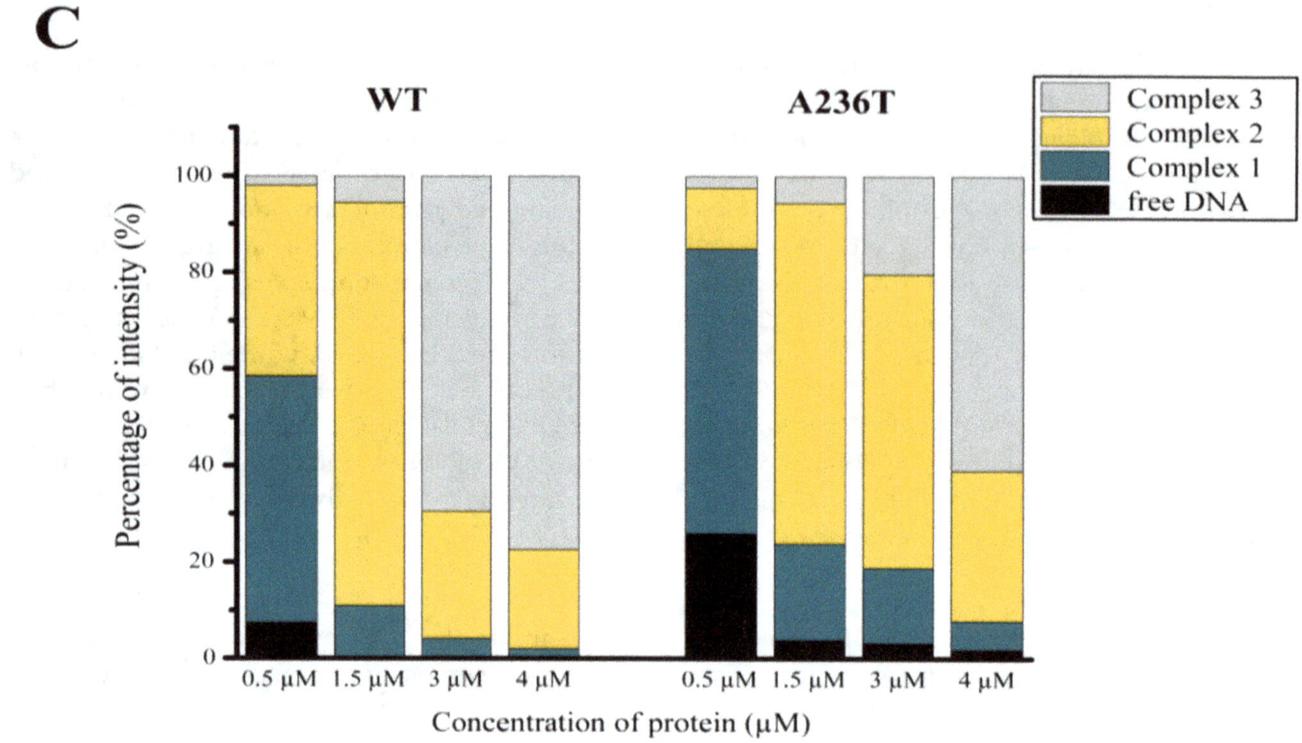

Figure 5. Characterization of the Stl^{A236T} mutant (**A**) Testing the Stl^{A236T} mutant in the Stl switch system. The structural model locates the A236T mutation at the *C*-terminal domain shown in blue. The HTH motif is colored yellow and the *N*-terminal domain is colored maroon. The expression of the Stl^{A236T} mutant resulted in blue colonies, verifying that it is defective in DNA binding in vivo; (**B**) Comparative electrophoretic mobility shift assay (EMSA) with Stl^{WT} and Stl^{A236T}. Stl^{A236T} is less prone to complex formation with DNA. Therefore, more of the free DNA is visible at the corresponding bands than that of Stl^{WT}. The comparative EMSA was repeated three times with same results; (**C**) Densitometry results of each of the species exhibiting different electrophoretic mobilities are shown by colored bands. The discrepancy between the wild-type and mutant Stl is quite noticeable, confirming our previous in vivo results. These data suggest that the DNA binding of Stl^{A236T} is compromised in vitro.

3.5. The Investigation of Protein:Protein Complexation Using the LacZ^{Str} Reporter System

As it is known from the literature, complexation between Stl and different phage proteins has immediate significance in spreading virulence factors. The LacZ^{Str} reporter system carries the opportunity to test the interaction of Stl with phage proteins that remove Stl from its cognate DNA binding site.

We probed a known interaction with the φ11 phage dUTPase to test this hypothesis [24,28]. Figure 6A shows the workflow of the double switch experiment. φ11 phage dUTPase expression was tuned up via a strong expression promoter, Phsp65, to ensure dUTPase excess relative to Stl. As a result, we observed 100% of blue colonies (Figure 6B), indicating that Stl was effectively sequestrated from the lacZ promoter. This experiment confirms that the switch system is applicable to the detection of protein:protein interactions. This feature also presents the opportunity to test phage protein libraries in search for new derepressing interactions.

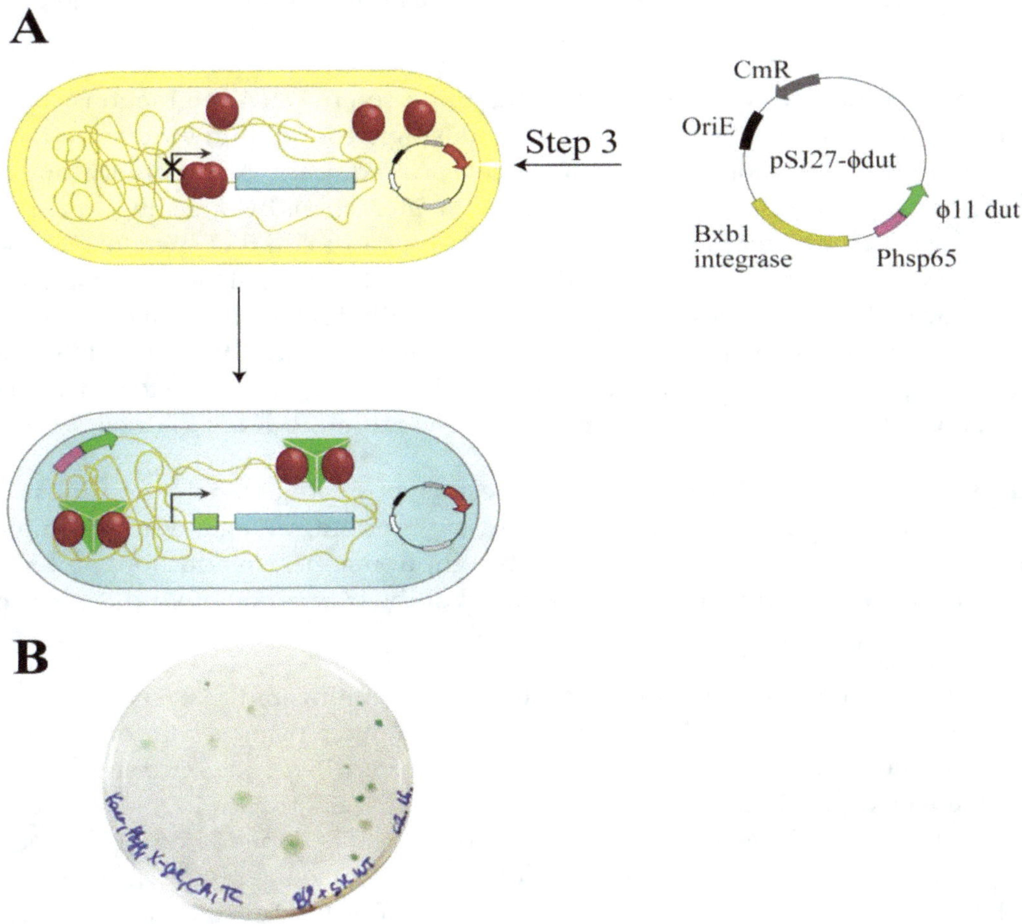

Figure 6. Workflow of the double switch system. (**A**) pSJ27-Φ11dut vector was electroporated into the Stl switch system. The relatively high expression level of Φ11 phage dUTPase enzyme is provided by a heat shock protein promoter (Phsp65). Φ11 phage dUTPase is a known partner of Stl, able to promote Stl to release the regulated DNA [24,28]; (**B**) Validation of the detection of protein:protein interaction using the $LacZ^{Str}$ reporter system. Stl in complex with Φ11 phage dUTPase releases DNA, therefore, blue colonies are formed. Note that the number of colonies represents the transformation efficiency.

4. Discussion

Switchable systems in molecular biology are associated with well-defined advantages and far-reaching applications in medical and industrial biotechnologies. These systems are based on cognate molecules that turn on/off metabolic pathways, signal transduction processes, and gene expression regulation. The actual benefit for each of these systems is defined by their specificity and applicability in various cellular models and/or organisms. In the present study, our aim was to exploit the phage-bacterium interaction and to design a molecular switch using the components regulating how phages modulate bacterial cell physiology.

The setup of our system is similar to that of the bacterial one-hybrid (B1H) system [29]; however, the two systems differ in their applicability and the actual molecular components. The B1H system was designed to identify target sites of a given DNA binding domain, while our system is constructed to identify protein segments crucial for DNA binding. In the B1H system, DNA binding activates the transcription of the reporter genes, whereas in the Stl-driven switch system, DNA binding blocks the transcription of the reporter gene. In addition, the transcription factor is expressed as a fusion to a subunit of RNA polymerase in the case of the B1H system, while the Stl repressor is able to bind and repress the promoter of the reporter gene at the same time and does not therefore need to be fused to another protein.

Our system has unique advantages in investigating protein:DNA interactions. First, the DNA binding segments of the protein can be identified in a single round of in vitro selection. Therefore, no additional protein purification steps or antibodies are required. Additionally, our method can potentially be used to discover phage-host protein:protein interactions by screening large DNA libraries. In our approach we exploited the capability of the Stl repressor to bind to specific DNA segments in a genetically engineered *M. smegmatis* strain that contained the marker β-galactosidase gene (*lacZ*) and the Stl-specific DNA binding segments in close proximity on the bacterial chromosome. Due to the integration vector relying on attP sites, both the *lacZ* gene and the Stl-specific DNA segment are present in just a single copy. Thus, the signal generated in this engineered system is stable among all cells present in the culture and there is no copy-number variation.

In contrast, Stl is expressed from an episomal vector. As an advantage, the amino acid sequence of the Stl repressor can be optionally modified, e.g., by random mutagenesis, and high variety of mutant repressors can be investigated in a high-throughput manner. In addition, protein expression from this vector is inducible, allowing efficient control of the expression of wild-type or mutated Stl proteins [20].

The application of the molecular switch system in the engineered *M. smegmatis* strain enabled us to focus on the Stl:DNA interaction in an unbiased manner using a random-mutagenized Stl library. Detailed characterization of protein:DNA interaction is of key current interest, with an additional biomedical aspect in our case, as we investigate the Stl protein involved in the regulation of pathogenicity factors in *S. aureus*. The virulence of *S. aureus* depends on helper phages which promote SaPI derepression by sequestering Stl. Interestingly, it is the phage dUTPase enzyme that interacts with Stl and conveys the derepression effect [4]. For the superfamily of dUTPases, Stl is the first well-established protein interacting partner; however, previous observations also suggested protein:protein interactions for dUTPases from other species [30,31].

In our engineered *M. smegmatis* strain, we exploit the β-Gal reporter system associated with a straightforward, easy readout suitable for high-throughput studies. This reporter system derived from the phage-bacterium interaction can be used to map DNA-binding protein surfaces in the living cell.

In this study, we demonstrated that the Stl switch system is capable of identifying side chain mutations in the Stl protein that are functionally involved in the Stl:DNA macromolecular complex formation. The random mutagenesis-generated mutations were projected on the 3D structural model of Stl. Interestingly, we observed that several residues in the C-terminal domain may also contribute to Stl DNA binding in the context of the full-length protein (cf. [8]). For some of these mutations, the actual phenotype was simple to rationalize due to major changes in the character of the residue involved in the mutation. We selected one specific mutant and subjected it to detailed in vivo and in vitro characterization. Results confirmed that this hit was fully valid. Hence, the herein presented system is capable of providing novel insights into macromolecular interactions involving Stl. As discussed earlier, Stl is an important repressor in the lifecycle of Staphylococcal pathogenicity islands and Staphylococcal phages [4]. Our test system is based on the perturbation of the Stl-induced switch in gene expression control. As such, our test system is also amenable to further biotechnological applications focusing on the identification of phage proteins that may perturb Stl-function. We have presented one such example for the Stl-interacting phage dUTPase protein. Phage libraries can also be tested for Stl-perturbing function in our switch system, where phage proteins can be investigated for their effect in a systematic approach. Previously, similar approaches were performed within the Staphylococcal cell and as such the possibility of identification of those proteins that may perturb Stl function and are essential for the phage lifecycle was practically excluded [4,7]. Importantly, this limitation is overcome in our system, which operates in the *M. smegmatis* environment and allows the investigation of complete phage libraries.

Acknowledgments: This work was supported by the National Research, Development and Innovation Office (K109486, K119493, NVKP_16-1-2016-0020 to BGV, K115993 and FK124527 to Judit Tóth), the Hungarian Academy of Sciences Medinprot program. VEKOP-2.3.2-16-2017-00013 was supported by the European Union and the State of Hungary, co-financed by the European Regional Development Fund, and ICGEB CRP/HUN14-01 to Beáta G. Vértessy. Judit Tóth is the recipient of the János Bolyai Research Scholarship of the Hungarian Academy of Sciences. This work was also supported by the ÚNKP-17-3-I New National Excellence Program of the Ministry of Human Capacities (Éva Viola Surányi).

Author Contributions: Éva Viola Surányi, Rita Hírmondó, Kinga Nyíri, Judit Tóth and Beáta G. Vértessy conceived and designed the experiments; Éva Viola Surányi, Rita Hírmondó, Szilvia Tarjányi and Bianka Kőhegyi performed the experiments; Éva Viola Surányi, Rita Hírmondó, Kinga Nyíri, Szilvia Tarjányi, Judit Tóth and Beáta G. Vértessy analyzed the data; Éva Viola Surányi, Rita Hírmondó, Kinga Nyíri, Judit Tóth and Beáta G. Vértessy wrote the paper. All authors reviewed the manuscript.

References

1. Malachowa, N.; DeLeo, F.R. Mobile genetic elements of *Staphylococcus aureus*. *Cell. Mol. Life Sci.* **2010**, *67*, 3057–3071. [CrossRef] [PubMed]
2. Mir-Sanchis, I.; Martínez-Rubio, R.; Martí, M.; Chen, J.; Lasa, Í.; Novick, R.P.; Tormo-Más, M.Á.; Penadés, J.R. Control of *Staphylococcus aureus* pathogenicity island excision. *Mol. Microbiol.* **2012**, *85*, 833–845. [CrossRef] [PubMed]
3. Papp-Kádár, V.; Szabó, J.E.; Nyíri, K.; Vertessy, B.G. In Vitro Analysis of Predicted DNA-Binding Sites for the Stl Repressor of the *Staphylococcus aureus* SaPIBov1 Pathogenicity Island. *PLoS ONE* **2016**, *11*, e0158793. [CrossRef] [PubMed]
4. Tormo-Más, M.A.; Mir, I.; Shrestha, A.; Tallent, S.M.; Campoy, S.; Lasa, I.; Barbé, J.; Novick, R.P.; Christie, G.E.; Penadés, J.R. Moonlighting bacteriophage proteins derepress staphylococcal pathogenicity islands. *Nature* **2010**, *465*, 779–782. [CrossRef] [PubMed]
5. Novick, R.P.; Christie, G.E.; Penadés, J.R. The phage-related chromosomal islands of *Gram-positive* bacteria. *Nat. Rev. Microbiol.* **2010**, *8*, 541–551. [CrossRef] [PubMed]
6. Hill, R.L.L.; Dokland, T. The Type 2 dUTPase of Bacteriophage PhiNM1 Initiates Mobilization of *Staphylococcus aureus* Bovine Pathogenicity Island 1. *J. Mol. Biol.* **2016**, *428*, 142–152. [CrossRef] [PubMed]
7. Donderis, J.; Bowring, J.; Maiques, E.; Ciges-Tomas, J.R.; Alite, C.; Mehmedov, I.; Tormo-Mas, M.A.; Penadés, J.R.; Marina, A. Convergent evolution involving dimeric and trimeric dUTPases in pathogenicity island mobilization. *PLoS Pathog.* **2017**, *13*, e1006581. [CrossRef] [PubMed]
8. Nyíri, K.; Kőhegyi, B.; Micsonai, A.; Kardos, J.; Vertessy, B.G. Evidence-Based Structural Model of the *Staphylococcal* Repressor Protein: Separation of Functions into Different Domains. *PLoS ONE* **2015**, *10*, e0139086. [CrossRef] [PubMed]
9. Brennan, R.G.; Matthews, B.W. The helix-turn-helix DNA binding motif. *J. Biol. Chem.* **1989**, *264*, 1903–1906. [PubMed]
10. Pabo, C.O.; Lewis, M. The operator-binding domain of λ repressor: Structure and DNA recognition. *Nature* **1982**, *298*, 443–447. [CrossRef] [PubMed]
11. Rohs, R.; Jin, X.; West, S.M.; Joshi, R.; Honig, B.; Mann, R.S. Origins of Specificity in Protein-DNA Recognition. *Annu. Rev. Biochem.* **2010**, *79*, 233–269. [CrossRef] [PubMed]
12. Jacob, F.; Monod, J. Genetic regulatory mechanisms in the synthesis of proteins. *J. Mol. Biol.* **1961**, *3*, 318–356. [CrossRef]
13. Hillen, W.; Berens, C. Mechanisms Underlying Expression of TN10 Encoded Tetracycline Resistance. *Annu. Rev. Microbiol.* **1994**, *48*, 345–369. [CrossRef] [PubMed]
14. Berens, C.; Hillen, W. Gene regulation by tetracyclines: Constraints of resistance regulation in bacteria shape TetR for application in eukaryotes. *Eur. J. Biochem.* **2003**, *270*, 3109–3121. [CrossRef] [PubMed]
15. Guzman, L.M.; Belin, D.; Carson, M.J.; Beckwith, J. Tight regulation, modulation, and high-level expression by vectors containing the arabinose PBAD promoter. *J. Bacteriol.* **1995**, *177*, 4121–4130. [CrossRef] [PubMed]
16. De Boer, H.A.; Comstock, L.J.; Vasser, M. The tac promoter: A functional hybrid derived from the trp and lac promoters. *Biochemistry* **1983**, *80*, 21–25. [CrossRef]
17. Brosius, J.; Erfle, M.; Storella, J. Spacing of the −10 and −35 Regions in the tac Promoter. *J. Biol. Chem.* **1985**, *260*, 3539–3541. [PubMed]

18. Tabor, S.; Richardson, C.C. A bacteriophage T7 RNA polymerase/promoter system for controlled exclusive expression of specific genes (T7 DNA polymerase/T7 gene 5 protein/proteolysis/13-lactamase/rifampicin). *Biochemistry* **1985**, *82*, 1074–1078.
19. Giesecke, A.V.; Joung, J.K. The bacterial two-hybrid system as a reporter system for analyzing protein-protein interactions. *Cold Spring Harb. Protoc.* **2007**. [CrossRef] [PubMed]
20. Hirmondó, R.; Szabó, J.E.; Nyíri, K.; Tarjányi, S.; Dobrotka, P.; Tóth, J.; Vértessy, B.G. Cross-species inhibition of dUTPase via the *Staphylococcal* Stl protein perturbs dNTP pool and colony formation in *Mycobacterium*. *DNA Repair* **2015**, *30*, 21–27. [CrossRef] [PubMed]
21. Parish, T.; Brown, A.C. (Eds.) *Mycobacteria Protocols*; Humana Press: Totowa, NJ, USA, 2009; ISBN 978-1-58829-889-8.
22. Parish, T.; Stoker, N.G. Use of a flexible cassette method to generate a double unmarked *Mycobacterium tuberculosis* tlyA plcABC mutant by gene replacement. *Microbiology* **2000**, *146*, 1969–1975. [CrossRef] [PubMed]
23. Mahenthiralingam, E.; Marklund, B.I.; Brooks, L.; Smith, D.; Bancroft, G.J.; Stokes, R.W. Site-directed mutagenesis of the 19-kilodalton lipoprotein antigen reveals No essential role for the protein in the growth and virulence of *Mycobacterium intracellulare*. *Infect. Immun.* **1998**, *66*, 3626–3634. [PubMed]
24. Szabó, J.E.; Németh, V.; Papp-Kádár, V.; Nyíri, K.; Leveles, I.; Bendes, A.Á.; Zagyva, I.; Róna, G.; Pálinkás, H.L.; Besztercei, B.; et al. Highly potent dUTPase inhibition by a bacterial repressor protein reveals a novel mechanism for gene expression control. *Nucleic Acids Res.* **2014**, *42*, 11912–11920. [CrossRef] [PubMed]
25. Kim, A.I.; Ghosh, P.; Aaron, M.A.; Bibb, L.A.; Jain, S.; Hatfull, G.F. *Mycobacteriophage* Bxb1 integrates into the *Mycobacterium* smegmatis groEL1 gene. *Mol. Microbiol.* **2003**, *50*, 463–473. [CrossRef] [PubMed]
26. Wilson, D.S.; Keefe, A.D. Random Mutagenesis by PCR. In *Current Protocols in Molecular Biology*; John Wiley & Sons, Inc.: Hoboken, NJ, USA, 2001.
27. Ubeda, C.; Maiques, E.; Barry, P.; Matthews, A.; Tormo, M.A.; Lasa, I.; Novick, R.P.; Penadés, J.R. SaPI mutations affecting replication and transfer and enabling autonomous replication in the absence of helper phage. *Mol. Microbiol.* **2008**, *67*, 493–503. [CrossRef] [PubMed]
28. Leveles, I.; Németh, V.; Szabó, J.E.; Harmat, V.; Nyíri, K.; Bendes, Á.Á.; Papp-Kádár, V.; Zagyva, I.; Róna, G.; Ozohanics, O.; et al. Structure and enzymatic mechanism of a moonlighting dUTPase. *Acta Crystallogr. Sect. D Biol. Crystallogr.* **2013**, *69*, 2298–2308. [CrossRef] [PubMed]
29. Xu, D.J.; Noyes, M.B. Understanding DNA-binding specificity by bacteria hybrid selection. *Brief. Funct. Genom.* **2015**, *14*, 3–16. [CrossRef] [PubMed]
30. Muha, V.; Horváth, A.; Békési, A.; Pukáncsik, M.; Hodoscsek, B.; Merényi, G.; Róna, G.; Batki, J.; Kiss, I.; Jankovics, F.; et al. Uracil-Containing DNA in Drosophila: Stability, Stage-Specific Accumulation, and Developmental Involvement. *PLoS Genet.* **2012**, *8*, e1002738. [CrossRef] [PubMed]
31. Békési, A.; Zagyva, I.; Hunyadi-Gulyás, E.; Pongrácz, V.; Kovári, J.; Nagy, A.O.; Erdei, A.; Medzihradszky, K.F.; Vértessy, B.G. Developmental regulation of dUTPase in *Drosophila melanogaster*. *J. Biol. Chem.* **2004**, *279*, 22362–22370. [CrossRef] [PubMed]

10

Broad Bactericidal Activity of the *Myoviridae* Bacteriophage Lysins LysAm24, LysECD7 and LysSi3 against Gram-Negative ESKAPE Pathogens

Nataliia P. Antonova [1,2], Daria V. Vasina [1], Anastasiya M. Lendel [2], Evgeny V. Usachev [1], Valentine V. Makarov [3], Alexander L. Gintsburg [1], Artem P. Tkachuk [1] and Vladimir A. Gushchin [1,2,*]

1 N.F. Gamaleya Federal Research Centre for Epidemiology and Microbiology, Ministry of Health of the Russian Federation, 123098 Moscow, Russia; northernnatalia@gmail.com (N.P.A.); d.v.vasina@gmail.com (D.V.V.); evgenyvusachev@gmail.com (E.V.U.); gintsburg@gamaleya.org (A.L.G.); artem.p.tkachuk@gmail.com (A.P.T.)
2 Lomonosov Moscow State University, 119991 Moscow, Russia; kazejosei@gmail.com
3 Center for Strategic Planning of the Ministry of Health of the Russian Federation, 119435 Moscow, Russia; makarovvalentine@gmail.com
* Correspondence: wowaniada@gmail.com or vladimir.a.gushchin@gamaleya.org

Abstract: The extremely rapid spread of multiple-antibiotic resistance among Gram-negative pathogens threatens to move humankind into the so-called "post-antibiotic era" in which the most efficient and safe antibiotics will not work. Bacteriophage lysins represent promising alternatives to antibiotics, as they are capable of digesting bacterial cell wall peptidoglycans to promote their osmotic lysis. However, relatively little is known regarding the spectrum of lysin bactericidal activity against Gram-negative bacteria. In this study, we present the results of in vitro activity assays of three putative and newly cloned *Myoviridae* bacteriophage endolysins (LysAm24, LysECD7, and LysSi3). The chosen proteins represent lysins with diverse domain organization (single-domain vs. two-domain) and different predicted mechanisms of action (lysozyme vs. peptidase). The enzymes were purified, and their properties were characterized. The enzymes were tested against a panel of Gram-negative clinical bacterial isolates comprising all Gram-negative representatives of the ESKAPE group. Despite exhibiting different structural organizations, all of the assayed lysins were shown to be capable of lysing *Pseudomonas aeruginosa*, *Acinetobacter baumannii*, *Klebsiella pneumoniae*, *Escherichia coli*, and *Salmonella typhi* strains. Less than 50 μg/mL was enough to eradicate growing cells over more than five orders of magnitude. Thus, LysAm24, LysECD7, and LysSi3 represent promising therapeutic agents for drug development.

Keywords: bacteriophages; *Myoviridae*; bacteriophage-derived lytic enzyme; enzybiotics; endolysin; in vitro activity; ESKAPE

---

1. Introduction

Antibiotic microbial resistance (AMR) is a natural aspect of microbe evolution under selective pressure. Because most antibiotics currently in use have natural analogues with similar native structures, AMR-associated genes of environmental bacteria can easily be distributed to clinically important strains through horizontal gene transfer [1,2]. The misuse, such as for the treatment of viral infections; underuse (premature antibiotic treatment termination); and overuse of antibiotics in agriculture leads to the catastrophic spread of AMR among bacteria surrounding human habitats [3–7]. According to a World Health Organization (WHO) report [3], the greatest attention needs to be paid to pathogens of the ESKAPE group (*Enterococcus faecium*, *Staphylococcus aureus*, *Acinetobacter baumannii*, *Pseudomonas aeruginosa*, *Klebsiella pneumoniae*, and other Enterobacteriaceae species). These pathogens represent the greatest threat among the so-called superbugs, which can rapidly acquire resistance to several classes of antibiotics and are able to cause a variety of nosocomial infections, such as bacteremia, pneumonia, and wound and skin infections [3,8].

Bacteriophages were used to control microbial populations long before penicillin was discovered [9]. The revived interest in bacteriophages and their enzymes as antibacterial agents is an expected consequence of entry into the so-called "post-antibiotic era," when most efficient and safe antibiotics will not provide a sufficient level of defense against bacterial pathogens. The use of bacteriophages has many advantages [10]. However, virus-based drugs also have significant drawbacks, including their labor-intensive acquisition and associated stringent safety requirements; also, they are unstable during storage and need constant pharmacokinetic monitoring [11,12]. Moreover, bacteriophages are highly immunogenic agents, which can reduce the effectiveness of their repeated use [9–11]. The primary problem in the application of bacteriophage-based therapeutics is their unpredictable action toward patient-specific infections, necessitating the performance of additional in vitro tests prior to treatment.

Drugs based on bacteriophage enzymes, such as lysins, rather than the viruses themselves, can have much more predictable results. Bacteriophage lysins are enzymes encoded by bacteriophages that cleave peptidoglycans (PGs) in the cell walls (CW) of bacteria. During their life cycle, bacteriophages secrete lysins to deliver phage DNA into cells during bacterial infection as well as to release new virions from the cell [13]. The catalytic domains of bacteriophage lysins can possess different classes of PG-degrading activities, including transglycosylase, glucosaminidase, lysozyme-like, amidase, and endopeptidase activities [14]. Among the benefits of lysins are their high rate of lytic action, their ability to act upon antibiotic-resistant bacterial strains (including bacteria growing under different metabolic conditions), as well as their degradation of bacterial biofilms [15] and the low probability of the development of bacterial resistance to lysins. A wide spectrum of endolysin activities, comparable to modern antibiotics, would solve one of the primary problems associated with bacteriophage use. It could allow therapies to be compatible with the condition of patients with acute infections, and the treatment outcome would be more predictable under conditions when there is a lack of time for preliminary in vitro testing.

In contrast, endolysins are traditionally believed to have lytic activity against the specific hosts of the phages from which they were isolated. Thus, individual endolysins are proposed to act against specific strains and species or, rarely, more broadly against the genera of Gram-negative bacteria [16,17]. Thus, most studies have focused on identifying the effects of these enzymes and their modified variants against the host species of parental phages [16,18,19]. However, the conserved PG structure of Gram-negative bacteria make the application endolysins potentially useful with respect to their breadth of action [20–22]. Recently, additional evidence has demonstrated the activity of specific endolysins toward Gram-negative bacteria [23,24], at least for prophages of *Escherichia coli* and *A. baumannii*. However, few studies have described the broad activity of bacteriophage-encoded lysins against Gram-negative bacteria, and growing evidence suggests that many bacteriophage lysins are capable of much more than just acting against parental phage-specific hosts.

To study the activity of lysins against a wide spectrum of Gram-negative bacteria, we cloned three putative endolysins from *Myoviridae* bacteriophage family members (LysAm24, LysECD7, and LysSi3) and studied their activity extensively in vitro. Bacteriophages encoding the selected lysins belong to different genera of the *Myoviridae* family. Importantly, the selected enzymes represent lysins with diverse domain organization (single-domain vs. two-domain) and different predicted mechanisms of action (lysozyme-like vs. peptidase). All of the enzymes were affinity purified and tested under different in vitro conditions against a panel of Gram-negative clinical bacterial isolates, including Gram-negative representatives of the ESKAPE group of pathogens (*P. aeruginosa*, *A. baumannii*, *K. pneumoniae*, *E. coli*, and *Salmonella typhi* strains). All three assayed lysins exhibited a wide spectrum of activity against the assayed bacteria, and less than 50 μg/mL of each enzyme was sufficient to eradicate growing cells of the tested bacterial strains over more than five orders of magnitude.

2. Materials and Methods

2.1. Bacterial Strains

The bacterial strains used in the study included both reference strains and clinical isolates of Gram-negative representatives of the ESKAPE group of pathogens, including *A. baumannii*, *P. aeruginosa*, *K. pneumoniae*, *E. coli*, and *S. typhi* as well as *Staphylococcus haemolyticus* and *S. aureus* from the collection of the N.F. Gamaleya Federal Research Center for Epidemiology and Microbiology, Ministry of Health of the Russian Federation (Table 1). Three strains of *K. pneumoniae* (B3060, Osh-2k, I6208) were from the state collection of pathogenic microorganisms and cell cultures "GKPM-Obolensk." All of the strains were stored at –80 °C and cultivated in LB broth at 37 °C and 240 rpm overnight before the assays were performed.

Table 1. Bacterial strains used in the study and the sources of their isolation.

Strain	Source	
Pseudomonas aeruginosa Ts 38-16	Patient's sputum, hospital strain	ICU [1]
Pseudomonas aeruginosa Ts 43-16	Patient's wound fluid, hospital strain	ICU
Pseudomonas aeruginosa Ts 44-16	Patient's urea, hospital strain	ICU
Acinetobacter baumannii Ts 50-16	Patient's sputum, hospital strain	ICU
Acinetobacter baumannii Ts 53-16	Patient's sputum, hospital strain	ICU
Acinetobacter baumannii Ts 58-16	Infusion pump, hospital strain	ICU
Acinetobacter baumannii Ts 54-16	Infusion pump, hospital strain	ICU
Klebsiella pneumoniae Ts 141-14	Patient's urea, hospital strain	Inpatient hospital
Klebsiella pneumoniae Ts 08-15	Patient's sputum, hospital strain	ICU
Klebsiella pneumoniae F 104-14	Patient's sputum, hospital strain	Outpatient hospital
Klebsiella pneumoniae B3060	Patient's liquor, hospital strain	Inpatient hospital
Klebsiella pneumoniae Osh-2k	Patient's blood, hospital strain	ICU
Klebsiella pneumoniae I6208	Patient's oral pharynx, hospital strain	Inpatient hospital
Escherichia coli M15	Reference laboratory strain	-
Salmonella typhi MVP 728	Reference laboratory strain	-
Staphylococcus aureus Z 73-14	Patient's nasal swab	Outpatient hospital
Staphylococcus haemolyticus G 58-0916	Patient's urethra	Outpatient hospital
Escherichia coli BL21(DE3) pLysS	Laboratory expression strain	-

[1] ICU: intensive care unit.

2.2. Recombinant Expression and Purification of Proteins

The coding sequences for the selected endolysins were PCR amplified from inactivated phage lysates (kindly provided by Dr. Eugenie O. Rubalskii) and were subsequently cloned into the expression vector pET42b(+) (kanamycin resistance) and checked for errors via Sanger sequencing (for primers, see Table S1). All of the proteins contained a C-terminal 8-His tag for affinity purification. The expression vectors were introduced into the competent *E. coli* cells, strain BL21(DE3) pLysS (chloramphenicol resistance) using a heat shock transformation protocol.

The *E. coli* cells were grown in LB broth (37 °C, 240 rpm) to an OD_{600} value of 0.55–0.65 and then induced with β-D-1-thiogalactopyranoside (1 mM IPTG) at 37 °C for 3 h. The cells were harvested by centrifugation (6000× *g* for 10 min at 4 °C) and resuspended in lysis buffer (20 mM Tris HCl, 250 mM NaCl, and 0.1 mM EDTA, pH 8.0), incubated with 100 μg/mL lysozyme at room temperature for 30 min, and disrupted by sonication. The cell debris was removed by centrifugation (10,000× *g* for 30 min at 4 °C) and the supernatant was filtered through a 0.2 μm filter. The proteins were purified on an NGC Discovery™ 10 FPLC system (Bio-Rad, Hercules, CA, USA) with a 5 mL HisTrap FF column (GE Healthcare, Chicago, IL, USA) pre-charged with Ni^{2+} ions. The filtered lysate was mixed with 30 mM imidazole and 1 mM $MgCl_2$ and loaded on the column that was pre-equilibrated with binding buffer (20 mM Tris HCl, 250 mM NaCl, and 30 mM imidazole, pH 8.0). The fractions were eluted using a linear gradient to 100% elution buffer (20 mM Tris HCl, 250 mM NaCl, and 500 mM imidazole pH 8.0). The collected protein fractions were dialyzed against 20 mM Tris HCl (pH 7.5). The purity of the proteins was determined by 16% SDS-PAGE.

The protein concentrations were measured using a spectrophotometer (Implen NanoPhotometer, IMPLEN, Munich, Germany) at 280 nm and calculated using a molar extinction coefficient of $E^{0.1\%}{}_{280nm}$.

2.3. Antibacterial Assay

Overnight bacterial cultures (OD_{600} = 1.4–1.6) were used as stationary phase cultures or were diluted 30-fold in LB broth and grown to exponential phase (OD_{600} = 0.6). Subsequently, the cells were harvested by centrifugation (3000× *g*, 10 min) and resuspended in the same volume of 20 mM Tris HCl (pH 7.5). Each suspension was diluted 100-fold in the same buffer to a final density of approximately 10^6 cells/mL. Afterwards, 100 µL of the bacterial suspension and 100 µL of the protein at the appropriate concentration were mixed in 96-well plate wells, and buffer without endolysins was used as a negative control. The mixtures were incubated at 37 °C for 30 min with shaking at 200 rpm and then were diluted 10-fold in PBS (pH 7.4). Subsequently, 100 µL of each dilution was plated onto LB agar, and bacterial colonies were counted after overnight incubation at 37 °C. All of the experiments were performed in triplicate, and the antibacterial activity is expressed as log10 of the number of surviving bacterial colonies.

The spectrum of antimicrobial activity was tested against a panel of sixteen Gram-negative bacterial strains and two Gram-positive bacterial strains (Table 1) using the conditions described above.

The effects of pH, salts, and buffers (Na or K phosphate buffers, pH 7.5) on the specific activity of endolysins were analyzed using the *A. baumannii* strain Ts 50-16 cultured to the logarithmic growth phase. The bacteria were incubated with the proteins in 20 mM Tris HCl buffer with different pH values (5.0 to 9.0); 5, 10, or 50 mM of Na or K phosphate buffer (pH 7.5) and 5 mM of Na or K phosphate buffer (pH 7.5) supplemented with different NaCl or KCl salts (0 to 500 mM). The effect of EDTA on the bactericidal activity of lysins at different pH values was assessed as mentioned above after the addition of 0.5 mM of permeabilizer during cell incubation with the lysins.

2.4. Storage Stability

To investigate the storage stability of the endolysins, the proteins were exposed to different temperature conditions (4, –20, and –80 °C) for one week or one, two, or three months. The protein samples were stored in 20 mM Tris HCl, 100 mM KCl, and 50% glycerin at pH 7.5. Antimicrobial activity was assessed as described above using the *A. baumannii* strain Ts 50-16.

2.5. Dynamic Light Scattering

The hydrodynamic diameters of the protein particles in solution were measured using a ZetaSizer Nano-ZS (Malvern Instruments LTD, Malvern, UK) in polystyrene cuvettes with an optical path length of 1 cm and a laser wavelength of 633 nm. The protein samples (approximately 1 mg/mL) were assayed in 10 mM sodium phosphate buffer (pH 7.5). The samples were heated from 25 to 70 °C with 2.5 °C step increases. The data were analyzed using Dispersion Technology version 5.10 (Malvern, PA, USA).

3. Results

3.1. Sequence Analysis, Expression, and Physicochemical Properties of Recombinant Enzymes

In this study, three putative endolysin-encoding genes from lytic phages of the family *Myoviridae* were assayed. The genes encoding LysAm24, LysECD7, and LysSi3 from Acinetobacter phage AM24 (NCBI: txid1913571), which infects *A. baumannii*, *Escherichia* phage ECD7 (NCBI: txid1981499, *E. coli*) and Enterobacteria phage UAB_Phi87 (NCBI: txid1197935, *S. typhi*), respectively, were cloned and expressed in *E. coli* (Table 2). None of these endolysins had been previously cloned or characterized.

Table 2. Endolysins used in the study.

Enzyme	Enzyme Source	Phage Host	GB Accession No.
LysAm24	Acinetobacter phage AM24	*A. baumannii*	APD20282.1
LysECD7	Escherichia phage ECD7	*E. coli*	ASJ80195.1
LysSi3	Enterobacteria phage UAB_Phi87	*S. typhi*	YP_009150069.1

BLAST searches for the deduced amino acid sequences derived from the cloned nucleotides showed a strong similarity to the phage-related lysozyme-like superfamily (muramidases, GH24 family) for the LysAm24 and LysSi3 proteins, whereas LysECD7 contained a D-alanyl-D-alanine carboxypeptidase domain (peptidase M15 family). Furthermore, whereas LysECD7 and LysSi3 contained only one enzymatic catalytic domain (ECD), the deduced amino acid sequence of LysAm24 included an additional CW-binding domain (CBD) at the N-terminus and an ECD at the C-terminus (Figure 1a). Such an inverted orientation of the CBD and ECD has been shown to be characteristic of Gram-negative bacteriophage endolysins [25,26].

To clone the selected lysins, their coding sequences were amplified from inactivated phage lysates. Subsequently, the amplified fragments were fused to an 8-His tag at the C-terminus of the encoded protein and were inserted into the *E. coli* expression vector pET42b(+). The proteins were purified using NiNTA affinity chromatography followed by SDS-PAGE gel analysis, which showed the monomeric form of the individual proteins with apparent molecular masses of 25.9, 16.1, and 18.5 kDa for LysAm24, LysECD7, and LysSi3, respectively (Figure 1b).

The use of dynamic light scattering allowed the hydrodynamic diameter as well as the thermal and structural stability of proteins molecules to be evaluated. The results showed that hydrodynamic diameters for all three enzymes had narrow peaks corresponding to 5.29 ± 0.97, 3.94 ± 1.33, and 3.69 ± 0.87 nm for LysAm24, LysECD7, and LysSi3, respectively (Figure 1c), indicating that all of the recombinant molecules are stable as primarily monomers in solution.

Figure 1d presents the results showing the dependence of the light scattering intensity parameters on temperature. The observed sharp increase in the total light scattering intensity and in the hydrodynamic diameter of the enzymes particles indicates the beginning of the aggregation of the proteins. The aggregation of LysAm24, LysECD7, and LysSi3 was observed to begin at 55, 50, and 42.5 °C, respectively. Thus, the results showed that LysAm24 possesses slightly increased thermostability compared to LysECD7 and LysSi3.

3.2. Bactericidal Activity and Biochemical Properties of Recombinant Lysins

The bactericidal activity of endolysins varies significantly depending on the protein of interest and the bacterial species and strains used. According previous studies, the optimal concentrations for endolysins vary from 10 to 500 μg/mL [24,27,28]. To initially evaluate bactericidal activity, the *A. baumannii* strain Ts 50-16 (for LysAm24 and LysSi3) and the *E. coli* strain M15 (for LysECD7) were used. The bactericidal activities of LysAm24, LysECD7, and LysSi3 against exponentially growing bacterial cells were shown to be concentration-dependent (Figure 2a). The minimal active concentration after 30 min of incubation in 20 mM Tris HCl buffer (pH 7.5) was observed to be 0.5 μg/mL for LysAm24 and LysECD7 and 10 μg/mL for LysSi3. Specifically, reductions of over 2 and 1.35 logs, respectively, were observed after treating the bacterial cells with 0.5 μg/mL of LysAm24 and LysECD7, and a reduction of approximately 3 logs was observed after the cells were treated with 10 μg/mL of LysSi3. In general, LysSi3 exhibited 10-fold lower activity than LysECD7 and LysAm24 against the selected bacterial strains. Lysin concentrations of more than 5 μg/mL (LysAm24 and LysECD7) and

50 µg/mL (LysSi3) eliminated bacterial growth completely, suggesting that recombinant LysAm24, LysECD7, and LysSi3 can exert bactericidal action without additional membrane permeabilization.

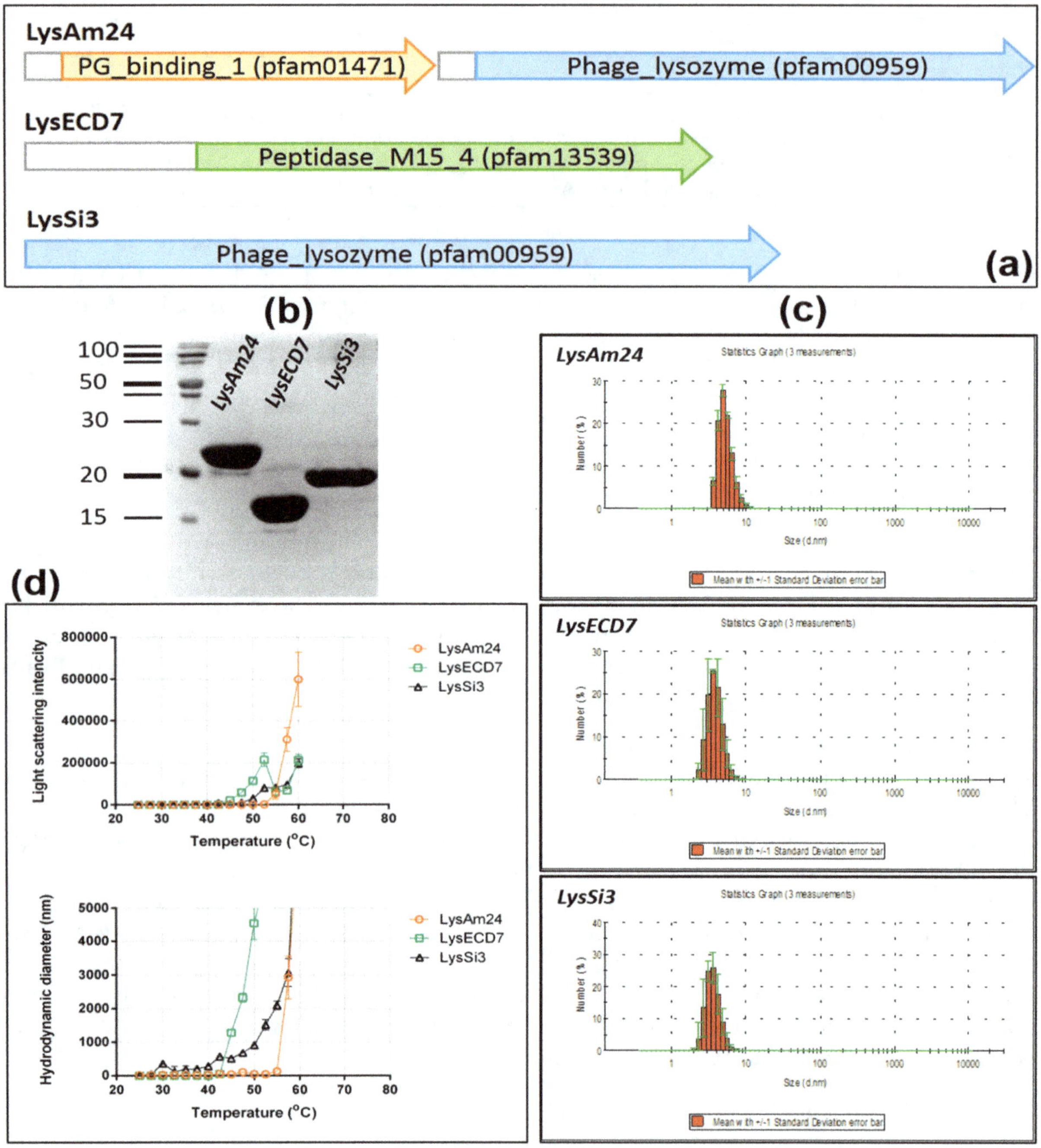

Figure 1. Domain organization, purification, and physicochemical properties of LysAm24, LysECD7, and LysSi3. (**a**) Putative domain organization of LysAm24, LysECD7, and LysSi3 predicted from the deduced amino acid sequences. The prediction was done with the protein BLAST search (https://blast.ncbi.nlm.nih.gov/Blast.cgi). (**b**) SDS-PAGE gel analysis of purified endolysins. The PageRuler Broad range Unstained Protein Ladder (Thermo Scientific, Vilnius, Lithuania) was used (**c**) Evaluation of the hydrodynamic diameter of *E. coli*-produced endolysins by DLS analysis. Statistical distribution of particle size by number. The data are presented as the mean values of three measurements ± SD. (**d**) Temperature dependence of the total light scattering intensity (upper panel) and the hydrodynamic diameters of the particles (lower panel). The data are presented as the mean values of three measurements made with an interval of 15 s ± SD.

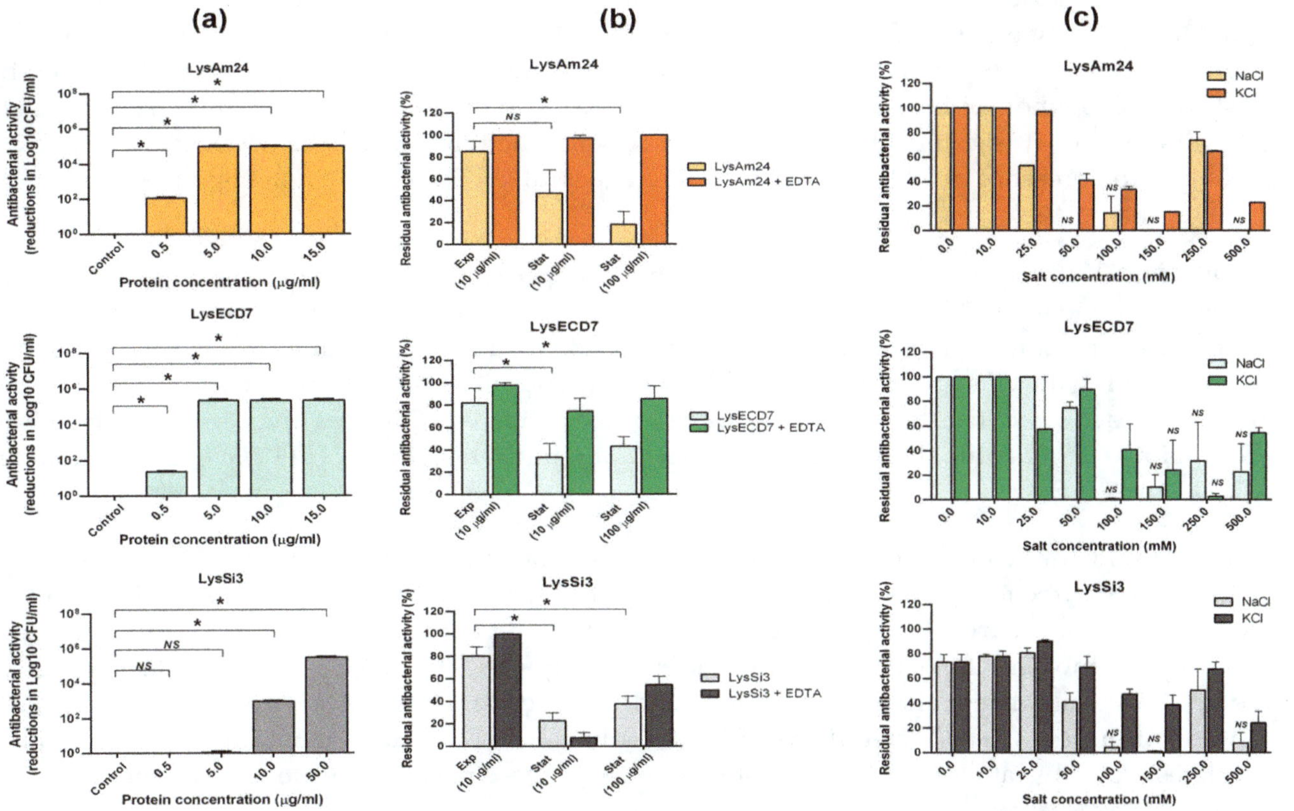

Figure 2. Bactericidal activity and biochemical properties of LysAm24, LysECD7, and LysSi3. (**a**) Bactericidal activity of different concentrations of LysAm24 against *A. baumannii* Ts 50-16, LysECD7 against *E. coli* M15 and LysSi3 against the *A. baumannii* Ts 50-16 in exponential phase. Cell cultures without incubation with endolysins were used as controls. The number of surviving cells after 18 h is expressed as reduction in log10 CFU/mL compared to the control. NS, no statistical significance of the data compared to the untreated culture is observed ($p > 0.05$, Mann–Whitney test). (**b**) Activity of lysins against both exponentially growing cells (Exp) and stationary-phase cells (Stat) of *A. baumannii* Ts 50-16 without and in the presence of EDTA. The residual activity after 18 h of growth compared to the untreated culture is shown. NS, no statistical significance of the data compared to the exponential growth phase is observed ($p > 0.05$, Mann–Whitney test). (**c**) Effects of salts on the bactericidal activity against *A. baumannii* Ts 50-16. The residual activity after 18 h of growth compared to the untreated culture is shown. NS, no statistical significance of the data compared to the untreated culture is observed ($p > 0.05$, Mann–Whitney test). For all experiments, the mean values are shown (± standard error of the mean (SEM)) from three independent experiments. Asterisk (*) indicates significant effect on bactericidal activity.

It is a known fact that lysins are generally less active against stationary-phase bacteria compared to exponentially growing bacterial cells. Indeed, for *A. baumanii* Ts 50-16, the addition of 10 µg/mL LysAm24, LysECD7, or LysSi3 reduced the viability of exponentially growing cells ($OD_{600} = 0.6$) by 76%, 70%, and 80% respectively, while reducing the viability of stationary-phase bacteria ($OD_{600} = 1.4–1.6$) by 47%, 33%, and 23% only. Even a concentration of endolysins of 100 µg/mL did not allow complete elimination of bacteria (Figure 2b). However, the addition of 0.5 mM EDTA restored the activity of lysins, indicating that the problem is related to the ability of the enzymes to overcome the bacterial outer membrane but not bactericidal activity itself.

For the effective use of endolysins, it is essential that they retain their activity under physiological conditions (high salt content, pH close to the physiological values present in the blood, oral cavity or on skin (7.4, 7.0, and 5.5, respectively), and exhibit resistance to proteolytic cleavage). Thus, the effect of the composition of the buffer system used and the salts present in the solution (Figure 2) on the preservation of the activity of endolysins was assessed.

We evaluated the effects of the presence of salts (NaCl or KCl) on the activity of the endolysins (Figure 2c). Both KCl and NaCl had an inhibitory effect on lysin activity, and the presence of both salts reduced the activity of all three assayed enzymes. The activity of LysAm24 was decreased by almost two-fold in the presence of 25 mM NaCl, while KCl had the same effect at 50 mM. For LysECD7, 100 mM of either salt significantly decreased bactericidal activity, while the activity of LysSi3 was inhibited in the presence of 50 mM NaCl or KCl. In general, all three lysins tolerated the presence of KCl better than NaCl.

Different buffer systems, specifically sodium or potassium phosphate buffers, at various concentrations and at pH 7.5 were also investigated, as these solutions are likely to be more suitable and cost-effective for future animal and human applications than buffers containing Tris HCl. The results of assays evaluating the effects of phosphate buffers at concentrations of 5, 10, and 50 mM on the activity of the enzymes showed that the most appropriate concentration was 5 mM, with potassium–phosphate buffer (pH 7.5) being optimal for LysAm24 and sodium–phosphate buffer (pH 7.5) being optimal for LysECD7 and LysSi3.

In addition, the stability of the enzymes during stock storage in the presence of 100 mM KCl and 50% glycerol at three temperatures (+4, −20, and −80 °C) for three months was tested. Bactericidal activity was tested in Tris HCl buffer pH 7.5. All of the preparations retained their initial activity after up to two months of storage. However, after three months, the activity of LysAm24 was almost undetectable, while that of LysECD7 and LysSi3 decreased significantly (for all three enzymes, the inhibition of the growth of bacterial colonies was only 1- and 0.5-orders of magnitude of that of the control, respectively, rather than 5-orders of magnitude). These results indicate that storage condition optimization is necessary for these enzymes.

3.3. *Effect of EDTA on the Bactericidal Activity of Lysins at Different pH Values*

A separate series of experiments was carried out to investigate the activity of LysAm24, LysECD7, and LysSi3 over a broad range of pH values in vitro (Figure 3), the results of which showed that all three enzymes had similar bactericidal activity in Tris HCl buffer at different pH values. The highest activity of LysAm24, LysECD7, and LysSi3 was observed under moderate acidic conditions (pH = 5.0–6.0). The activity of all three enzymes decreased when the pH was increased to 6.5–7.0, with LysAm24, LysSi3, and LysAm24 exhibiting 45%, 30%, and 60% of their initial activity, respectively. At higher pH values (7.5–8.0), the activities of all three enzymes recovered, albeit to different degrees.

The effect of EDTA, which is used as a membrane permeabilizer to enhance the bacteriolytic activity of lysins, was also evaluated (Figure 3). The synergy between EDTA and the enzymes was observed under neutral and alkaline conditions (pH 6.5–8.5). The most pronounced effect of the presence of EDTA was observed for LysAm24 and LysSi3, for which almost the complete recovery of activity was observed at pH values of 6.5–7.0, whereas this effect occurred to lesser degree for LysECD7.

3.4. *Broad Substrate Specificity and Antibacterial Properties of Recombinant Phage Endolysins*

The spectrum of activity of the three endolysins was tested against a panel of eighteen bacterial isolates from patients, including *A. baumannii*, *P. aeruginosa*, and *K. pneumoniae* as well as a collection of reference strains (*E. coli* and *S. typhi*) (Table 1). The results of the antimicrobial assays of the three endolysins towards these bacterial species are presented in Figure 4. All of the enzymes showed a wide but diverse range of bactericidal activity and efficiently inhibited the growth of several strains each from *P. aeruginosa*, *A. baumannii*, and *K. pneumoniae* (Figure 4). When the proteins were added to exponentially growing cultures of bacteria, no growth of the host cells was observed within 18 h of the treatment. However, the strains *P. aeruginosa* Ts 44-16 and *K. pneumoniae* F 104-14 were only moderately susceptible to the lysins. The strain *E. coli* BL21(DE3) pLysS, when used for the recombinant expression of lysins, was partly resistant to action of LysAm24 and LysSi3. Additionally, LysSi3 had no activity against the *K. pneumoniae* B3060, Osh-2k, and I6208 strains. Among the Gram-positive bacterial strains, no activity was found against *S. aureus* 73-14, but LysAm24 and LysSi3 partially inhibited the growth of *S. haemolyticus* G58 (to 45% and 10%, respectively).

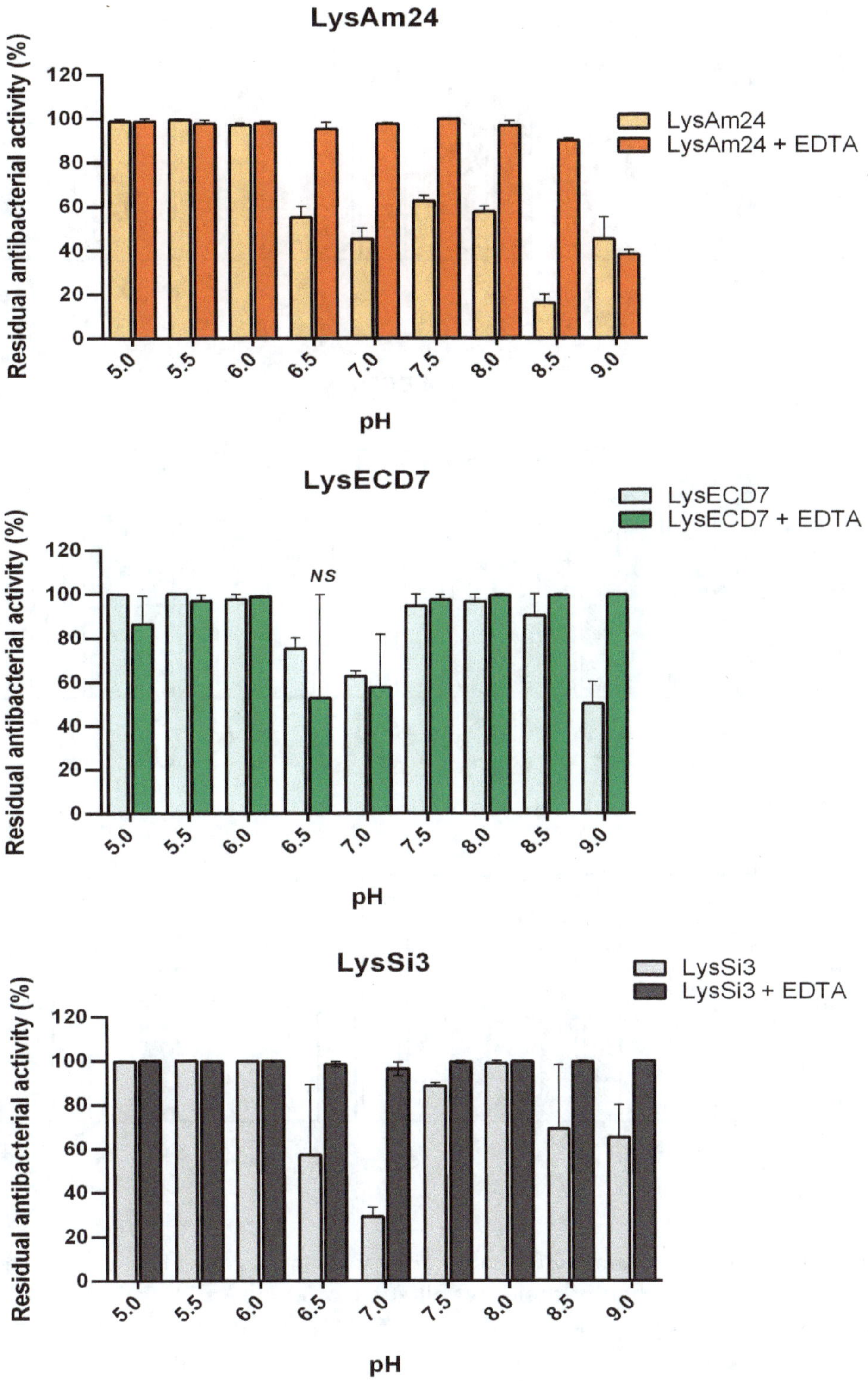

Figure 3. Effects of pH and EDTA on the antibacterial activity of LysAm24, LysECD7, and LysSi3 against exponentially growing cells. The residual activity after 18 h of growth compared to the untreated culture is shown. For all experiments, the mean values are shown (±SEM) from three independent experiments. NS, no statistical significance of the data compared to the untreated control culture ($p > 0.05$, Mann–Whitney test). All other data points were statistically significant compared to the untreated control culture.

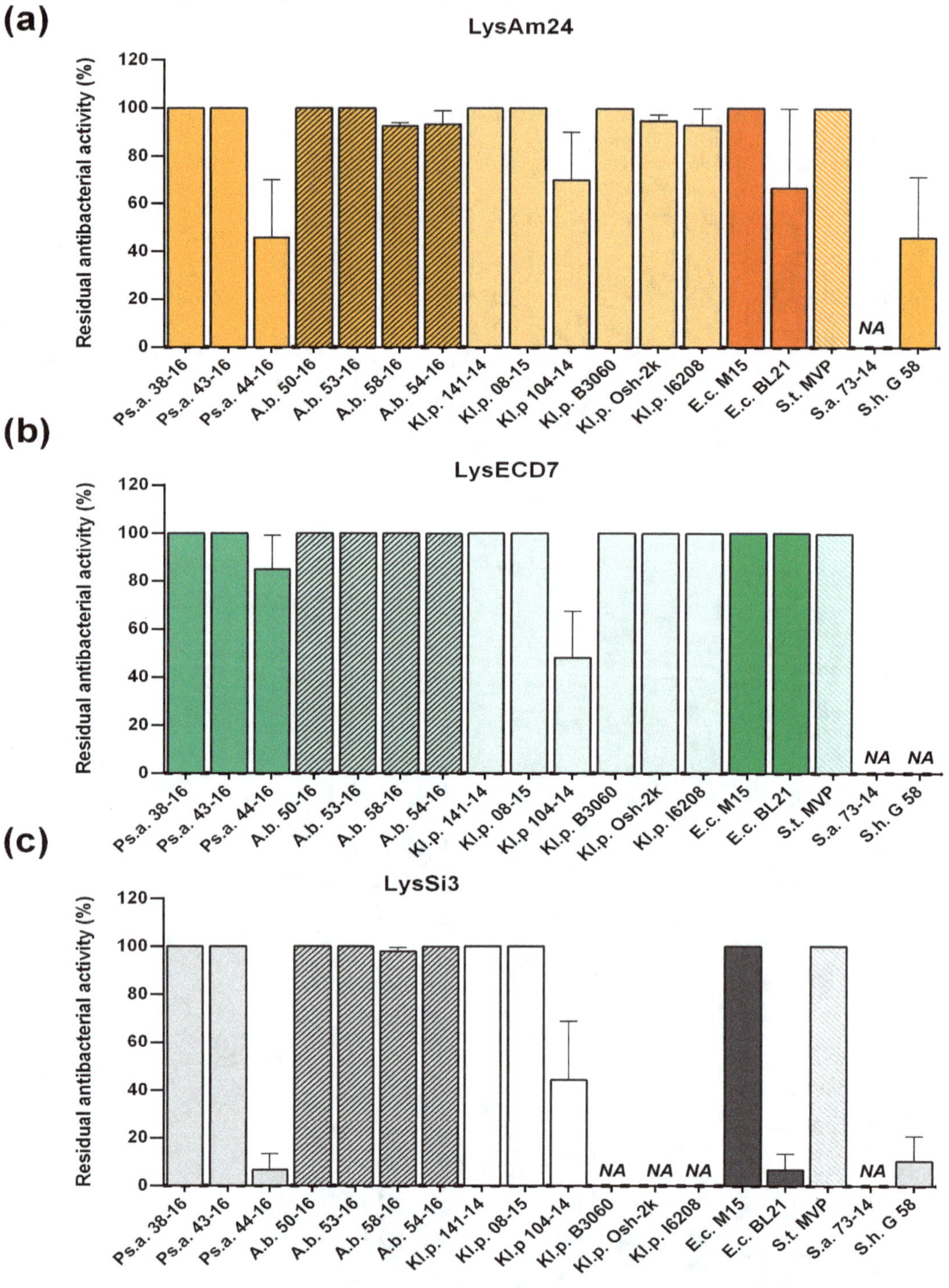

Figure 4. Endolysin activity against different strains of Gram-negative bacteria. Bacterial count (cfu/mL) was used as the method of choice for all strains: (**a**) LysAm24, (**b**) LysECD7, (**c**) LysSi3. For all experiments, the mean values are shown (±SEM) from three independent experiments. NA, no bactericidal activity was detected. All other data points were statistically significant compared to the untreated control culture ($p < 0.05$, Mann–Whitney test).

Interestingly, although all three assayed lysins were active or moderately active against the *E. coli* BL21(DE3) pLysS expression strain, extensive cells lysis of these cells was not detected during the expression and purification procedures, indicating that either internal cell membrane penetration did not take place or that lysis of the expression strain occurred to some degree with no significant effect on the efficiency of protein production.

4. Discussion

All three phage endolysins assayed in this study were obtained from members of the family *Myoviridae* that differ in their taxonomic assignments: Escherichia phage ECD7, Tevenvirinae subfamily; Enterobacteriaphage UAB_Phi87, Ounavirinae; and Acinetobacter phage AM24, different unclassified genus. The significance of the multiple-drug resistance of these phage hosts is reflected in the recent classification of these species as "Priority 1: Critical" pathogens on the WHO priority pathogens list for the R&D of new antibiotics [3]. The genomes of the selected bacteriophages are highly variable in terms of both size and organization. Since most bacteriophages of this family are lytic, they have potential use as antibacterial agents by themselves. Taking into account the size of the cloned lysins in this study (25.9, 16.1, and 18.5 kDa for LysAm24, LysECD7 and LysSi3, respectively), we expected them to be endolysins, because virion-associated lysins (VALs) have molecular weights of greater than 40 kDa [13]. Small endolysins are compatible with cost-effective biotechnological production, and our results show that LysAm24, LysECD7, and LysSi3 can be efficiently and cheaply produced in *E. coli*. Furthermore, it was shown that the assayed enzymes have good stability profiles and possess wide bactericidal activity, as they were capable of inhibiting the growth of all assayed Gram-negative representatives of the ESKAPE group. The bactericidal effect against Gram-positive bacterial strains was not pronounced. No activity was found against *S. aureus* 73-14, while LysAm24 and LysSi3 partially inhibited the growth of *S. haemolyticus* G58 (to 45% and 10% respectively). More strains of Gram-positive bacteria need to be tested to assess the activity of the studied lysins on this group more carefully.

The endolysins chosen for this study represent proteins with diverse domain organizations, including those with the most common catalytic domains; glycosidases, which hydrolyze glycosidic bonds in glycan strands; and endopeptidases, which cleave the cross-bridges between strands and are proposed to have distinct mechanisms of bactericidal action. This set of endolysins allowed us to study the general patterns of activity in this class of proteins.

Endolysins from phages infecting Gram-positive bacteria are believed to not exhibit broad lytic activity and are typically limited to particular bacterial genera, species, or even strains [29–31]. This is probably due to the highly diverse structures of the CW peptidoglycans of Gram-positive bacteria needed to evade the host immune system [20–22]. Thus, the bacteriophage lysins from a Gram-positive background may require tailoring to the CW, leading to a narrowing of their spectrum of specificity. In addition, bacteriophages infecting Gram-positive bacteria need to control the activity of lysins released from newly lysed cells to prevent the premature lysis of neighboring cells, which are suitable for infection by the released bacteriophage progeny. The firm fixation of endolysins on the CW debris of lysed Gram-positive bacterial cells could prevent their activity toward the CW surface of neighboring cells. This explanation seems to be feasible [13] and is supported by CBD domains being most commonly present in the endolysins of phages that infect Gram-positive bacteria and mycobacteria [14,32]. It was previously shown that the omission of the CBD in Gram-positive endolysins significantly improves their activity and broadens the range of their bacterial targets [33,34]. Thus, the presence of specific CBD and diverse PG structures makes the activity of Gram-positive endolysins narrowly targeted [35,36].

In contrast, Gram-negative bacteria possess an outer membrane (OM) which explains the conservative nature of Gram-negative bacterial PG structures [20–22]. Endolysins of Gram-negative bacteria are small molecules that commonly have a single-domain globular structure. With some exclusions, lysins that act against Gram-negative bacteria do not have a CBD domain [25,37]. Thus, because the absence of a CBD domain in targeting endolysins is the norm for Gram-negative bacteria, such lysins should generally be much less specific.

Some experimental evidence of the broad lytic activity of endolysins against Gram-negative bacteria has come from studies of *E. coli* and *A. baumannii* prophages [23,24,27]. An *E. coli* prophage lysin (PlyE146) has been shown to lyse a wide range of Gram-negative bacteria (*K. pneumoniae*, *A. baumannii*, *P. aeruginosa*, *E. coli*, and *S. enterica*) [24], and *A. baumannii* prophage lysin (AcLys) exhibits a similar range of bactericidal activity [23]. Our results show that the wide bactericidal activity of endolysins from Gram-negative-infecting bacteriophages is a much more general phenomenon than was previously appreciated. In the present study, we demonstrated that three different lysins (LysAm24, LysECD7, and LysSi3) from *Myoviridae* bacteriophage family members that exhibit diverse domain organization (single-domain vs. two-domain) and different putative mechanisms of action (lysozyme vs. peptidase) have equally wide activity profiles. Notably, even LysAm24, which contains an additional glycan-binding domain that is typically associated with strain-specific lytic activity, showed a broad spectrum of activity with respect to the assayed bacteria. A quantity of 50 μg/mL of lysins was enough to eradicate growing bacterial cells by over more than five orders of magnitude for the assayed *P. aeruginosa*, *A. baumannii*, *K. pneumoniae*, *E. coli*, and *S. typhi* strains. In contrast, PlyE146 [24] only displayed wide specificity at a concentration of 400 μg/mL. Moreover, similar results were achieved at lower concentrations of this enzyme (below 1 μg/mL) only after cells were partially permeabilized by hypotonic conditions, freezing during storage or after the cells were washed with water [23].

Importantly, none of the recombinant enzymes assayed in our study (LysAm24, LysECD7, and LysSi3) required a specialized OM penetration approach to exert full bactericidal activity at an acidic pH and low salt concentration on exponentially growing cells. We previously showed that the recombinant endolysin L-KPP10 only exhibits pronounced activity in the presence of 0.5 mM EDTA [28] and that L-KPP10 alone does not possess significant lytic activity even on exponentially growing cells. Thus, EDTA and organic acids that displace divalent cations and destabilize the bacterial cell membranes could be an optional solution for the use of such endolysins [13]. Alternatively, the addition of a positively charged peptide (SMAP-29) at the C-terminus of the endolysin L-KPP10 (generating AL-KPP10) eliminated the need for EDTA permeabilization [28]. The effectiveness of this approach has been demonstrated in several other studies [16,17], suggesting that the use of positively charged peptide fusion may be a more convenient way to ensure efficient outer membrane permeabilization.

Interestingly all three assayed lysins were active or moderately active against the *E. coli* BL21(DE3) pLysS expression strain used for the production of recombinant enzymes (Figure 4). The strain was fully sensitive to LysECD7 (100%) and partly resistant to the action of LysAm24 (70%) and LysSi3 (10%). Moreover, all proteins were shown to be soluble when expressed intracellularly. In our previous studies of L-KPP10 endolysin fused with the strong cell penetration peptide SMAP-29 [28], we were able to purify it from inclusion bodies (insoluble condensed matter). All attempts to produce it in a soluble form led to either a lack of protein expression or the production of a soluble protein with no bactericidal activity. Lysins described in the current paper do not possess specific membrane penetration peptides. It is not clear whether studied endolysins can be obtained from the cytoplasm, where recombinant proteins are expressed, into periplasmic space where they can exert their bactericidal action. On the one hand, there is evidence that, for efficient permeabilization, some lysins require additional phage-encoded proteins for the control of inner membrane permeabilization [14]. For the lysins under study, the necessity of additional membrane permeabilizing proteins is not known and requires further investigation. On the other hand, as we mentioned above, buffer conditions, salts, pH, and membrane permeabilizers significantly affect the bactericidal activity of lysins. Thus, the intracellular conditions might be not optimal for enzymatic action, allowing soluble endolysins to be produced intracellularly in the strain that is sensitive to its action from outside.

The finding that none of our enzymes required a chemical permeabilizer under some conditions was unexpected. Some lysins are known to possess a natural ability to penetrate the bacterial OM due to positively charged segments that can disorganize or disrupt the OM [13,23], and the recombinant enzymes assayed in this study likely possess such sequences. Specifically, the C-terminal sequences of LysAm24 and LysSi3 contain a positively charged α-helix with nine lysine and arginine residues in 30-aa stretch. For proteins with homologous domain structures, this region was speculated to be an OM penetration sequence [24]. In contrast, no such positively charged stretch of amino acids is present in LysECD7. In ChiX, an L-Ala-D-Glu endopeptidase from *Serratia marcescens* that is homologous to LysECD7, the formation of a sugar-binding site presumably occurs through three tryptophan residues

(Trp87, Trp89, and Trp116) that occupy the outer edge of the active site cleft [38]. However, LysECD7 only contains two of these residues (Trp80 and Trp105), and whether these amino acids possess permeabilization activity is not yet clear. An alternative explanation for the cell permeabilization activity of the recombinant enzymes assayed in this study is that they contain an 8-His tag sequence at the C-terminus with adjacent sequences that form a positively charged cluster that is capable of improving membrane penetration (KLEHHHHHHHH). To test this hypothesis, we performed assays over a range of pH values with and without added permeabilizer (EDTA).

The observed pH-dependent changes in the dynamics of the assayed endolysin activities indirectly confirmed this hypothesis. Decreased activity in response to increasing pH values is known to affect other endolysins [24]. The additional protonation of C-terminal cationic peptides of LysAm24 and LysSi3 increase their destabilizing activity toward the bacterial OM at low pH values. Indeed, the charges of LysAm24 and LysSi3 C-termini changed from +12.8/13.8 at pH 5.0 to 1.8/2.5 at pH 9.0. Since LysECD7 does not contain similar peptides, with the exception of the His-Tag sequence at the C-terminus, and the change of its C-terminus charge from +8.0 at pH 5.0 to −1.0 at pH 9, this effect was less pronounced for LysECD7, and it retained up to 65% of its activity. Under pH values with C-termini charges of less than 10.0, LysAm24 and LysSi3 activity decreased significantly in the absence of EDTA. Detailed information on the influence of different pH values on the protein charge and activity is summarized in Table 3.

Table 3. Heatmap of estimated endolysin C-termini charges over the pH range and the residual antibacterial activity within this range (http://protcalc.sourceforge.net/). Min to Max pKa and residual activity is indicated with the color scale from red to green background. For all experiments, the mean values of three independent experiments are shown.

pH	C-Termini Charge (pKa)			Residual Antibacterial Activity, %		
	LysAm24	LysECD7	LysSi3	LysAm24	LysECD7	LysSi3
5.00	12.80	8.00	13.80	98.8	100.0	99.8
5.50	11.60	7.30	12.60	99.,4	100.0	100.0
6.00	10.20	6.10	11.20	97.2	97.6	99.9
6.50	8.00	4.00	9.00	55.0	75.0	57.2
7.00	5.80	1.80	6.80	45.0	62.5	29.2
7.50	4.40	0.50	5.40	62.5	94.5	88.5
8.00	3.40	−0.30	4.40	57.5	96.5	98.8
8.50	2.60	−0.70	3.50	16.0	90.0	69.0
9.00	1.80	−1.00	2.50	45.0	50.0	65.0

To differentiate the permeabilization ability from the bactericidal activity of the endolysins, the effect of pH in the presence of EDTA, which was used as a membrane permeabilizer, was evaluated. In the presence of EDTA, the activity of all lysins increased under neutral and basic pH conditions (Figure 3). These results indicate that the decrease in activity of the proteins associated with changes in pH is primarily associated with the ability of the lysins to overcome the cell membrane rather than by changes in their bactericidal activity. Our results suggest that there is a correlation between the structural changes in proteins under different pH values and their endogenous ability to penetrate the bacterial membrane (Spearman's rank correlation coefficients $r = 0.86$, 0.73, and 0.53 for LysAm24, LysECD7, and LysSi3 correspondingly). The charges of the C-termini can explain the observed changes but not the recovery of their activity over a narrow range of physiological pH values (approximately pH 7.5). Further detailed structural studies of these proteins will shed light on this issue.

The bactericidal activity of all studied lysins was concentration-dependent with one exception. At the highest concentration point (100 µg/mL), LysAm24 had two times weaker activity on the stationary phase bacterial cells (Figure 2) compared to at the lower concentration (10 µg/mL). For LysSi3 with putative muramidase activity, as well as for LysECD7 with putative peptidase activity, this effect was not observed. This is an interesting fact that we suggest is characteristic of some endolysins. Previously, we have shown the same effect for endolysin with putative

N-acetylmuramidase activity—artificial lysin AL-KPP10 fused with antimicrobial peptide SMAP-29 and tested against *P. aeruginosa* [28]. It was significantly less active at a concentration of 100 μg/mL than at concentrations of 25 and 50 μg/mL. The possible explanation for this lysin activity drop off under high concentrations is that some lysins might be prone to self-aggregation (e.g., dimerization of molecules). Interestingly, the presence of EDTA fully compensated for this drop in activity.

The presence of potassium and sodium chlorides negatively affected the activity of all studied endolysins. Interestingly, the bactericidal activity of LysAm24 and LysSi3 was restored to up to 60–80% that of the control at a concentration of 250 mM of either salt. For LysECD7, this effect was less pronounced and seemed to move to a higher salt concentration, at least for potassium chloride. Proteins usually have a particular optimal salt concentration because salt affects the intermolecular interactions. Thus, this effect can hardly be explained by the properties of the recombinant proteins alone. The bactericidal activity was tested on a complex system that included live bacterial cells. To exclude the factors pertaining to the live bacterial cells, experiments with pure peptidoglycan would be very useful. The specificity of bacteriophages and their encoded endolysins is considered to be a benefit for their practical use and preparation [16,17]. However, we believe that endolysins with activity toward specific bacteria are disadvantageous, as this would make patient treatment outcomes less predictable and would likely require preliminary sensitivity evaluations. Our results show that a large number of bacteriophage lytic enzymes can have broad bactericidal activity toward Gram-negative bacteria, demonstrating their potential in the development of medicines with reliable and predictable effectiveness.

Author Contributions: Conceptualization, V.A.G., N.P.A. and D.V.V.; Methodology, N.P.A., A.M.L. and E.V.U.; Formal Analysis, D.V.V., N.P.A., V.V.M. and V.A.G.; Investigation, N.P.A., V.V.M. and E.V.U.; Writing—Original Draft Preparation, Review and Editing, V.A.G., N.P.A. and D.V.V.; Project Administration, V.A.G., A.L.G.; Funding Acquisition, A.P.T. and V.A.G.

Acknowledgments: The authors wish to thank Eugenie O. Rubalskii for providing bacteriophage DNA and Roman S. Ovchinnikov and Mikhail V. Fursov for providing the clinical strain isolates collection. We also would like to thank Pavel I. Semenyuk for assistance with the DLS experiments and Vadim U. Balabanyan and Alexander V. Grishin for fruitful discussions of the results and careful manuscript reading.

References

1. Davies, J.; Davies, D. Origins and Evolution of Antibiotic Resistance. *Microbiol. Mol. Biol. Rev.* **2010**, *74*, 417–433. [CrossRef]
2. Furusawa, C.; Horinouchi, T.; Maeda, T. Toward prediction and control of antibiotic-resistance evolution. *Curr. Opin. Biotechnol.* **2018**, *54*, 45–49. [CrossRef]
3. Shrivastava, S.; Shrivastava, P.; Ramasamy, J. World health organization releases global priority list of antibiotic-resistant bacteria to guide research, discovery, and development of new antibiotics. *J. Med. Soc.* **2018**, *32*, 76–77. [CrossRef]
4. Singer, A.C.; Shaw, H.; Rhodes, V.; Hart, A. Review of antimicrobial resistance in the environment and its relevance to environmental regulators. *Front. Microbiol.* **2016**, *7*, 1–22. [CrossRef]
5. Woolhouse, M.; Ward, M.; van Bunnik, B.; Farrar, J. Antimicrobial resistance in humans, livestock and the wider environment. *Phil. Trans. R. Soc.* **2015**, *370*, 7. [CrossRef]
6. Venter, H.; Henningsen L., M.; Stephanie L., B. Antimicrobial resistance in healthcare, agriculture and the environment: the biochemistry behind the headlines. *Essays Biochem.* **2017**, *61*, 1–10. [CrossRef]
7. Afshinnekoo, E.; Meydan, C.; Chowdhury, S.; Jaroudi, D.; Boyer, C.; Bernstein, N.; Maritz, J.M.; Reeves, D.; Gandara, J.; Chhangawala, S.; et al. Geospatial Resolution of Human and Bacterial Diversity with City-Scale Metagenomics. *Cell Syst.* **2015**, *1*, 72–87. [CrossRef]
8. Magill, S.S.; Edwards, J.R.; Bamberg, W.; Beldavs, Z.G.; Dumyati, G.; Kainer, M.A.; Lynfield, R.; Maloney, M.; McAllister-Hollod, L.; Nadle, J. Microwave assisted synthesis of some novel benzimidazole substituted fluoroquinolones and their antimicrobial evaluation. *J. Pharm. Sci. Res.* **2010**, *2*, 69–76.
9. Myelnikov, D. An Alternative Cure: The Adoption and Survival of Bacteriophage Therapy in the USSR, 1922–1955. *J. Hist. Med. Allied Sci.* **2018**, *73*, 385–411. [CrossRef]

10. Loc-Carrillo, C.; Abedon, S.T. Pros and cons of phage therapy. *Bacteriophage* **2011**, *1*, 111–114. [CrossRef]
11. Drulis-Kawa, Z.; Majkowska-Skrobek, G.; Maciejewska, B.; Delattre, A.-S.; Lavigne, R. Learning from Bacteriophages—Advantages and Limitations of Phage and Phage-Encoded Protein Applications. *Curr. Protein Pept. Sci.* **2012**, *13*, 699–722. [CrossRef]
12. Nilsson, A.S. Phage therapy-constraints and possibilities. *Ups. J. Med. Sci.* **2014**, *119*, 192–198. [CrossRef]
13. Oliveira, H.; São-José, C.; Azeredo, J. Phage-Derived Peptidoglycan Degrading Enzymes: Challenges and Future Prospects for In Vivo Therapy. *Viruses* **2018**, *10*, 292. [CrossRef]
14. Oliveira, H.; Melo, L.D.R.; Santos, S.B.; Nobrega, F.L.; Ferreira, E.C.; Cerca, N.; Azeredo, J.; Kluskens, L.D. Molecular Aspects and Comparative Genomics of Bacteriophage Endolysins. *J. Virol.* **2013**, *87*, 4558–4570. [CrossRef]
15. Fischetti A., V. Lysin Therapy for *Staphylococcus aureus* and Other Bacterial Pathogens. *Curr. Top. Microbiol. Immunol.* **2015**, *6*, 23–27.
16. Briers, Y.; Walmagh, M.; Grymonprez, B.; Biebl, M.; Pirnay, J.P.; Defraine, V.; Michiels, J.; Cenens, W.; Aertsen, A.; Miller, S.; et al. Art-175 is a highly efficient antibacterial against multidrug-resistant strains and persisters of *Pseudomonas aeruginosa*. *Antimicrob. Agents Chemother.* **2014**, *58*, 3774–3784. [CrossRef]
17. Defraine, V.; Schuermans, J.; Grymonprez, B.; Govers, S.K.; Aertsen, A.; Fauvart, M.; Michiels, J.; Lavigne, R.; Briers, Y. Efficacy of Artilysin® Art-175 against resistant and persistent. *Antimicrob. Agents Chemother.* **2016**, *60*, 3480–3488. [CrossRef]
18. Lood, R.; Winer, B.Y.; Pelzek, A.J.; Diez-Martinez, R.; Thandar, M.; Euler, C.W.; Schuch, R.; Fischettia, V.A. Novel Phage Lysin Capable of Killing the Multidrug-Resistant Gram- Negative Bacterium *Acinetobacter baumannii* in a Mouse Bacteremia Model. *Antimicrob. Agents Chemother.* **2015**, *59*, 1983–1991. [CrossRef]
19. Briers, Y.; Walmagh, M.; Lavigne, R. Use of bacteriophage endolysin EL188 and outer membrane permeabilizers against *Pseudomonas aeruginosa*. *J. Appl. Microbiol.* **2011**, *110*, 778–785. [CrossRef]
20. Vollmer, W.; Blanot, D.; de Pedro, M.A. Peptidoglycan structure and architecture. *FEMS Microbiol. Rev.* **2008**, *32*, 149–167. [CrossRef]
21. Vollmer, W.; Seligman, S.J. Architecture of peptidoglycan: more data and more models. *Trends Microbiol.* **2010**, *18*, 59–66. [CrossRef]
22. Schleifer, K.H.; Kandler, O. Peptidoglycan types of bacterial cell walls and their taxonomic implications. *Bacteriol. Rev.* **1972**, *36*, 407–477. [PubMed]
23. Sykilinda, N.N.; Nikolaeva, A.Y.; Shneider, M.M.; Mishkin, D.V.; Patutin, A.A.; Popov, V.O.; Boyko, K.M.; Klyachko, N.L.; Miroshnikov, K.A. Structure of an Acinetobacter broad-range prophage endolysin reveals a C-terminal α-helix with the proposed role in activity against live bacterial cells. *Viruses* **2018**, *10*, 309. [CrossRef] [PubMed]
24. Larpin, Y.; Oechslin, F.; Moreillon, P.; Resch, G.; Entenza, J.M.; Mancini, S. In vitro characterization of PlyE146, a novel phage lysin that targets Gram-negative bacteria. *PLoS ONE* **2018**, *13*, e0192507. [CrossRef] [PubMed]
25. Walmagh, M.; Briers, Y.; dos Santos, S.B.; Azeredo, J.; Lavigne, R. Characterization of modular bacteriophage endolysins from myoviridae phages OBP, 201Q2-1 and PVP-SE1. *PLoS ONE* **2012**, *7*, e36991. [CrossRef] [PubMed]
26. Schuch, R.; Nelson, D.; Fischetti, V.A. A bacteriolytic agent that detects and kills *Bacillus anthracis*. *Nature* **2002**, *418*, 884. [CrossRef] [PubMed]
27. Lai, M.J.; Lin, N.T.; Hu, A.; Soo, P.C.; Chen, L.K.; Chen, L.H.; Chang, K.C. Antibacterial activity of *Acinetobacter baumannii* phage ΦaB2 endolysin (LysAB2) against both Gram-positive and Gram-negative bacteria. *Appl. Microbiol. Biotechnol.* **2011**, *90*, 529–539. [CrossRef] [PubMed]
28. Antonova, N.P.; Balabanyan, V.Y.; Tkachuk, A.P.; Makarov, V.V.; Gushchin, V.A. Physical and chemical properties of recombinant KPP10 phage lysins and their antimicrobial activity against *Pseudomonas aeruginosa*. *Bull. Russ. State Med. Univ.* **2018**, *1*. [CrossRef]
29. Ha, E.; Son, B.; Ryu, S. Clostridium perfringens virulent bacteriophage CPS2 and its thermostable endolysin lysCPS2. *Viruses* **2018**, *10*, 251. [CrossRef]
30. Loeffler, J.M.; Nelson, D.; Fischetti, V.A.; Clin, J. Rapid Killing of *Streptococcus pneumoniae* with a Bacteriophage Cell Wall Hydrolase. *Science* **2016**, *294*, 2170–2172. [CrossRef] [PubMed]
31. Daniel, A.; Euler, C.; Collin, M.; Chahales, P.; Gorelick, K.J.; Fischetti, V.A. Synergism between a novel chimeric lysin and oxacillin protects against infection by methicillin-resistant *Staphylococcus aureus*. *Antimicrob. Agents Chemother.* **2010**, *54*, 1603–1612. [CrossRef]

32. Payne, K.M.; Hatfull, G.F. Mycobacteriophage endolysins: Diverse and modular enzymes with multiple catalytic activities. *PLoS ONE* **2012**, *7*, e34052. [CrossRef]
33. Low, L.Y.; Yang, C.; Perego, M.; Osterman, A.; Liddington, R. Role of net charge on catalytic domain and influence of cell wall binding domain on bactericidal activity, specificity, and host range of phage lysins. *J. Biol. Chem.* **2011**, *286*, 34391–34403. [CrossRef]
34. Mayer, M.J.; Garefalaki, V.; Spoerl, R.; Narbad, A.; Meijers, R. Structure-based modification of a Clostridium difficile-targeting endolysin affects activity and host range. *J. Bacteriol.* **2011**, *193*, 5477–5486. [CrossRef]
35. Pastagia, M.; Schuch, R.; Fischetti, V.; Huang, D. Lysins: The arrival of pathogen-directed anti-infectives. *Artforum Int.* **2003**, *41*, 211–219. [CrossRef]
36. Fischetti, V.A.; Nelson, D.; Schuch, R.; Fischetti, V.A.; Nelson, D. Reinventing phage therapy: are the parts greater than the sum? *Nature Biotechnol.* **2006**, *24*, 1508–1511. [CrossRef]
37. Rodríguez-Rubio, L.; Martínez, B.; Donovan, D.M.; Rodríguez, A.; García, P. Bacteriophage virion-associated peptidoglycan hydrolases: Potential new enzybiotics. *Crit. Rev. Microbiol.* **2013**, *39*, 427–434. [CrossRef]
38. Owen, R.A.; Fyfe, P.K.; Lodge, A.; Biboy, J.; Vollmer, W.; Hunter, W.N.; Sargent, F. Structure and activity of ChiX: A peptidoglycan hydrolase required for chitinase secretion by *Serratia marcescens*. *Biochem. J.* **2018**, *475*, 415–428. [CrossRef]

11

Potential Application of Bacteriophages in Enrichment Culture for Improved Prenatal *Streptococcus agalactiae* Screening

Jumpei Uchiyama [1,*], Hidehito Matsui [2], Hironobu Murakami [1], Shin-ichiro Kato [3], Naoki Watanabe [1], Tadahiro Nasukawa [1], Keijiro Mizukami [1], Masaya Ogata [1], Masahiro Sakaguchi [1], Shigenobu Matsuzaki [4] and Hideaki Hanaki [2]

1 School of Veterinary Medicine, Azabu University, Kanagawa 252-5201, Japan; h-murakami@azabu-u.ac.jp (H.M.); n.watanabe1011@gmail.com (N.W.); dv1804@azabu-u.ac.jp (T.N.); mizukami@azabu-u.ac.jp (K.M.); a15121@azabu-u.ac.jp (M.O.); sakagum@azabu-u.ac.jp (M.S.)
2 Kitasato Institute for Life Sciences, Kitasato University, Tokyo 108-8641, Japan; m_hidehito@yahoo.co.jp (H.M.); hanakihideaki@yahoo.co.jp (H.H.)
3 Research Institute of Molecular Genetics, Kochi University, Kochi 783-8502, Japan; katoshin@kochi-u.ac.jp
4 Kochi Medical School, Kochi University, Kochi 783-8505, Japan; matuzaki@kochi-u.ac.jp
* Correspondence: uchiyama@azabu-u.ac.jp

Abstract: Vertical transmission of *Streptococcus agalactiae* can cause neonatal infections. A culture test in the late stage of pregnancy is used to screen for the presence of maternal *S. agalactiae* for intrapartum antibiotic prophylaxis. For the test, a vaginal–rectal sample is recommended to be enriched, followed by bacterial identification. In some cases, *Enterococcus faecalis* overgrows in the enrichment culture. Consequently, the identification test yields false-negative results. Bacteriophages (phages) can be used as antimicrobial materials. Here, we explored the feasibility of using phages to minimize false-negative results in an experimental setting. Phage mixture was prepared using three phages that specifically infect *E. faecalis*: phiEF24C, phiEF17H, and phiM1EF22. The mixture inhibited the growth of 86.7% (26/30) of vaginal *E. faecalis* strains. The simple coculture of *E. faecalis* and *S. agalactiae* was used as an experimental enrichment model. Phage mixture treatment led to suppression of *E. faecalis* growth and facilitation of *S. agalactiae* growth. In addition, testing several sets of *S. agalactiae* and *E. faecalis* strains, the treatment with phage mixture in the enrichment improved *S. agalactiae* detection on chromogenic agar. Our results suggest that the phage mixture can be usefully employed in the *S. agalactiae* culture test to increase test accuracy.

Keywords: phage; *Enterococcus faecalis*; *Streptococcus agalactiae*; culture enrichment

---

1. Introduction

Streptococcus agalactiae (also called group B *Streptococcus*) is vertically transmitted to the newborn during delivery, and can cause neonatal infections [1,2]. Common early-onset diseases caused by this organism in infants include sepsis and pneumonia, and (rarely) meningitis [1,2]. To prevent such infections, a prenatal *S. agalactiae* culture test is recommended in the late stage of pregnancy [1,2]. In the case of a positive test result, the pregnant carrier is prophylactically treated with antibiotics to prevent vertical transmission of *S. agalactiae* during the intrapartum period [1,2].

For the *S. agalactiae* culture test, the Centers for Disease Control and Prevention highly recommend an enrichment culture, followed by conventional *S. agalactiae* identification [3,4]. In the culture test, a swab is taken from the vaginal and anorectal areas, and the samples are inoculated and cultured in an enrichment culture broth selective for *S. agalactiae*. After the enrichment culture, bacterial identification is performed, e.g., using the Christie–Atkins–Munch-Petersen test, serologic identification, growth on chromogenic agar, and nucleic acid amplification [4]. However, although a selective culture broth is used for the enrichment culture, *S. agalactiae* is poorly recovered along with overgrowth of *Enterococcus faecalis* in some cases [5–8]. This may lead to false-negative results in the subsequent identification tests [5–8]. To address this problem, selective antimicrobial agents to be included in the enrichment broth should be reevaluated.

Bacteriophages (phages), i.e., bacterial viruses, infect specific bacteria. Some phages infect and lyse bacteria at the specificity level of species and strains. These phage characteristics were used to eliminate most cells in a bacterial population and facilitate the isolation of less prevalent environmental bacteria that produce novel bioactive compounds [9]. Phage applicability for the isolation of food-poisoning microbes in the food microbiology field was also examined [10]. Hence, potentially, phage application might also be used to reduce the unwanted growth of *E. faecalis* in an *S. agalactiae* enrichment culture and to facilitate *S. agalactiae* detection in clinical microbiology. Indeed, phages that specifically infect *E. faecalis* were isolated from environmental samples, such as sewage and canal water [11–13]. In the current study, we examined the applicability of *E. faecalis*-specific phages to suppress *E. faecalis* growth in an *S. agalactiae* enrichment culture in an experimental setting.

2. Materials and Methods

2.1. Bacteria, Phages, and Culture Media

Strains of *E. faecalis* (n = 30), *S. agalactiae* (n = 7), *Enterococcus avium* (n = 5), and *Enterococcus faecium* (n = 5) were isolated from vaginal swabs using the Chrom-ID Strepto B test (bioMérieux, Marcy-l'Étoile, France). The swabs were obtained after random sampling at local hospitals in eastern Japan (Table S1). Bacteria were cultured at 37 °C under aerobic or microaerobic (i.e., 5% CO_2) conditions, as appropriate, based on their specific growth requirements (Table S1).

Phage phiEF24C was isolated and characterized, as described elsewhere [12,14,15]. Phage phiEF17H was newly isolated from canal water in Kochi (Japan). Phage phiM1EF22 was newly isolated from sewage water in Tokyo (Japan) (Table S2). The isolation procedures are described in Reference [12]. *E. faecalis* strains KUEF01, KUEF25, and KUEF27, described in Table S1, were used as host bacteria for phages phiEF24C, phiEF17H, and phiM1EF22, respectively, for phage amplification and plaque assay. Bacterial-phage suspensions were cultured aerobically at 37 °C.

Enterococcus spp. and phages were cultured in tryptic soy broth or agar (TSA), and *S. agalactiae* was cultured in Todd–Hewitt broth (THB), unless stated otherwise. Granada-type broth (GtB; 25.0 g/L proteose peptone no. 3, 14.0 g/L soluble starch, 2.5 g/L glucose, 1.0 g/L pyruvic acid sodium salt, 0.1 g/L cysteine hydrochloride, 0.3 g/L magnesium sulfate, 11.0 g/L 3-(*N*-morpholino)propane sulfonic acid, 10.7 g/L disodium hydrogen phosphate, 0.5 mg/L crystal violet, 10 mg/L colistin sulfate, 10 mg/L metronidazole, and 15 mg/L nalidixic acid, pH 7.4) was originally prepared as the *S. agalactiae* enrichment broth [16,17]. Alternatively, the pigmented enrichment Lim broth (modified Lim broth; Kyokuto Pharmaceutical Industrial, Tokyo, Japan) was used as an *S. agalactiae* enrichment broth. Unless stated otherwise, all culture media were purchased from Becton, Dickinson, and Co. (Franklin Lakes, NJ, USA). All chemicals and reagents were purchased from Nacalai Tesque (Kyoto, Japan) and FUJIFILM Wako Pure Chemical (Osaka, Japan).

2.2. Phage Genome Sequencing and Analysis

After phage amplification, phage particles were purified from 500 mL of phage lysate by CsCl density-gradient centrifugation, as described elsewhere [18]. Phage genomic DNA was then prepared by phenol–chloroform extraction of the collected purified phage band [18]. A shotgun library was prepared for each phage DNA using the GS FLX Titanium rapid library preparation kit (Roche Diagnostics, Indianapolis, IN, USA), according to the manufacturer's instructions. The libraries were analyzed using a GS Junior 454 sequencer (Roche Diagnostics, Risch-Rotkreuz, Switzerland).

The sequence reads were assembled using the 454 Newbler software (version 3.0; 454 Life Sciences, Branford, CT, USA). The genomes were annotated using a prokaryotic genome annotation pipeline, DFAST (https://dfast.nig.ac.jp/) [19,20]. The phiEF17H and phiM1EF22 genome sequences were deposited in GenBank under the accession numbers AP018714 and AP018715, respectively.

The genome sequences were analyzed using nucleotide Basic Local Alignment Search Tool (BLASTn) at the National Center for Biotechnology Information (NCBI; https://blast.ncbi.nlm.nih.gov/Blast.cgi?PROGRAM=blastn&PAGE_TYPE=BlastSearch&LINK_LOC=blasthome; last accessed: 5 May 2018). Moreover, the genomes of phages belonging to the family *Myoviridae* subfamily *Spounavirinae* were downloaded from GenBank (last accessed: 20 September 2018), and the viral phylogeny was analyzed using a proteomic tree analysis tool, ViPTree version 1.0 [21].

2.3. *Multilocus Sequence Typing (MLST) of* E. faecalis *Strains*

E. faecalis strains were cultured overnight, bacterial DNA was extracted, and MLST analysis was performed, according to the procedures described elsewhere [22]. The sequence alleles were analyzed using the *E. faecalis* MLST database (https://pubmlst.org/efaecalis/; last accessed: 5 January 2018) to designate sequence types (STs) [23]. The concatenating allele sequences were analyzed using MEGA 7.0.18, and sequence alignment implemented in ClustalW was followed by phylogenetic tree construction using the unweighted pair group with arithmetic mean (UPGMA) method [24].

2.4. *Examination of Antibacterial Activity of* E. faecalis *to* S. agalactiae

The anti-*S. agalactiae* activity of *E. faecalis* was examined using a spot-on-lawn assay, as described elsewhere [25]. Briefly, 200 μL of overnight bacterial culture of a single *S. agalactiae* strain was mixed with a melted 0.5% (*w*/*v*) soft agar and plated onto 1.5% (*w*/*v*) agar. One microliter of *E. faecalis* overnight culture was spotted on the solidified top agar. After incubation overnight at 37 °C in a microaerophilic condition, *S. agalactiae* growth around the spotted *E. faecalis* was examined.

2.5. *Analysis of Phage Lytic Activity*

The phage host range was determined by a streak test, as described elsewhere [12,15]. Briefly, 200 μL of overnight bacterial culture of a single bacterial strain was mixed with a melted 0.5% (*w*/*v*) soft agar and plated onto 1.5% (*w*/*v*) agar. The phage suspension (ca. $1.0 \times 10^{8–9}$ plaque-forming units (PFU)/mL) was streaked onto the solidified top agar. After incubation overnight at 37 °C, bacterial lysis, with or without plaque formation, was examined.

2.6. *Analysis of Bacterial Densities in* S. agalactiae *and* E. faecalis *Coculture with Phage Mixtures*

A rifampicin-resistant mutant clone of *S. agalactiae* was isolated by aerobically culturing *S. agalactiae* strain KUGBS2 on TSA containing 20 μg/mL rifampicin at 37 °C for two days. The putative mutant clones were repurified at least three times; each repurification round was repeated for one day under the same incubation conditions. One resultant rifampicin-resistant mutant clone of strain KUGBS2 was obtained and was tentatively designated as strain KUGBS2rif. *S. agalactiae* strain KUGBS2rif and *E. faecalis* strain KUEF08 were cultured individually until an optical density of 0.4–0.6 at 600 nm was attained. After diluting with the enrichment broth, suspensions of 3.0×10^4 colony-forming units (CFU)/mL *S. agalactiae* strain KUGBS2rif and 3.0×10^7 CFU/mL *E. faecalis* strain KUEF08 were prepared. Each phage suspension was diluted with THB to ca. 3.0×10^6 PFU/mL or 3.0×10^4 PFU/mL. By mixing equal volumes of phage suspensions at the same dilution, mixtures of two different dilutions of phages were prepared.

For the experiment, 100 μL each of *S. agalactiae* strain KUGBS2rif and *E. faecalis* strain KUEF08, and 300 μL of phage mixture were added to 10 mL of the enrichment broth. As negative controls, the same volume of THB was added instead of bacterial suspensions and/or phage suspensions. The mixtures were microaerobically incubated at 37 °C for 24 h. Total bacterial density and *S. agalactiae* strain KUGBS2rif and *E. faecalis* strain KUEF08 densities were determined.

Total bacterial densities were determined on TSA. TSA supplemented with 20 μg/mL rifampicin and *Enterococcus*-selective agar (EF agar base "Nissui"; Nissui Pharmaceutical Co., Tokyo, Japan) were used to determine the densities of *S. agalactiae* strain KUGBS2rif and *E. faecalis* strain KUEF08, respectively. *S. agalactiae* strain KUGBS2rif did not grow on the *Enterococcus*-selective agar; *E. faecalis* strain KUEF08 did not grow on TSA containing 20 μg/mL rifampicin.

2.7. Detection of Bacteria on Chromogenic Selective Agar after S. agalactiae *and* E. faecalis *Coculture with Phage Mixtures*

S. agalactiae and *E. faecalis* were cultured individually until an optical density of 0.4–0.6 at 600 nm was obtained. *S. agalactiae* and *E. faecalis* cultures were diluted with THB to ca. $3.0–5.0 \times 10^4$ CFU/mL and ca. 3.0×10^7 CFU/mL, respectively. After dilution of individual phage suspensions in THB to ca. 1.0×10^7 PFU/mL, the phage mixture was prepared by mixing equal volumes of the diluted phage suspensions.

For the experiment, 30 μL each of bacterial suspensions of *S. agalactiae* and *E. faecalis*, and 30 μL of phage mixture were added to 3 mL of the enrichment broth. As a negative control, the same volume of THB was added instead of the phage mixture. After 24-h incubation at 37 °C, a loopful of the suspension was inoculated on the Chrom-ID Strepto B agar (BioMérieux, Marcy-l'Étoile, France). After 24-h incubation at 37 °C in darkness, colony color and appearance on agar plates were examined. All incubations were carried out under microaerophilic conditions.

2.8. Statistical Analysis

The data were statistically analyzed using EZR (Saitama Medical Center, Jichi Medical University, Saitama, Japan), which is a graphical user interface for R (The R Foundation for Statistical Computing, Vienna, Austria) [26]. Student's *t*-tests were used to analyze differences between bacterial densities in different treatments. A p-value < 0.01 was considered to indicate a statistically significant difference.

3. Results and Discussion

3.1. Phage Characteristics

Phages phiEF24C, phiEF17H, and phiM1EF22 were used in the current study. Phage phiEF24C, one of the best-studied *Enterococcus* phages, is classified into the family *Myoviridae* subfamily *Spounavirinae* [12,14,15]. The other two phages, phiEF17H and phiM1EF22, were newly isolated and their whole genomes were sequenced. The whole-genome sequence similarity analysis using the BLASTn showed that phages phiEF17H and phiM1EF22 are similar to phage phiEF24C (Table S3). Thus, they were considered to be classified into the family *Myoviridae* subfamily *Spounavirinae* [27].

Moreover, the phylogenetic relationship of these three phages used in this study was analyzed with 33 other phages of the family *Myoviridae* subfamily *Spounavirinae*. Phages sharing this particular viral taxonomy of the family *Myoviridae* subfamily *Spounavirinae* are highly virulent toward host bacteria [28]. The tree showed that phages phiEF24C, phiEF17H, and phiM1EF22 were phylogenetically clustered but slightly different from each other among the *Enterococcus* phages belonging to this viral subfamily (Figure 1A). These *Enterococcus* phages may be categorized into a new virus genus in this subfamily.

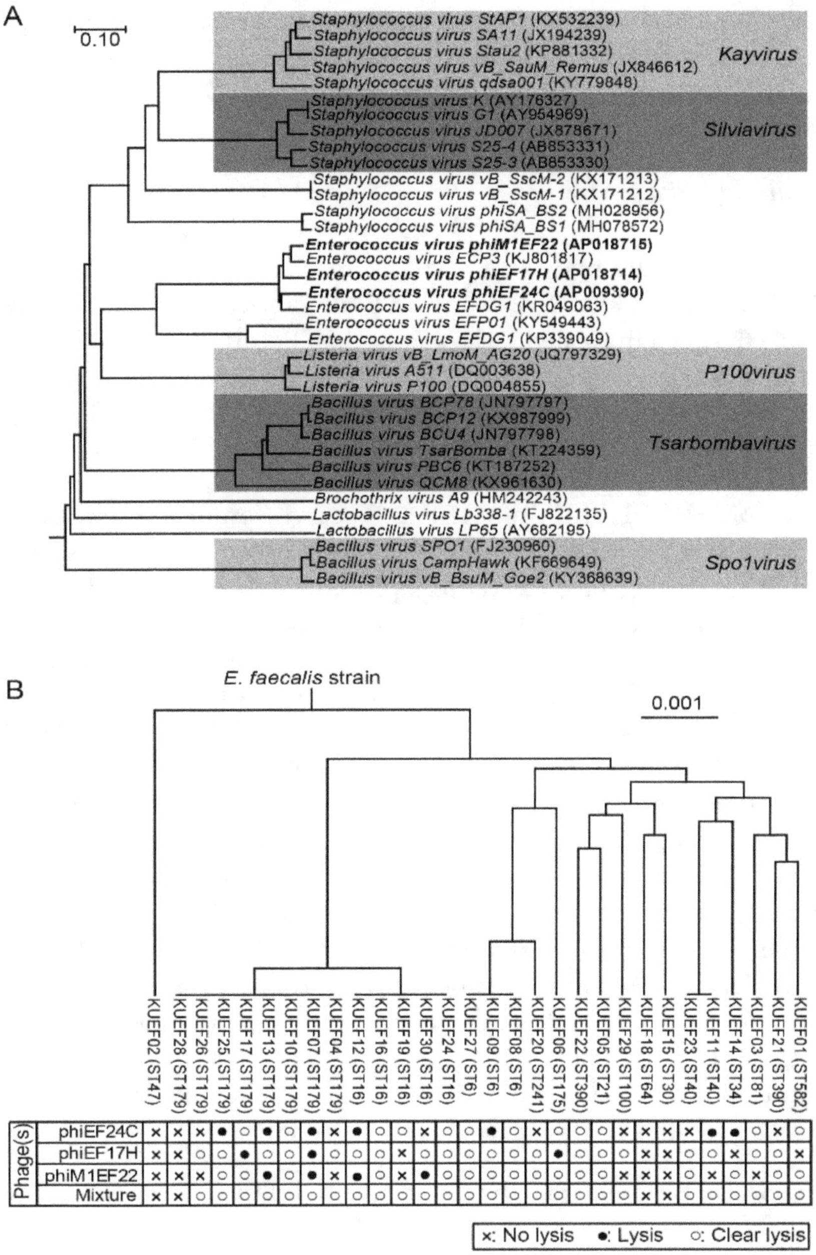

Figure 1. *Enterococcus* phages phiEF24C, phiEF17H, and phiM1EF22 and their lytic activity to various *E. faecalis* strains. (**A**) Viral proteomic trees of *Enterococcus* phages phiEF24C, phiEF17H, and phiM1EF22 in the family *Myoviridae* subfamily *Spounavirinae*. *Enterococcus* phages phiEF24C, phiEF17H, and phiM1EF22 are shown in bold. Phage taxonomical names are shown followed by the GenBank accession number in parentheses. The phages belonging to a certain viral genus are shown in grey highlight, on which the viral genus names are indicated. (**B**) *E. faecalis* strains isolated from vaginal swabs and their sensitivity to phages. Phylogenetic tree of *E. faecalis* strains was constructed based on the concatenated multilocus sequence typing (MLST) alleles. In the phylogenetic tree, *E. faecalis* strain names are followed by sequence types (STs) in brackets. Phage sensitivities to each phage and phage mixture are shown below the phylogenetic tree.

3.2. *Characteristics of* E. faecalis *Strains Isolated from Vaginal Swabs*

The genetic background of *E. faecalis* strains isolated from the vaginal swabs in this study was examined. MLST analysis of the *E. faecalis* vaginal swab isolates revealed that 43.3% (13/30) of *E. faecalis* strains were phylogenetically closely related, representing either ST16 or ST179 (Figure 1B).

The remaining strains (56.7% (17/30)) were genetically diverse. Moreover, *E. faecalis* strains do not seem to show antibacterial activity (e.g., bacteriocin production) to *S. agalactiae*, while some of them are able to show antibacterial activity to a variety of bacteria [5,29]. Testing the antibacterial activity to *S. agalactiae* strains using the spot-on-lawn assay, no anti-*S. agalactiae* activity was observed among these *E. faecalis* strains.

3.3. Phage Lytic Spectrum and Phage Mixture

Phage lytic activity was examined with a streak test using these *E. faecalis* strains (Figure 1B). Phages phiEF24C, phiEF17H, and phiM1EF22 showed lytic activity toward 63.3% (19/30), 76.7% (23/30), and 66.7% (20/30), respectively, of the tested *E. faecalis* strains. Because these phages have slightly different genomes, they have different host ranges to *E. faecalis* strains. Moreover, lytic activities of the three phages with other bacterial vaginal swab isolates (*E. avium*, *E. faecium*, and *S. agalactiae* strains) were also examined, but no lytic activity was observed.

Phages phiEF24C, phiEF17H, and phiM1EF22 lysed different *E. faecalis* strains and also some common strains. Theoretically, a combination of these three phages lysed a broader range of *E. faecalis* strains than any single phage tested. A phage mixture containing the three phages was prepared by mixing phage particles in a 1:1:1 ratio, which was also used in the following experiments. The lytic spectrum of the phage mixture was then examined using the streak test described above. The phage mixture showed lytic activity toward 86.7% (26/30) of *E. faecalis* strains tested (Figure 1B). Four *E. faecalis* strains were not lysed by the phages, namely, KUEF02 (MLST ST47), KUEF18 (ST64), KUEF15 (ST30), and KUEF28 (ST179). Because these bacterial strains may have several phage-resistant mechanisms [30], they cannot be lysed with these phages in the phage mixture. In addition, the phage mixture did not show any lytic activity with the other tested bacteria, i.e., *E. avium*, *E. faecium*, and *S. agalactiae*, as seen in the assessments of lytic activity of individual phages as above.

The phage mixture may have been contaminated with anti-*S. agalactiae* agents during phage mixture preparation (i.e., during phage propagation on the *E. faecalis* host). The antimicrobial activity of the phage mixture to *S. agalactiae* was examined. Incubation of the phage mixture with *S. agalactiae* for 30 min and 24 h did not influence bacterial viability, compared with a THB-treated negative control (Figure S1), excluding the possibility of phage mixture contamination with anti-*S. agalactiae* substances. Thus, the phage mixture could be used as anti-*E. faecalis* agents in the enrichment broth to examine the suppression of *E. faecalis* growth in *S. agalactiae* enrichment culture.

3.4. *Effect of Phage Mixture on* S. agalactiae *and* E. faecalis *Cell Densities in Experimental Enrichment Cultures*

The vaginal microbiome of pregnant women is majorly composed of *Lactobacillus* spp., together with the bacteria of phyla *Actinobacteria*, *Firmicutes*, and others in minor quantities [31–33]. During the culture of swab seed in *S. agalactiae*-selective enrichment broth, the growth of the majority of vaginal microflora is inhibited and a minority, including *S. agalactiae*, *E. faecalis*, and other bacteria, are selectively grown when they are presented in the swab. In this study, a situation in which *E. faecalis* thrives and *S. agalactiae* grows poorly was mimicked in an experimental setting, using a simple coculture of *E. faecalis* and *S. agalactiae* in *S. agalactiae*-selective enrichment broth. Using the model, the effects of the phage mixture were then examined.

The simple coculture of *E. faecalis* and *S. agalactiae* in *S. agalactiae*-selective enrichment broth was used as an experimental enrichment model, which mimicked the situation of poor recovery of *S. agalactiae*. In the experimental enrichment model, *S. agalactiae* and *E. faecalis* cell densities were monitored in the presence and absence of the phage mixture over time. To mimic the situation in which *S. agalactiae* was poorly recovered, *S. agalactiae* strain KUGBS2rif and *E. faecalis* strain KUEF08 were inoculated into GtB at 3.0×10^2 CFU/mL and 3.0×10^5 CFU/mL, respectively. Either of the two dilutions of phage mixture (at a multiplicity of infection (MOI) of each phage of 10^{-1} and 10^{-3} to *E. faecalis*) was added. As a negative control, THB was used instead of the phage mixture.

Bacterial cell densities (total bacteria, *S. agalactiae*, and *E. faecalis*) were then monitored over time (Figure 2). Firstly, based on the determined total bacteria numbers in both the negative control and phage treatment groups, bacteria grew exponentially for up to 12 h, following which the cultures entered the stationary phase of growth (Figure 2). Hence, incubation for 12–24 h was sufficient to achieve bacterial enrichment in that particular experimental setting. Moreover, *S. agalactiae* cell densities were compared with *E. faecalis* cell densities at different time points in the negative control and phage treatment groups. In the negative control group (i.e., no phage treatment), *E. faecalis* densities were significantly higher than *S. agalactiae* at incubation for 6, 12, and 24 h ($p < 0.01$) (Figure 2A). In contrast, in the phage treatment groups at MOIs of 10^{-1} and 10^{-3} of each phage, the cell density of *E. faecalis* was significantly lower than that of *S. agalactiae* at incubation for 6, 12, and 24 h in the phage treatment groups ($p < 0.01$) (Figure 2B,C).

We also evaluated the effectiveness of phage treatment in the experimental enrichment model using the commercially available pigmented enrichment Lim broth (modified Lim broth) (Figure S2). The experiments were performed using the same method described above. Inhibition of *E. faecalis* growth, compared with the untreated group, was observed in the phage treatment groups at both MOIs of each phage tested (i.e., 1 and 10^{-2}) (Figure S3). These results indicate that the phage mixture inhibited *E. faecalis* growth and facilitated *S. agalactiae* growth in both *S. agalactiae* enrichment broths.

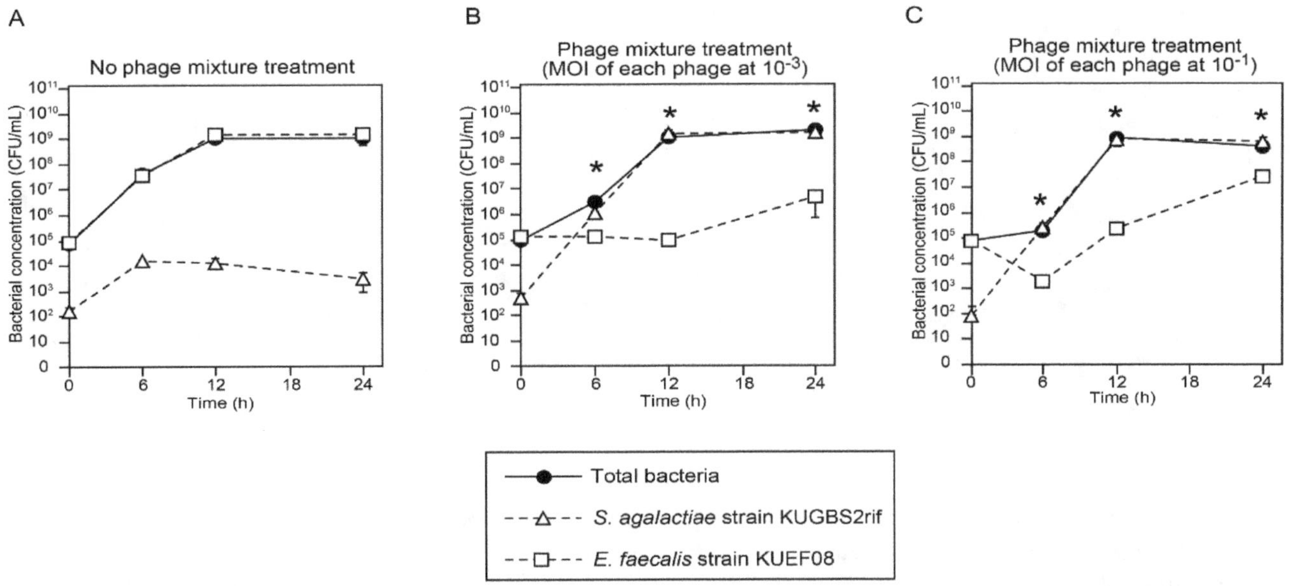

Figure 2. Growth of *E. faecalis* and *Streptococcus agalactiae* cocultures in the presence or absence of phage mixtures in Granada-type broth (GtB). No phage treatment (**A**); or treatment with phages at 10^{-3} (**B**) or at 10^{-1} (**C**) multiplicities of infection (MOIs) of each phage to *E. faecalis*. The means with standard deviations were calculated from triplicate experiments, and are plotted as points with error bars. Time points at which *S. agalactiae* density was significantly higher than that of *E. faecalis* are indicated by asterisks ($p < 0.01$; Student's *t*-test).

3.5. *Efficient Detection of* S. agalactiae *after Experimental Enrichment Culture in the Presence of the Phage Mixture*

In the *S. agalactiae* culture test, bacteria are generally identified in a culture aliquot after enrichment culture. Consequently, we then evaluated the efficiency of *S. agalactiae* identification after the experimental enrichment culture. As the identification assay, we used growth on the *S. agalactiae* chromogenic agar, in which *S. agalactiae* colonies are distinguished from *E. faecalis* colonies based on color. Five *S. agalactiae–E. faecalis* sets were tested: KUGBS2–KUEF08, KUGBS1–KUEF24, KUGBS6–KUEF26, KUGBS4–KUEF21, and KUGBS7–KUEF29. *E. faecalis* and *S. agalactiae* strains were first inoculated at a 100:1 ratio in the Granada-type enrichment broth; then, the phage mixture was added (at an MOI of each phage at 10^{-1} to *E. faecalis*), and the cultures were incubated. Enrichment culture aliquots were plated on chromogenic agar, and the resultant colony appearance was evaluated (Figure 3). After enrichment of all phage-treated *S. agalactiae*–*E. faecalis* sets, *S. agalactiae* colonies were dominant on the agar plates. By contrast, in enrichment cultures without phage treatment, only a few *S. agalactiae* colonies were observed on the chromogenic agar, while *E. faecalis* colonies were dominant. The same experiment was performed using the modified Lim broth and several *S. agalactiae*–*E. faecalis* combinations (Figure S4). The data were in agreement with observations made using the GtB. Thus, the phage mixture treatment is believed to improve the *S. agalactiae* culture test by inhibiting the undesirable growth of *E. faecalis*.

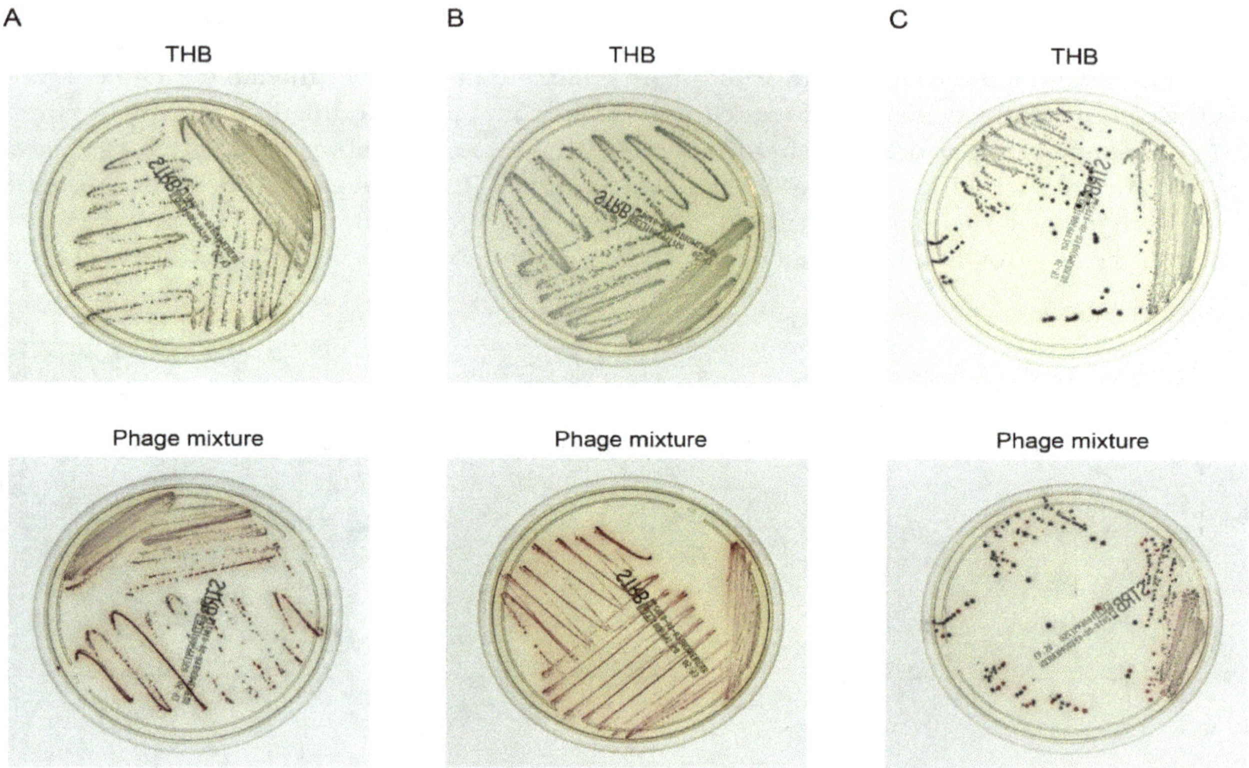

Figure 3. Bacterial identification on chromogenic agar after experimental enrichment coculture of *S. agalactiae* and *E. faecalis*. Combinations of single strains of *E. faecalis* and *S. agalactiae* were used to inoculate GtB, and were cultured in the presence of the phage mixture (MOI of each phage to *E. faecalis*: 10^{-1}) or Todd–Hewitt broth (THB). After enrichment culture, aliquots were spread on the chromogenic agar, and the resultant bacterial colonies were evaluated. Colonies of *S. agalactiae* and *E. faecalis* are red and blue, respectively. Top and bottom panels show photographs of chromogenic agar plates inoculated with enriched cultures treated with THB or phage mixtures, respectively. Representative data for three out of five *S. agalactiae*–*E. faecalis* sets are shown, namely, KUGBS2–KUEF08 (**A**), KUGBS1–KUEF24 (**B**), and KUGBS6–KUEF26 (**C**).

3.6. Phage Application Potential in the Clinical Setting

In the current study, we showed that a specific phage mixture effectively inhibited the growth of *E. faecalis* in an *S. agalactiae* culture test in the experimental setting. Before clinical application and reagent manufacture, it is necessary to discuss issues such as the phage composition in the phage mixture, the usage per assay (i.e., volume and phage density), storage of the phage mixture, and production cost.

As a first consideration, strains insensitive to the phage mixture may occur at a higher than expected rate in the clinical setting. To address this issue, phage sensitivity of *E. faecalis* strains should be constantly examined. The effective spectrum of phages in a mixture can be modified by replacing and/or adding other phages, including newly isolated phages, naturally evolved phages, and/or genetically modified phages [34,35]. For example, by adding six newly isolated phages to the phage mixture developed in the current study, we increased the inhibition efficiency among the tested *E. faecalis* strains to 96.7% (29/30). Hence, updating the composition of the phage mixture will help address the problem associated with the phage-insensitive strains.

Next, we considered the usage per assay. Assuming that the volume of the enrichment broth is ca. 3–4 mL, a drop of phage mixture suspension (i.e., 30–50 μL of a solution of ca. 1.0×10^7 PFU/mL of each phage) would suffice for an assay. Such usage per assay is in line with the usage of the phage mixture in the experiments performed in the current study (Figure 3). Thus, the phage mixture complies with the usage expected in a clinical setting.

Importantly, the quality of the phage mixture should be guaranteed to allow its commercialization. Phages are generally stable at 4 °C in culture media for a certain period of time [36,37], which suggests that phage products can be distributed on a market scale using the cold chain. Accordingly, we examined the infectious density of each phage solution during storage at 4 °C for 270 days; we did not observe any substantial reduction in phage particle density during this period (Figure S5). The shelf life of culture medium is generally up to six months [38]; no loss in sensitivity in *S. agalactiae* enrichment broths was observed after at least four months [39]. Thus, the stability of the phage mixture appears to be in line with the storage of the enrichment broths.

Finally, the use of phage mixture should be cost-effective. Phage production on a bioreactor scale was recently investigated because of increased interest in phage therapy. Consequently, the cost of phage production is estimated to be $\$4.4 \times 10^{-13}$/phage particle [40–42]. Hence, the calculated cost of the phage mixture per assay in the current study is only $\$2.2 \times 10^{-7}$ ($\$4.4 \times 10^{-13}$/phage particle multiplied by the number of phage particles in the mixture, i.e., 1.0×10^7 PFU/mL × 50 μL). A number of commercial phage companies were established worldwide [43]. Thus, in theory, phage mixtures may be produced for the *S. agalactiae* culture test in a cost-effective manner.

After commercialization of the phage mixture, special attention should be devoted to the biological characteristics of phages to avoid laboratory accidents, because clinical tests are at high risk of interference by phage contamination via aerosols. For example, *Enterococcus* phages may contaminate clinical *E. faecalis* culture tests and interfere with the test results in a clinical laboratory. To avoid such accidents, appropriate good laboratory hygiene and sterilization practices should be instated when preparing to use phage products. Such measures can include aseptic handling at a specified bench (e.g., safety cabinet), change of gloves and ultraviolet (UV) irradiation of the bench after usage, and storage in a specific cabinet [44]. We believe that the phage mixture is appropriate for use in the clinical laboratory, as long as the knowledge about phage products is disseminated and implemented. Thus, we anticipate that the phage mixture will become commercially available for *S. agalactiae* enrichment broths in the future and will be used in the clinical setting.

Supplementary Materials

Table S1: Bacterial strains used in the current study; Table S2: Phages used in the current study; Table S3: Genome sequences of *Enterococcus* phages used in this study; Figure S1: Examination of the anti-*S. agalactiae* effects of the phage mixture; Figure S2: Color change of the pigmented modified Lim broth with bacterial growth; Figure S3: Effects of the phage mixture on bacteria inoculated in a commercially available *S. agalactiae* enrichment broth; Figure S4: Bacterial identification on chromogenic agar after experimental enrichment coculture of *S. agalactiae* and *E. faecalis*; Figure S5: Phage stability.

Author Contributions: Conceptualization, J.U., H.M. (Hidehito Matsui), and H.H. Methodology, J.U. and H.M. (Hidehito Matsui). Software, J.U. and K.M. Validation, J.U. and T.N. Formal analysis, J.U. and S.K. Investigation, J.U., S.K., N.W., and M.O. Resources, J.U., H.M. (Hidehito Matsui), S.M., and H.H. Data curation, M.S. Writing—original draft preparation, J.U. and W.N. Writing—review and editing, J.U., H.M. (Hidehito Matsui), H.M. (Hironobu Murakami), T.N., K.M., and S.M. Visualization, J.U. and T.N. Supervision, J.U. Project administration, J.U. Funding acquisition, J.U. and M.S.

Acknowledgments: We thank Takuya Nakajima and Mio Sasaki (School of Veterinary Medicine, Azabu University, Kanagawa, Japan) for support in performing the experiments.

References

1. Morgan, J.A.; Cooper, D.B. Pregnancy, group b *streptococcus*. In *StatPearls*; StatPearls Publishing: Treasure Island, FL, USA, 2018; (updated on 11 February 2018).
2. Edmond, K.M.; Kortsalioudaki, C.; Scott, S.; Schrag, S.J.; Zaidi, A.K.; Cousens, S.; Heath, P.T. Group b streptococcal disease in infants aged younger than 3 months: Systematic review and meta-analysis. *Lancet* **2012**, *379*, 547–556. [CrossRef]
3. Cagno, C.K.; Pettit, J.M.; Weiss, B.D. Prevention of perinatal group b streptococcal disease: Updated cdc guideline. *Am. Fam. Physician* **2012**, *86*, 59–65. [PubMed]
4. Rosa-Fraile, M.; Spellerberg, B. Reliable detection of group b *streptococcus* in the clinical laboratory. *J. Clin. Microbiol.* **2017**, *55*, 2590–2598. [CrossRef] [PubMed]
5. Dunne, W.M., Jr.; Holland-Staley, C.A. Comparison of nna agar culture and selective broth culture for detection of group b streptococcal colonization in women. *J. Clin. Microbiol.* **1998**, *36*, 2298–2300. [PubMed]
6. Park, C.J.; Vandel, N.M.; Ruprai, D.K.; Martin, E.A.; Gates, K.M.; Coker, D. Detection of group b streptococcal colonization in pregnant women using direct latex agglutination testing of selective broth. *J. Clin. Microbiol.* **2001**, *39*, 408–409. [CrossRef] [PubMed]
7. Binghuai, L.; Yanli, S.; Shuchen, Z.; Fengxia, Z.; Dong, L.; Yanchao, C. Use of maldi-tof mass spectrometry for rapid identification of group b *streptococcus* on chromid strepto b agar. *Int. J. Infect. Dis.* **2014**, *27*, 44–48. [CrossRef] [PubMed]
8. Baden, M.; Higashiyama, T.; Ikemoto, T.; Okada, Y. Evaluation of direct latex agglutination of selective broth for detection of group b streptococcal carriage in pregnant women. *J. Jpn. Soc. Clin. Microbiol.* **2016**, *26*, 7–13.
9. Kurtböke, D. Actinophages as indicators of actinomycete taxa in marine environments. *Antonie Van Leeuwenhoek* **2005**, *87*, 19–28. [CrossRef] [PubMed]
10. Muldoon, M.T.; Teaney, G.; Li, J.; Onisk, D.V.; Stave, J.W. Bacteriophage-based enrichment coupled to immunochromatographic strip-based detection for the determination of *salmonella* in meat and poultry. *J. Food Prot.* **2007**, *70*, 2235–2242. [CrossRef] [PubMed]
11. Khalifa, L.; Coppenhagen-Glazer, S.; Shlezinger, M.; Kott-Gutkowski, M.; Adini, O.; Beyth, N.; Hazan, R. Complete genome sequence of *enterococcus* bacteriophage eflk1. *Genome Announ.* **2015**, *3*, e01308–e01315. [CrossRef] [PubMed]
12. Uchiyama, J.; Rashel, M.; Maeda, Y.; Takemura, I.; Sugihara, S.; Akechi, K.; Muraoka, A.; Wakiguchi, H.; Matsuzaki, S. Isolation and characterization of a novel *enterococcus faecalis* bacteriophage phief24c as a therapeutic candidate. *FEMS Microbiol. Lett.* **2008**, *278*, 200–206. [CrossRef] [PubMed]
13. Khalifa, L.; Gelman, D.; Shlezinger, M.; Dessal, A.L.; Coppenhagen-Glazer, S.; Beyth, N.; Hazan, R. Defeating antibiotic- and phage-resistant *enterococcus faecalis* using a phage cocktail in vitro and in a clot model. *Front. Microbiol.* **2018**, *9*, 326. [CrossRef] [PubMed]
14. Uchiyama, J.; Rashel, M.; Takemura, I.; Wakiguchi, H.; Matsuzaki, S. In silico and in vivo evaluation of bacteriophage phief24c, a candidate for treatment of *Enterococcus faecalis* infections. *Appl. Environ. Microbiol.* **2008**, *74*, 4149–4163. [CrossRef] [PubMed]
15. Uchiyama, J.; Takemura, I.; Satoh, M.; Kato, S.; Ujihara, T.; Akechi, K.; Matsuzaki, S.; Daibata, M. Improved adsorption of an *Enterococcus faecalis* bacteriophage phief24c with a spontaneous point mutation. *PLoS ONE* **2011**, *6*, e26648. [CrossRef] [PubMed]
16. de la Rosa, M.; Perez, M.; Carazo, C.; Pareja, L.; Peis, J.I.; Hernandez, F. New granada medium for detection and identification of group b streptococci. *J. Clin. Microbiol.* **1992**, *30*, 1019–1021. [PubMed]
17. Heelan, J.S.; Struminsky, J.; Lauro, P.; Sung, C.J. Evaluation of a new selective enrichment broth for detection of group b streptococci in pregnant women. *J. Clin. Microbiol.* **2005**, *43*, 896–897. [CrossRef] [PubMed]
18. Nasukawa, T.; Uchiyama, J.; Taharaguchi, S.; Ota, S.; Ujihara, T.; Matsuzaki, S.; Murakami, H.; Mizukami, K.; Sakaguchi, M. Virus purification by cscl density gradient using general centrifugation. *Arch. Virol.* **2017**, *162*, 3523–3528. [CrossRef] [PubMed]

19. Tanizawa, Y.; Fujisawa, T.; Nakamura, Y. Dfast: A flexible prokaryotic genome annotation pipeline for faster genome publication. *Bioinformatics* **2018**, *34*, 1037–1039. [CrossRef] [PubMed]
20. Tanizawa, Y.; Fujisawa, T.; Kaminuma, E.; Nakamura, Y.; Arita, M. Dfast and daga: Web-based integrated genome annotation tools and resources. *Biosci. Microbiota Food Health* **2016**, *35*, 173–184. [CrossRef] [PubMed]
21. Nishimura, Y.; Yoshida, T.; Kuronishi, M.; Uehara, H.; Ogata, H.; Goto, S. Viptree: The viral proteomic tree server. *Bioinformatics* **2017**, *33*, 2379–2380. [CrossRef] [PubMed]
22. Ruiz-Garbajosa, P.; Bonten, M.J.; Robinson, D.A.; Top, J.; Nallapareddy, S.R.; Torres, C.; Coque, T.M.; Cantón, R.; Baquero, F.; Murray, B.E.; et al. Multilocus sequence typing scheme for *Enterococcus faecalis* reveals hospital-adapted genetic complexes in a background of high rates of recombination. *J. Clin. Microbiol.* **2006**, *44*, 2220–2228. [CrossRef] [PubMed]
23. Jolley, K.A.; Maiden, M.C. Bigsdb: Scalable analysis of bacterial genome variation at the population level. *BMC Bioinform.* **2010**, *11*, 595. [CrossRef] [PubMed]
24. Kumar, S.; Stecher, G.; Tamura, K. Mega7: Molecular evolutionary genetics analysis version 7.0 for bigger datasets. *Mol. Biol. Evol.* **2016**, *33*, 1870–1874. [CrossRef] [PubMed]
25. Vijayakumar, P.P.; Muriana, P.M. A microplate growth inhibition assay for screening bacteriocins against *Listeria monocytogenes* to differentiate their mode-of-action. *Biomolecules* **2015**, *5*, 1178–1194. [CrossRef] [PubMed]
26. Kanda, Y. Investigation of the freely available easy-to-use software "ezr" for medical statistics. *Bone Marrow Transplant.* **2013**, *48*, 452–458. [CrossRef] [PubMed]
27. Lefkowitz, E.J.; Dempsey, D.M.; Hendrickson, R.C.; Orton, R.J.; Siddell, S.G.; Smith, D.B. Virus taxonomy: The database of the international committee on taxonomy of viruses (ictv). *Nucleic Acids Res.* **2018**, *46*, D708–D717. [CrossRef] [PubMed]
28. Klumpp, J.; Lavigne, R.; Loessner, M.J.; Ackermann, H.W. The spo1-related bacteriophages. *Arch. Virol.* **2010**, *155*, 1547–1561. [CrossRef] [PubMed]
29. Nes, I.F.; Diep, D.B.; Holo, H. Bacteriocin diversity in *streptococcus* and *enterococcus*. *J. Bacteriol.* **2007**, *189*, 1189–1198. [CrossRef] [PubMed]
30. van Houte, S.; Buckling, A.; Westra, E.R. Evolutionary ecology of prokaryotic immune mechanisms. *Microbiol. Mol. Biol. Rev.* **2016**, *80*, 745–763. [CrossRef] [PubMed]
31. Rosen, G.H.; Randis, T.M.; Desai, P.V.; Sapra, K.J.; Ma, B.; Gajer, P.; Humphrys, M.S.; Ravel, J.; Gelber, S.E.; Ratner, A.J. Group b *streptococcus* and the vaginal microbiota. *J. Infect. Dis.* **2017**, *216*, 744–751. [CrossRef] [PubMed]
32. Rick, A.M.; Aguilar, A.; Cortes, R.; Gordillo, R.; Melgar, M.; Samayoa-Reyes, G.; Frank, D.N.; Asturias, E.J. Group b streptococci colonization in pregnant guatemalan women: Prevalence, risk factors, and vaginal microbiome. *Open Forum Infect. Dis.* **2017**, *4*, ofx020. [CrossRef] [PubMed]
33. Nuriel-Ohayon, M.; Neuman, H.; Koren, O. Microbial changes during pregnancy, birth, and infancy. *Front Microbiol.* **2016**, *7*, 1031. [CrossRef] [PubMed]
34. Ando, H.; Lemire, S.; Pires, D.P.; Lu, T.K. Engineering modular viral scaffolds for targeted bacterial population editing. *Cell Syst.* **2015**, *1*, 187–196. [CrossRef] [PubMed]
35. Koskella, B.; Brockhurst, M.A. Bacteria-phage coevolution as a driver of ecological and evolutionary processes in microbial communities. *FEMS Microbiol. Rev.* **2014**, *38*, 916–931. [CrossRef] [PubMed]
36. Ackermann, H.; Tremblay, D.; Moineau, S. Long-term bacteriophage preservation. *WFCC Newsl.* **2004**, *38*, 35–40.
37. Lobocka, M.B.; Glowacka, A.; Golec, P. Methods for bacteriophage preservation. *Methods Mol. Biol.* **2018**, *1693*, 219–230. [PubMed]
38. Ulisse, S.; Peccio, A.; Orsini, G.; Di Emidio, B. A study of the shelf-life of critical culture media. *Vet. Ital.* **2006**, *42*, 237–247. [PubMed]
39. Carvalho Mda, G.; Facklam, R.; Jackson, D.; Beall, B.; McGee, L. Evaluation of three commercial broth media for pigment detection and identification of a group b *Streptococcus* (*Streptococcus agalactiae*). *J. Clin. Microbiol.* **2009**, *47*, 4161–4163. [CrossRef] [PubMed]
40. Agboluaje, M.; Sauvageau, D. Bacteriophage production in bioreactors. *Methods Mol. Biol.* **2018**, *1693*, 173–193. [PubMed]
41. Krysiak-Baltyn, K.; Martin, G.J.O.; Gras, S.L. Computational modeling of bacteriophage production for process optimization. *Methods Mol. Biol.* **2018**, *1693*, 195–218. [PubMed]
42. Krysiak-Baltyn, K.; Martin, G.J.O.; Gras, S.L. Computational modelling of large scale phage production using a two-stage batch process. *Pharmaceuticals (Basel)* **2018**, *11*, 31. [CrossRef] [PubMed]

43. Forde, A.; Hill, C. Phages of life-the path to pharma. *Br. J. Pharmacol.* **2018**, *175*, 412–418. [CrossRef] [PubMed]
44. Los, M.; Czyz, A.; Sell, E.; Wegrzyn, A.; Neubauer, P.; Wegrzyn, G. Bacteriophage contamination: Is there a simple method to reduce its deleterious effects in laboratory cultures and biotechnological factories? *J. Appl. Genet.* **2004**, *45*, 111–120. [PubMed]

12

Directed in Vitro Evolution of Therapeutic Bacteriophages: The Appelmans Protocol

Ben H. Burrowes [1,2], Ian J. Molineux [3,*] and Joe A. Fralick [1,*]

1 Department of Immunology and Molecular Microbiology, Texas Tech University Health Sciences Center, 3601 4th Street, Lubbock, TX 79430, USA; benburrowes@live.com
2 Roche Molecular Systems, 983 University Avenue B200, Los Gatos, CA 95032, USA
3 Center for Infectious Disease, Department of Molecular Biosciences, The University of Texas at Austin, 1 University Station A5000, Austin, TX 78712, USA
* Correspondence: molineux@austin.utexas.edu (I.J.M.); joe.fralick@ttuhsc.edu (J.A.F.)

Abstract: The 'Appelmans protocol' is used by Eastern European researchers to generate therapeutic phages with novel lytic host ranges. Phage cocktails are iteratively grown on a suite of mostly refractory bacterial isolates until the evolved cocktail can lyse the phage-resistant strains. To study this process, we developed a modified protocol using a cocktail of three *Pseudomonas* phages and a suite of eight phage-resistant (including a common laboratory strain) and two phage-sensitive *Pseudomona aeruginosa* strains. After 30 rounds of selection, phages were isolated from the evolved cocktail with greatly increased host range. Control experiments with individual phages showed little host-range expansion, and genomic analysis of one of the broad-host-range output phages showed its recombinatorial origin, suggesting that the protocol works predominantly via recombination between phages. The Appelmans protocol may be useful for evolving therapeutic phage cocktails as required from well-defined precursor phages.

Keywords: Appelmans; bacteriophage evolution; bacteriophage recombination; phage therapy; *Pseudomonas aeruginosa*; antibiotic resistance

---

1. Introduction

The emergence and increasing prevalence of bacterial strains that are resistant to available antibiotics poses a serious threat to world health [1], which, according to the World Health Organization (WHO), is heading toward a post-antibiotic era when many common infections will no longer have a cure [2]. This imminent threat demands an evaluation of novel approaches toward treating antibiotic-refractory infections. One such approach that is being re-examined is bacteriophage (phage) therapy, the use of bacterial viruses to specifically target and clear bacterial infections [3–6]. The National Institute of Allergy and Infectious Diseases included phage therapy as one of seven strategic approaches to combat antimicrobial resistance [7]. Bacteria and their phages have been evolving for over three billion years, and although bacteria can become resistant to phages, the latter, unlike antibiotics, can also evolve to infect otherwise resistant bacteria. For phages to be successful therapeutically, it may be necessary to continually isolate new phages that infect refractory bacterial strains. Therapeutic phages are usually isolated from the environment using sources such as sewage, marine and fresh water, or even patient samples [8]. While such sources can offer a great diversity of phages from which to select, it is not always possible or efficient to isolate phages with the most clinically relevant host range.

One approach successfully used in the Republic of Georgia is to prepare large cocktails of phages that target a broad range of pathogenic bacteria [9]. When a pathogen is encountered that is refractory to the cocktail, a new phage that can grow on the resistant pathogen is added, or what is referred to as the "Appelmans protocol" is employed to "invigorate" the cocktail (pers. comm. Dr Z. Alavidze, see Appendix A). This protocol is based on an empirical liquid method of phage titration developed in the 1920s by Appelmans [9,10]. Despite its history and common use by several laboratories in countries of the former Soviet Union and Europe, the protocol has only recently been reported in Western journals [11], and the mechanism(s) by which it works has not been determined.

For a mixture of phages, two major genetic mechanisms by which one or more phages adapt to a bacterium that was refractory to the original cocktail are spontaneous mutation and recombination. Phage genomes are architecturally mosaics, with each individual genome representing a unique assemblage of individual exchangeable modules [12–15]. Mechanisms for generating such mosaics include homologous recombination at shared boundary sequences of modular junctions, illegitimate recombination in a non-specific sequence-directed process, and site-specific recombination [13,16,17]. A cocktail of phages applied to host strains susceptible to several members of the mixture offers an opportunity for recombination events that can generate more genetic diversity in the phage pool.

In the contemporary laboratory, host-range mutants are usually isolated using a single phage, either through spontaneous or induced mutations by selecting for those that can grow on the resistant bacterial strain, which often map to the tail fiber genes (e.g. [16,18–23]). However, in the Appelmans protocol, a phage cocktail consisting of multiple phages is grown iteratively on multiple separate hosts, including resistant bacteria, until the cocktail evolves the ability to lyse the entire culture. The bacterial strains are maintained from their original stocks and are kept separate throughout. Any mutant phage with the ability to propagate on a strain that is refractory to other members of the mixture will be selected as phages are pooled at the end of each round of selection.

As a proof-of-principle experiment aimed at examining the mechanism of host-range expansion, we designed a 96-well plate format Appelmans protocol using a three-phage cocktail to which a suite of seven clinical isolates were initially refractory by plaque analysis. After 30 rounds of selection, an individual phage was isolated that could grow on all seven clinical isolates. Multiple recombination events between two phages in the cocktail were seen, at least some of which likely conferred an expanded host range. Our results describe a relatively simple, straightforward method by which a phage cocktail can expand its host range without the addition of new genetic information.

2. Materials and Methods

2.1. Bacterial and Bacteriophage Strains

All bacterial and bacteriophage strains used are listed in Table 1. The phages chosen for the protocol were ΦKZ, a large, virulent member of the *Myoviridae* family [24,25], Pa2, shown here to be an *N4virus* and used in our laboratory in previous in vivo studies of phage therapy [26], and phage RWG, also shown here to be an *N4virus,* which was isolated in our laboratory from a patient wound swab.

The bacterial strains included three common laboratory strains, PAO1, PA14 [27], and PAK [28]. PAO1 and PAK were kind gifts from Dr A. N. Hamood, Texas Tech University Health Sciences Center (TTUHSC); PA14 was generously donated by Dr K. Rumbaugh, TTUHSC. Seven clinical *Pseudomonas aeruginosa* strains were isolated from wound swabs within a six month period from patients at the Southwest Regional Wound Care Center (WCC), Lubbock, Texas. All isolates were from separate patients and showed high degrees of antibiotic resistance but with distinct antibiograms. These isolates are identified by the prefix WCC accompanied by a three-digit identifier (see Table 1). PAO1 was the fully permissive host used to propagate all *P. aeruginosa* phages.

Bacterial strains were grown in LB (5 g/L NaCl (Sigma Aldrich, St. Louis, MO, USA), 5 g/L yeast extract (BD, Franklin Lakes, NJ, USA), 10 g/L tryptone (BD, Franklin Lakes, NJ, USA)) supplemented with 1 mM $CaCl_2$ (Sigma Aldrich, St. Louis, MO, USA) and 1 mM $MgSO_4$ (Sigma Aldrich, St. Louis, MO, USA). Phages RWG and Pa2 were grown by infecting PAO1 at an A_{600} ~0.2, using an MOI ~0.1. After 3–5 h incubation at 37 °C, 250 rpm, the cultures were seen to lyse, and the phage was harvested by adding 1:100 $CHCl_3$ (Sigma Aldrich, St. Louis, MO, USA), vortexing vigorously, and then centrifuging (≥18,000× *g*, 5 min, 4 °C) to remove cellular debris. The supernatants were stored over $CHCl_3$ at 4 °C. ΦKZ was grown on semi-solid media as described elsewhere [25].

Table 1. *Pseudomonas aeruginosa* and bacteriophage strains used for the Appelmans protocol. Plaque morphologies are described for growth on PAO1.

Bacterial/Phage Strain	Source	Reference	Notes
PAO1	Dr A. N. Hamood (TTUHSC)	[1]	Host strains used for the Appelmans protocol
PAK	Dr A. N. Hamood (TTUHSC)	[2]	Host strains used for the Appelmans protocol
PA14	Dr K. Rumbaugh (TTUHSC)	[3]	Host strains used for the Appelmans protocol
WCC176	Dr R. D. Wolcott MD Southwest Regional Wound Care Center. (Lubbock, TX)	This study	Host strains used for the Appelmans protocol
WCC199	Dr R. D. Wolcott MD Southwest Regional Wound Care Center. (Lubbock, TX)	This study	Host strains used for the Appelmans protocol
WCC201	Dr R. D. Wolcott MD Southwest Regional Wound Care Center. (Lubbock, TX)	This study	Host strains used for the Appelmans protocol
WCC205	Dr R. D. Wolcott MD Southwest Regional Wound Care Center. (Lubbock, TX)	This study	Host strains used for the Appelmans protocol
WCC222	Dr R. D. Wolcott MD Southwest Regional Wound Care Center. (Lubbock, TX)	This study	Host strains used for the Appelmans protocol
WCC229	Dr R. D. Wolcott MD Southwest Regional Wound Care Center. (Lubbock, TX)	This study	Host strains used for the Appelmans protocol
WCC232	Dr R. D. Wolcott MD Southwest Regional Wound Care Center. (Lubbock, TX)	This study	Host strains used for the Appelmans protocol
Pa2	ATCC [14203-B1]	[1]	73,008 bp. Plaques 1–2 mm, clear
ΦKZ	Felix d'Herelle Reference Center for Viruses	[4]	280,334 bp. Plaques <1 mm, clear
RWG	This laboratory	This study	72,646 bp. Plaques 2–3 mm, clear

Where required, phage stocks were concentrated by the addition of 1/5 volume of 25% polyethylene glycol (PEG) M.W. 8000 (Sigma Aldrich, St. Louis, MO, USA), 2.5 M NaCl after incubation overnight at 4 °C. After centrifugation at 15,000× *g* for 15 min, the pellet was suspended in SM buffer (50 mM Tris-HCl [pH 7.5] (Sigma Aldrich, St. Louis, MO, USA), 0.1 M NaCl, 8 mM $MgSO_4$, 0.01% *w*/*v* gelatin (Sigma Aldrich, St. Louis, MO, USA)). PEG precipitation was performed twice to give a final purified phage suspension [29]. Phage titers were assayed by the agar overlay method [30].

2.2. Analysis of Host Range

The Appelmans protocol was carried out as described in Results. Every 10 rounds, the pooled lysates were checked for novel phages by streaking a loopful (~1 μL) onto an LB agar plate and overlaying with ~10^7 exponential-phase bacteria in top agar (LB plus 5 g/L agar). Five plaques were picked on each bacterial strain, choosing distinct plaque morphologies where possible, and were purified on the same strain. After at least three rounds of purification, the plaques were picked with a sterile Pasteur pipet and added to 2 mL early log-phase cells in supplemented LB. After lysis, or after 5 h incubation (whichever came first), the bacteria were killed by the addition of 20 μL $CHCl_3$, and cell debris was removed by centrifugation at 15,000× *g* for 10 min. Most (38/50) phages isolated after the final round of Appelmans selection gave only faint, diffuse plaques and poor titers after purification and amplification on either the isolation strain or on PAO1; these phages were not studied further.

Host range was initially assessed by placing a 10 μL drop of phage suspension (≥10^6 plaque-forming units, pfu) onto a pre-seeded lawn of host cells in top agar [30]. After overnight incubation at 37 °C, zones of lysis or the presence of plaques indicated that the bacterial strain was susceptible to the phage. The lytic host range of phage suspensions was further confirmed by titering to observe individual plaques.

2.3. DNA Sequencing and Annotation

The phages Pa2, RWG, and phi176 were purified by equilibrium density gradient centrifugation in CsCl [29]. Genomic DNA was isolated by extraction with phenol [31] and was subjected to 454 pyrosequencing by the University of Texas at Austin genomics core facility. Assembly used Newbler 2.6 and DNAStar software, followed by manual inspection of read frequency that revealed the presence of terminal repeats. These were taken as the physical genome ends but they were not formally established as such experimentally. Alignment of the Pa2, RWG, and phi176 genomes was performed using the NCBI Blast suite (https://blast.ncbi.nlm.nih.gov/Blast.cgi) and DNAStar. Genome sequences have been deposited in GenBank with Accession Numbers KM411958 (RWG), KM411959 (Pa2), and KM411960 (phi176).

The DNA sequences of phi176, Pa2, and RWG were annotated using the RAST (Rapid Annotation using Subsystem Technology) annotation service available at http://rast.nmpdr.org/ and manually. Differences were resolved by comparison to the previously annotated, closely related phage genomes PA26 (JX194238) and LIT1 (NC_013692) and, post facto, by internal comparisons. Open reading frames (ORFs) were analyzed for putative function using the Translated BLAST (blastx) and Protein BLAST (blastp) algorithms, searching the non-redundant protein sequences database.

2.4. Mitomycin C Induction and Southern Blot Analysis of Temperate Phages

Bacterial strains were grown to an A_{600} ~0.5 in LB before adding 4 μg/mL mitomycin C and incubating overnight. The cultures were then centrifuged and filtered through a 0.2 μM membrane. The resulting lysate was DNase I-treated prior to DNA extraction. Output phage DNA was digested with HincII and EcoRV (NEB), electrophoresed, and hybridized to a nitrocellulose membrane. Probes were prepared with temperate phage DNA using the ThermoFisher North2South Random Prime DNA Biotinylation Kit. Hybridization was detected using the ThermoFisher North2South Chemiluminescent Hybridization and Detection Kit.

3. Results

3.1. Appelmans ProtocolUusing a Phage Cocktail Expands Bacteriophage Host Ranges

In order to generate novel phages from our laboratory strains, we developed a protocol based on that used by the George Eliava Institute of Bacteriophage, Microbiology and Virology (IBMV), Tbilisi, Georgia (see Appendix A). The protocol is represented schematically in Figure 1. Phages Pa2, RWG, and ΦKZ were combined 1:1:1 by titer to yield an input cocktail of 1×10^{10} pfu/mL. Using a 96-well microtiter plate, 100 μL of serial 10-fold dilutions (10^0 to 10^{-9}) of the cocktail were added to 100 μL of double-strength LB containing 1 μL of an overnight bacterial culture of a single strain. Eight bacterial strains can be tested on each plate. One well was used for each dilution plus a control well with no phage. After overnight incubation at 37 °C on a shaking platform at 200 rpm, the plates were visually inspected. Wells showing complete lysis, plus the first turbid well, were pooled. If no lysis was seen, the well containing the undiluted phage cocktail was harvested. Pooled lysates were cleared by vortex mixing with 1:100 $CHCl_3$, followed by centrifugation (15,000× *g* for 15 min). The lysate was termed the round 1 cocktail and was used to initiate the next round of the protocol using the same set of bacterial strains.

In the original Appelmans protocol the endpoint is reached when the output cocktail lyses >80% of the host strains at a minimum dilution of 10^{-7} (Appendix A). This cocktail is then usually used therapeutically. However, we were interested in analyzing the phages in the developing cocktail and we added additional steps in order to allow the analysis of individual phages present every 10 rounds of evolution. Table 2 shows that after 30 rounds, the evolved phage population generated plaques on all 10 *P. aeruginosa* strains, whereas the parent phages were only able to plaque on two laboratory strains.

The round-30 cocktail was plated onto each of the 10 bacterial strains used for the Appelmans protocol; five phages were isolated on each bacterial strain. For each plaque on a particular host, the same strain was used for further phage purification. Of the 15 plaques isolated on the laboratory *P. aeruginosa* strains PAO1, PA14, and PAK, eight phages were successfully amplified to high titer ($\geq 10^8$ pfu/mL) on their isolation strain. Only 4 of the 35 plaques isolated on the WCC clinical strains contained phages that grew to high titers in liquid culture, either on their isolation strain or on PAO1, so only these were used for further study. Table 3 shows the host range of individual phages isolated from the round-30 cocktail. Phi229.2, phi229.1, and phi176 had a relatively broad host range that encompass, respectively, 80%, 90%, and 100% of the tested strains. Because of resource constraints, we obtained sequence data only for the parental strains and phi176; the genetic origins of phages phi229.1 and phi229.2 are therefore unknown.

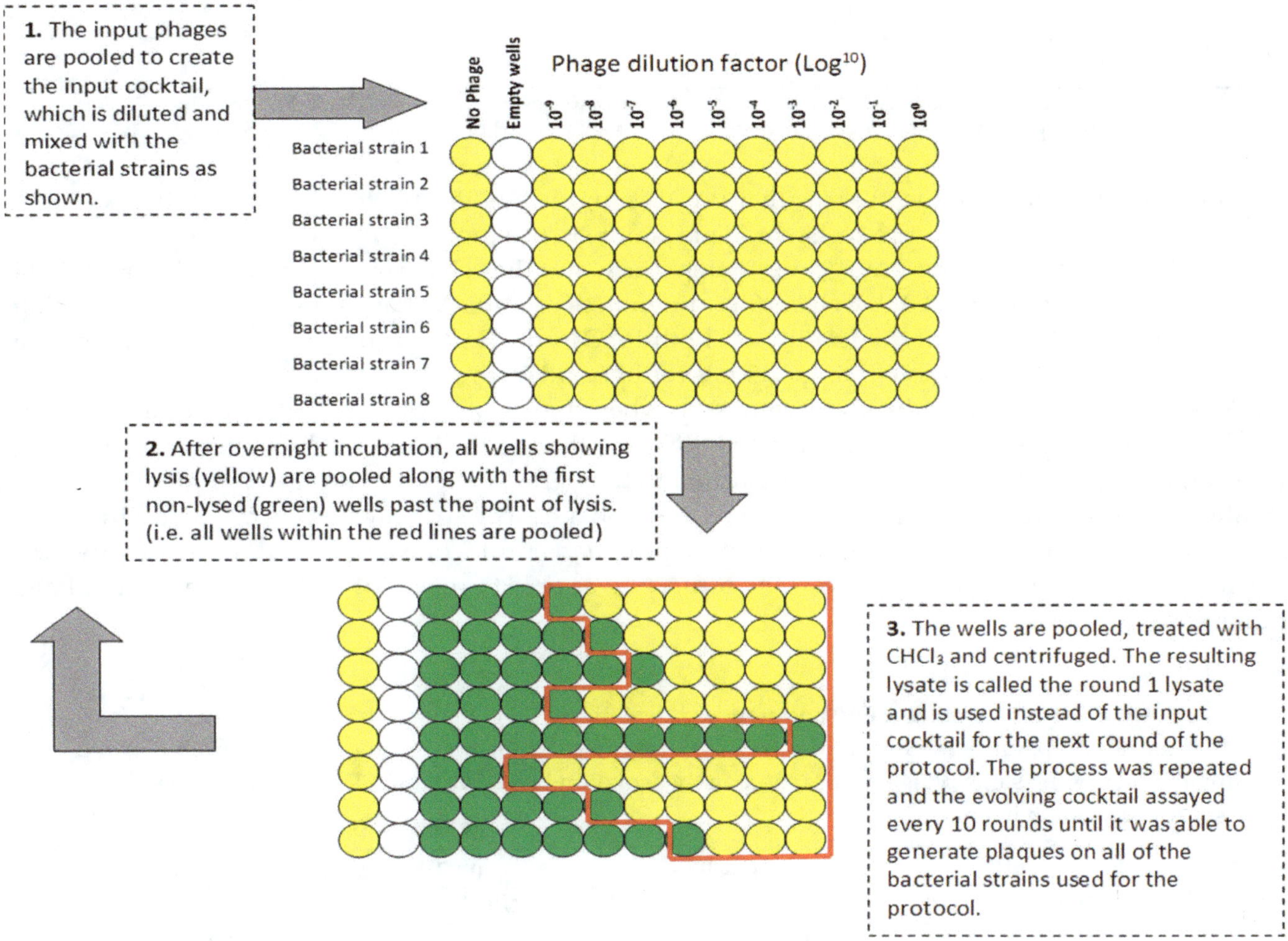

Figure 1. Schematic representation of the 96-well-plate-format Appelmans protocol.

Table 2. Phage host range. A 10 µL spot of a lysate containing $\geq 10^6$ plaque-forming units (pfu) was placed onto an overlay lawn seeded with ~10^7 colony-forming units (cfu) of *P. aeruginosa*. "+" indicates visible lysis after 16 h incubation at 37 °C, no entry indicates no visible lysis.

Test Strain	Phage					
	Pa2	ΦKZ	RWG	Round 10 Cocktail	Round 20 Cocktail	Round 30 Cocktail
PAO1	+	+	+	+	+	+
PA14				+	+	+
PAK	+	+		+	+	+
WCC176					+	+
WCC199				+	+	+
WCC201				+	+	+
WCC205				+	+	+
WCC222						+
WCC229						+
WCC232						+

Table 3. Host range of individual phages isolated from the round-30 cocktail.

Test Strain	Phage [1]											
	phiPAO1.1	**phiPAO1.2**	**phiPAO1.3**	**phiPA14.1**	**phiPA14.2**	**phiPAK.1**	**phiPAK.2**	**phiPAK.3**	**phi176**	**phi201**	**phi229.1**	**phi229.2**
PAO1	+	+	+	+		+	+	+	+	+	+	+
PA14				+	+				+	+	+	+
PAK	+	+	+	+		+	+	+	+	+	+	+
WCC176									+		+	+
WCC199									+		+	
WCC201			+	+					+	+		
WCC205									+		+	+
WCC222									+		+	+
WCC229									+		+	+
WCC232									+		+	+

[1] Phages are named with the prefix 'phi' followed by the isolation strain (see Table 1) and by the phage isolate number where appropriate (e.g., phi229.2 was the second phage isolated on strain WCC229).

Plating efficiencies (Table 4) show that PAO1 is highly susceptible to all phages, whereas PA14, PAK, and WCC201 are far more resistant. This is not surprising, as Pa2 and RWG (and ΦKZ) had been routinely propagated on PAO1 prior to this study, and the cell surfaces of PAO1, PAK, and PA14 are known to be different [32,33]. Cell surface differences can affect phage sensitivity. It is interesting that phi201 plated reasonably efficiently on both PAO1 and PAK but poorly on PA14, suggesting that the phages within the developing cocktail could be adapting to different hosts. In several cases, efficiency of plating (EOP) values >>1 were seen when the isolated phages were titered on non-host strains. These phages were better adapted to the hosts than to the host on which they were isolated, yielding the highest EOP. In other words, although they grew on a given strain, they grew better on an alternative host (usually PAO1). Although these phages were propagated primarily on the more permissive host, their maintenance during the Appelmans protocol was also selected for by the less permissive host(s) present. In this way, the protocol was able to select for and propagate rare phage mutants as they arose.

Table 4. Efficiency of plating (EOP) and standard deviations (SD) of phages isolated on clinical isolates. '-' indicates no plaques (<10 pfu/mL). EOP is determined as the phage titer on a test strain divided by the titer of the same phage preparation on the isolation host. EOP and SD values on the isolation strains are shown in bold.

Test Strain	phi176		phi201		phi229.1		phi229.2	
	EOP	SD	EOP	SD	EOP	SD	EOP	SD
PAO1	73.6	5.9	0.7	0.1	6.7	2.0	51.2	11.0
PA14	9.6×10^{-5}	5.4×10^{-4}	1.0×10^{-9}	2.4×10^{-8}	8.3×10^{-5}	4.6×10^{-5}	1.1×10^{-3}	4.2×10^{-4}
PAK	2.7×10^{-5}	2.0×10^{-6}	0.1	1.8×10^{-2}	1.3×10^{-6}	4.6×10^{-7}	2.0×10^{-4}	4.6×10^{-5}
WCC176	**1.0**	**0.1**	-	-	1.5	0.5	7.9	2.7
WCC199	4.3	1.1	-	-	3.3×10^{-2}	1.2×10^{-2}	-	-
WCC201	1.8×10^{-6}	1.2×10^{-6}	**1.0**	**0.2**	-	-	-	-
WCC205	3.3	0.4	-	-	1.4	0.5	2.6	0.6
WCC222	18.2	1.6	-	-	1.1	0.3	2.6	0.8
WCC229	50.0	6.9	-	-	**1.0**	**0.4**	**1.0**	**0.3**
WCC232	39.3	6.7	-	-	0.1	4.7×10^{-2}	7.9	1.9

3.2. *Appelmans Protocol Using Single Phages Generates Little Host-Range Expansion*

When a single phage was used for the Appelmans protocol, little expansion in host range was seen (Supplementary Table S1A–C). Only phage RWG adapted to utilize a new host, and the total host-range expansion of all three phages separately was far less than that generated when the phages were combined into a cocktail. We saw no plaques from the single-phage Appelmans protocol experiments on most clinical *P. aeruginosa* isolates. Thus, although each individual phage was under continuous selection to grow on a mixture of different hosts, spontaneous mutations alone were insufficient to generate much change in host range.

3.3. Phi176 is a Recombinant Derivative of Phages Pa2 and RWG

At the outset of this study, we had no information on phage RWG and limited data on Pa2. As a preliminary check to determine whether recombination with endogenous genetic material (e.g., prophage or bacterial genes) was likely to contribute to phage evolution, we used Southern analyses of Pa2, RWG, and ΦKZ DNA with phage DNA isolated after mitomycin C induction of all bacterial strains. No hybridization was detected.

The genome of ΦKZ is known (GenBank: AF399011.1), but the genomes of phi176, Pa2, and RWG were sequenced for this study. Phages Pa2 and RWG are closely related, showing an overall nucleotide sequence identity of >99%. Both phages are also closely related to phages PA26 (JX194238, 97% identity), LIT1 (NC_013692, 97% identity), vB_PaeP_C2-10_Ab09 (GenBank: NC_024140.1, 98% identity), and PEV2 (NC_031063.1, 98% identity). Pa2 and RWG are therefore *N4virus Podoviridae* [34,35].

The genome of phi176 (GenBank: KM411960.1) is derived exclusively from phages Pa2 (GenBank: KM411959.1) and RWG (GenBank: KM411958.1), with no ΦKZ-derived sequences. Remarkably, only a single missense mutation—a residue not found at the corresponding position in either Pa2 or RWG—was definitively identified across the phi176 genome. The absence of point mutations was unexpected because of the extensive phage growth (~150 cycles) that took place during the Appelmans protocol—a conservative estimate is five rounds of phage replication for each round of the protocol (assuming a burst size of 50 pfu/cell, a single phage will be amplified to $\sim 3 \times 10^8$ pfu in five rounds).

The Pa2, RWG, and phi176 genome sequences were aligned, allowing a visual determination that at least 48 crossovers occurred during the overall Appelmans procedure to yield phi176 (Figure 2 and Table 5). The terminal repeats of Pa2 and phi176 are the same, and the left (and right) genetic end of phi176 is thus derived from Pa2. Working in base-pair order (i.e., left to right), the first crossover from Pa2 DNA into RWG in generating phi176 is predicted by the last identical nucleotide to Pa2; this process was then repeated using the last position of identity on each respective genome over the complete sequence. Overall, about 2/3 of the phi176 genome is of Pa2 heritage. In Table 5, we named the gene by the crossover site predicted, but obviously, it is only known that recombination includes that site and not exactly where it was initiated or terminated. The phi176 sequence differed from both Pa2 and RWG at only five sites. One, at position 30,110 (phi176 ORF52) was a spontaneous missense mutation, but the remaining four (after phi176 positions 9355, 28671, 29,415, and 59342) occurred at homopolymeric runs and were judged more likely to be sequencing errors than mutational events. Two of the four affect intergenic regions, whereas the 151 amino acid phi176 ORF48 is only homologous to Pa2 and RWG for its initial 71 residues, and the 82 amino acid phi176 ORF50 differs in its last eight residues from both of its parents. The corresponding protein pairs from Pa2 and RWG were 100% identical and had no known or predicted function, supporting our premise of sequencing errors.

Forty-eight recombination events is a lower bound because, unless the recombination led to a change in sequence, it would go undetected; in addition, some recombination events might not have survived through all 30 rounds of the protocol. The majority (63/92) of ORFs underwent no crossovers between Pa2 and RWG in generating phi176. This was not unexpected because most genes are not directly involved in determining the host range. One recombination event was detected in 22 of the 29 recombinant ORFs, and one altered the ORF44/45 intergenic region (Table 5). Three ORFs underwent two crossovers, and a further three ORFs underwent three, whereas five and six recombination events occurred in ORF43 and ORF52, respectively. ORF52 also harbors the spontaneous missense mutation.

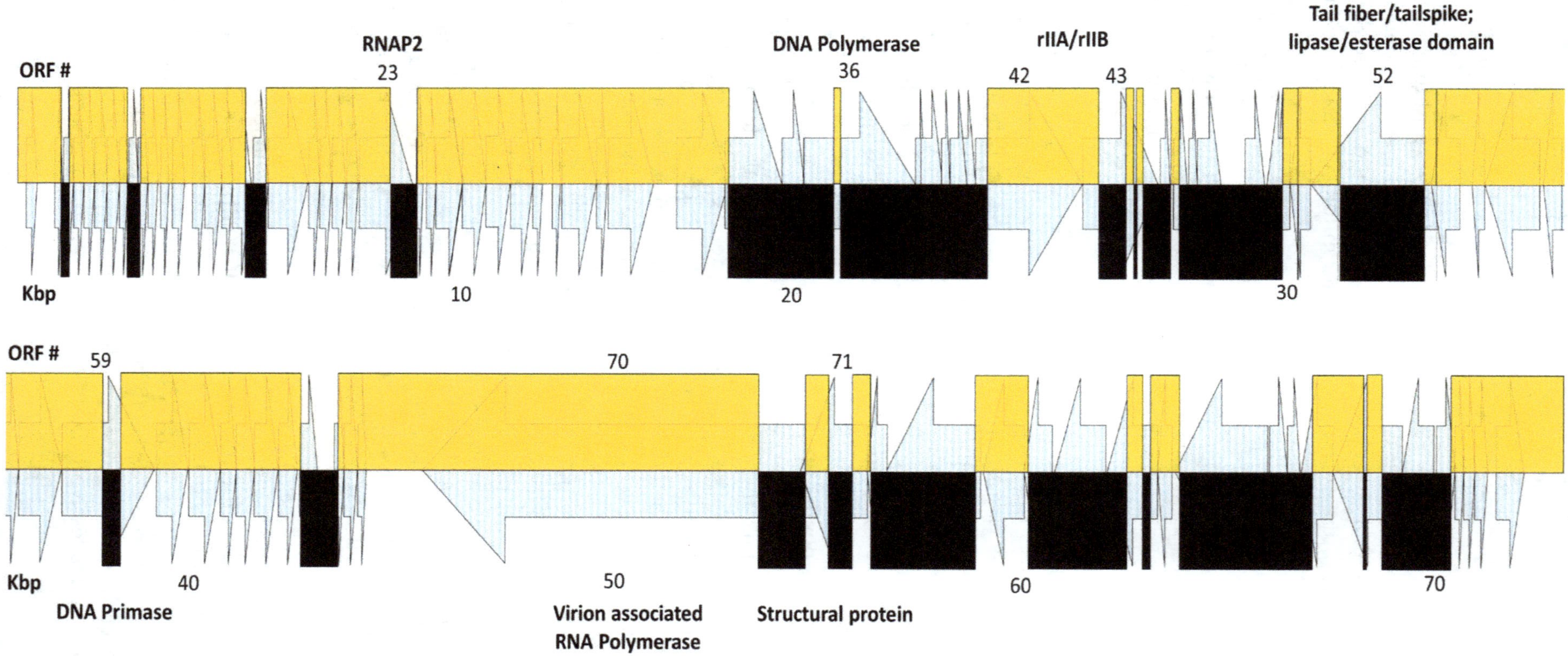

Figure 2. Schematic representation of the phi176 genome and its recombinatorial derivation. Open reading frame (ORF) numbers are shown at the top, and genome position (Kbp) at the bottom. Arrows represent individual ORFs. Yellow boxes denote genetic material of Pa2 origin, and black boxes represent RWG-derived sequences in the recombinant phi176 genome. The putative function of key recombinant genes (Table 5) are shown in bold above and below the map.

Table 5. Recombinant phi176 ORFs.

Phi176 ORF	Start	Stop	ORF Size (bp)	Cross-Overs	Crossover Location(s) in phi176 Genome (bp) [1]	Putative Gene Function
02	872	1063	192	1	1020	
03	1078	1308	231	1	1210	
08	2394	2597	204	1	2581	
10	2832	3065	234	1	2884	
17	5053	5493	441	1	5352	
18	5533	5880	348	1	5840	
23	8042	9283	1242	1	8764	RNA polymerase 2
23a	9380	9532	153	1	9384	
35	16,639	17,976	1167	1	16,721	DNA helicase
37	18,503	21,118	2616	2	19,203 19,359	DNA polymerase
42	22,534	25,050	2517	1	22,811	rIIA-like
43	25,062	26,840	1779	5	25,434 26,071 26,254 26,321 26,473	rIIB-like
IG44-45	N/A	N/A	N/A	1	27,137	N/A
45	27,196	27,513	318	1	27,326	
51	29,943	29,464	480	1	29,758	
52	33,215	29,943	3273	6 (+ 1 missense)	30,115 30,122 31,070 31,073 31,118 33,095	Tail fiber/tailspike; lipase/esterase domain
53	33,925	33,254	672	2	33,375 33,379	
59	37,720	39,888	2169	2	38,669 39,095	DNA primase P4 type
66	43,308	43,712	405	1	43,324	
67	44,118	44,354	237	1	44,210	
70	55,093	44,897	10,197	1	54,100	Virion RNA polymerase
71	56,659	55,094	1566	3	55,210 55,751 56,313	Structural protein
72	57,126	56,659	468	1	56,747	
73	59,329	57,107	2223	1	59,203	
75	61,019	60,354	666	1	60,455	
77	63,503	62,310	1194	3	62,768 63,143 63,333	
79	66,091	63,911	2181	1	64,015	
82	67,614	66,880	735	1	67,134	
83	69,263	67,611	1653	3	68,333 68,417 68,774	
85	70,055	70,486	432	1	70,398	

[1] The starting 1–1020 bp and final 70,399–73,050 bp (including the terminal repeats) are derived from parental phage Pa2.

The frequencies of recombination within a given gene are not stochastic. Most ORFs undergoing multiple crossovers are somewhat larger than the average gene length of ~740 bp, but the putative phi176 virion RNA polymerase (ORF70), which is by far the largest gene at 10,197 bp, almost completely retained its Pa2 ancestry, undergoing only a single crossover across its length, despite ~275 polymorphisms with respect to its RWG counterpart. Retention of Pa2 DNA sequences in ORF70 is unlikely to be due to incompatibility of amino acid substitutions, as only 29 polymorphisms result in amino acid substitutions. Figure 2 shows the composition of the phi176 genome as recombinatorial crossovers between the Pa2 and RWG genomes, and Table 5 lists the Pa2 genes in which recombination occurred, generating phi176. Note that the phi176 rIIB-like protein ORF43 contained only two distinct amino acid substitutions, whereas six recombination events were detected within the gene; similarly, only four of six cross-overs in the phi176 ORF52 tailspike gene gave rise to amino acid substitutions. Recombination itself is of course totally distinct from any selective change conferred on an ORF by an amino acid substitution. However, it cannot be determined whether recombination events that did not lead to a change in an ORF were simply that, or whether they were fossils from prior events during the evolution of phi176.

4. Discussion

A cocktail comprised of three distinct phages was evolved over 30 rounds of the Appelmans protocol on a suite of seven clinical and three laboratory strains of *P. aeruginosa*. Of these 10 isolates, only two of the laboratory strains were sensitive to the original phages. When the protocol was carried out with each phage separately, very little expansion of host range was seen. However, after 30 rounds of the protocol with three input phages, progenies were isolated with expanded host ranges that included most (and in one case all) of the 10 bacterial strains. Genetic analysis of the phage with broadest host range showed it to be a recombinant derivative of the two most closely related members of the input cocktail: RWG and Pa2. These results support our premise that the Appelmans protocol is able to rapidly extend the host range of a cocktail of phages via recombination between the members of the cocktail.

The recombinant phage phi176 was essentially a result of multiple ($\geq$48) recombination events between Pa2 and RWG, with only a single spontaneous point mutation. Given that ΦKZ is unrelated to Pa2 and that Pa2 and RWG share >99% DNA sequence identity, it is not surprising that no ΦKZ sequences were found in the phi176 recombinant. Although recombination between very different phages has clearly occurred in the environment (e.g., between the λ, P2, and T4 side-tail fibers [36], also see [12–15]), a short-term laboratory experiment provides little opportunity for recombination that does not involve significant DNA homology.

With so many recombination events occurring in genes of unknown function, it is impossible to elucidate exactly which of those event(s) were crucial in generating the host range of phi176. However, recombination clearly occurred in some genes likely to affect host range. The first step in infection is adsorption, which must involve structural gene products (e.g., tail fibers or tailspikes) that recognize a receptor on the surface of a bacterial cell. Two possible such gene products are encoded by phi176 ORF71 and ORF52. Although ORF71, which is a homologue of LIT1 gp72 [34], underwent three recombination events, it only differed from its Pa2 and RWG parents by a single amino acid residue. However, that was enough to confer host-range differences in both phages and eukaryotic viruses [19,21]. Phi176 ORF52, on the other hand, underwent six recombination events and additionally acquired a single point mutation. Furthermore, two of these recombinations resulted in a template switch for only three or seven contiguous nucleotides and hence were likely the end result of multiple recombination events in this region. Phi176 ORF52 differed from its parental sequences by 17 (RWG) and 75 (Pa2) amino acids. ORF52 is similar to LUZ7 gp56, or a hybrid of LIT1 gp52 and gp53, which have been tentatively identified as phage tail fibers [34]. However, as the protein sequences contain lipase and esterase motifs, they are more likely to be tailspikes. Phi176 ORF52 was therefore under strong selection for adaptation during growth on different host strains in the Appelmans protocol.

Interestingly, phi176 ORF55, which putatively encodes a tail fiber, is not altered from its Pa2 ancestor. Perhaps, ORF55 is not required for phage adsorption on the strains tested here.

Resistance to phage is often achieved by cell surface receptor changes. Phages often respond by altering their receptor-binding protein(s), usually a tail fiber or tailspike. The most dramatic example is the accumulation of mutations found in Ox2 when selected to grow on a series of resistant strains. The tail fiber changed its receptor from OmpA, through other Omps, to LPS [22,23]. However, bacteria have several anti-phage defense mechanisms that operate after phage adsorption. These include CRISPR/Cas, restriction–modification systems, and other abortive infection mechanisms [37–39]. Some of these defenses are coded by prophage genes, including the exclusion of T4 *r*II mutants after infection of a *rex*+ λ lysogen [40]. T4 *r* mutants were originally isolated as plaque morphology variants on *Escherichia coli* B strains [41], which had lost the lysis inhibition characteristic of $T4^{+}$ and thus exhibited a rapid-lysis phenotype [42]. However, this phenotype is not maintained in the non-lysogenic *E. coli* K-12 strains where plaques of *r*II mutants have wild-type morphology, and lysis inhibition is close to normal [40,43]. The rII proteins were originally only associated with T-even phages, but DNA sequencing has revealed similar proteins in the genomes of many different phage types, including the *N4virus* lineage and thus Pa2, RWG, and phi176, the phages studied here. It is important to note that only the roles of T4 *r*IIA and *r*IIB in overcoming the inhibitory effects of λ prophage *rexA* and *rexB* on T4 growth and the association of these genes with rapid lysis in *E. coli* B have been studied experimentally. Lysis inhibition, overcoming prophage- or plasmid-mediated exclusion or other functions of sequence homologues in other phages has not been directly demonstrated. Nevertheless, effects on lysis inhibition can alter plaquing host range because mechanisms that increase the rate of viral propagation compared to bacterial propagation, such as increasing burst size or reducing lysis time [37,44], can all increase fitness in vitro [45] and can lead to plaques from lysis-inhibition mutants when compared to their non-plaquing lysis-inhibited parents.

A single recombination event was seen in ORF42, a putative *r*IIA gene. The N-terminal halves of Pa2 and RWG ORF42 are very similar, and recombination between the two respective genes resulted in the phi176 ORF42 having just two amino acid changes from Pa2. The C-terminal half of phi176 ORF42 is derived entirely from Pa2, with 120 amino acids differing from its RWG counterpart. However, five recombination events took place in generating phi176 ORF43, a putative *r*IIB-like gene. The chimeric phi176 rIIB-like protein has 11 and 27 amino acid changes from ORF43 of RWG and Pa2, respectively. ORF43 was clearly under strong selection during adaptation in the Appelmans experiment to expand phage host range. By analogy with the T4 rII proteins, we suggest that phi176 ORFs 42 and 43 could counteract unknown inhibitory genes in the various clinical isolates that allowed for a phi176 host-range expansion, relative to either of its parents.

Directed evolution of phages is a common laboratory approach that has been used, among other things, to develop phages with enhanced therapeutic potential [46] and to study and modify phage host range [20,47–49]. We suggest that several factors appear to be crucial for the rapid generation of expanded host range phages with the Appelmans protocol. First, the use of a cocktail of virulent phages is critical to allow recombination to generate diversity within the protocol. When we isolated individual phages from the round-30 cocktail, those selected on our laboratory strains showed very little expansion of host range. The three phages with broadest host range we identified were isolated using bacterial strains refractory to the round-10 cocktail. However, when these broad-host-range phages were tested against a new set of urinary tract infection isolates of *P. aeruginosa*, isolated in a separate laboratory three years after the WCC isolates were obtained, no collateral expansion of host range was seen. Therefore, it appears that the use of both clinically relevant and up–to-date bacterial isolates that are resistant to the phages of the cocktail is important in generating therapeutically useful phages. It is noteworthy that the Eliava IBMV regularly adds new bacterial isolates to their Appelmans panel, and the phages generated are added to their complex and evolving therapeutic preparations (Pers. Comm., Z. Alavidze). We also believe it is important to include a bacterial host that can support lytic growth of all or most of the phages of the cocktail, thereby enabling recombination after co-infection. In contrast, collateral host-range expansion was maintained when a similar protocol was carried out by Mapes et al. [11], who saw an increase in host range on a set bacterial isolates that had not been used for phage evolution. It is not clear why the two studies arrived at different conclusions, and clearly, further work is needed both to optimize the Appelmans protocol and to better understand how to generate clinically useful phages efficiently.

Phage evolution is postulated to occur principally through recombination [12,13,15,17,50–54], and recombination has been demonstrated between the members of an experimental therapeutic cocktail of virulent *E. coli* phages in an anaerobic, continuous culture system [55]. Therefore, we consider it likely that phage evolution itself occurs predominantly through what is essentially the Appelmans protocol writ large. Indeed, the Appelmans protocol could already be occurring naturally, in situ, during active phage therapy [56–58]. Multiple rounds of infection and lysis in a dense and relatively localized bacterial population would provide many opportunities for horizontal exchange of genetic elements between therapeutic phages and even non-therapeutic phages present as prophages in the bacterial population or phages that are present in the local environment.

Our results are in keeping with this idea and suggest that recombination events between the members of a phage cocktail during the Appelmans protocol or during therapy generate far more genetic diversity than non-recombinant mutations alone. The total phenotypic diversity generated by all three single phage experiments was far less than that generated by evolving a cocktail of those same three phages. In the absence of recombination, we would have expected the total diversity generated from all three input phages separately to be similar to that generated by the three phages simultaneously. Moreover, only one point mutation was seen in the phi176 genome, whereas at least 48 recombination events occurred, corroborating the predominantly recombinatorial nature of phage evolution in the Appelmans protocol.

Author Contributions: B.H.B, J.A.F., and I.J.M. conceptualized and helped design the methods. B.H.B. performed the Appelmans protocol and all phage work. I.J.M. sequenced and identified crossovers, I.J.M. and B.H.B. annotated the phage genomes. B.H.B. and I.J.M. prepared the manuscript.

Acknowledgments: The authors would like to thank Zemphira Alavidze (IBMV) for providing details of the Appelmans protocol shown in Appendix A. Randall Wolcott (WCC), the University Medical Center Clinical Microbiology Department (Lubbock, TX), Kendra Rumbaugh (Dept. of Surgery, TTUHSC), and Abdul Hamood (Dept. of Immunology and Molecular Microbiology, TTUHSC) supplied the bacterial strains used in this work.

Appendix A

The Appelmans protocol, as supplied by Dr Z. Alavidze, IBMV. (In Georgia the protocol is often called the 'Appelman' technique.)

"The Appelman technique is a method for developing phage isolates into strains that are much more active against a wide range of bacterial strains of the same family. This long-standing technique is a main method used in the development of phage cocktails in the Republic of Georgia. The phage cocktail, generally active on at least 60% of the host collection, is diluted serially, with one dilution series for each of the 10–12 strains of interest, adding 0.1 mL of bacteria, diluted 10-fold from a slant tube, to each tube of 5 mL of broth. The tubes are incubated at 37 °C overnight. The following day, each dilution is checked for lysis compared to a control tube. The lysed tubes are combined across all the strains and filtered. The new filtrate is considered Appelman 1 and is used for the next dilution series, with the addition of one dilution past the previous point of lysis. Once a phage is able to lyse at a dilution of 10^{-7} across 10 or more strains, it is considered a "mother" phage that can be used in combination with other Appelman phages for a cocktail.

Qualitative analysis of the Appelman phages by spot testing shows an increase in virulence of the phage between the first and last Appelman. The efficiency of plating of each phage on the various strains is necessary to determine quantitatively how much change occurs as a result of Appelman passaging. The important question raised by this method is: how are the phages becoming more active? It is probable that a small change in the tail fibers of the phages allows for a better infection."

References

1. Fauci, A.S.; Marston, H.D. The Perpetual Challenge of Antimicrobial Resistance. *JAMA* **2014**, *311*, 1853–1854. [PubMed]
2. Smith, R.D.; Coast, J. Antimicrobial resistance: A global response. *Bull. World Health Organ.* **2002**, *80*, 126–133. [PubMed]
3. Burrowes, B.; Harper, D.R.; Anderson, J.; McConville, M.; Enright, M.C. Bacteriophage therapy: Potential uses in the control of antibiotic-resistant pathogens. *Expert Rev. Anti-Infect. Ther.* **2011**, *9*, 775–785. [PubMed]
4. Kutter, E.; De Vos, D.; Gvasalia, G.; Alavidze, Z.; Gogokhia, L.; Kuhl, S.; Abedon, S.T. Phage therapy in clinical practice: Treatment of human infections. *Curr. Pharm. Biotechnol.* **2010**, *11*, 69–86. [CrossRef] [PubMed]
5. Abedon, S.T. Hot Topic: The 'Nuts and Bolts' of Phage Therapy. *Curr. Pharm. Biotechnol.* **2010**, *11*, 1. [PubMed]
6. Sulakvelidze, A.; Alavidze, Z.; Morris, J.G. Bacteriophage therapy. *Antimicrob. Agents Chemother.* **2001**, *45*, 649–659. [PubMed]
7. National Institutes of Health. *NIAID's Antibacterial Resistance Program: Current Status and Future Directions*; National Institutes of Health: Washington, DC, USA, 2014.
8. Gill, J.J.; Hyman, P. Phage choice, isolation, and preparation for phage therapy. *Curr. Pharm. Biotechnol.* **2010**, *11*, 2–14. [CrossRef] [PubMed]
9. Chanishvili, N.; Sharp, R. *Eliava Institute of Bacteriophage, Microbiology and Virology, Tbilisi, Georgia. A Literature Review of the Practical Application of Bacteriophage Research*; Eliava Foundation: Tbilisi, Georgia, 2009.
10. Appelmans, R. Le dosage du Bacteriophage. *Compt. Rend. Soc. Biol.* **1921**, *85*, 1098.
11. Mapes, A.C.; Trautner, B.W.; Liao, K.S.; Ramig, R.F. Development of expanded host range phage active on biofilms of multi-drug resistant Pseudomonas aeruginosa. *Bacteriophage* **2016**, *6*, e1096995. [PubMed]
12. Hendrix, R.W. Bacteriophage genomics. *Curr. Opin. Microbiol.* **2003**, *6*, 506–511. [CrossRef] [PubMed]
13. Hendrix, R.W. Bacteriophages: Evolution of the majority. *Theor. Popul. Biol.* **2002**, *61*, 471–480.
14. Krylov, V.; Pleteneva, E.; Bourkaltseva, M.; Shaburova, O.; Volckaert, G.; Sykilinda, N.; Kurochkina, L.; Mesyanzhinov, V. Myoviridae bacteriophages of Pseudomonas aeruginosa: A long and complex evolutionary pathway. *Res. Microbiol.* **2003**, *154*, 269–275. [PubMed]
15. Silander, O.K.; Weinreich, D.M.; Wright, K.M.; O'Keefe, K.J.; Rang, C.U.; Turner, P.E.; Chao, L. Widespread genetic exchange among terrestrial bacteriophages. *Proc. Natl. Acad. Sci. USA* **2005**, *102*, 19009–19014. [CrossRef] [PubMed]
16. Tetart, F.; Desplats, C.; Krisch, H. Genome plasticity in the distal tail fiber locus of the T-even bacteriophage: Recombination between conserved motifs swaps adhesin specificity. *J. Mol. Biol.* **1998**, *282*, 543–556. [CrossRef] [PubMed]
17. Morris, P.; Marinelli, L.J.; Jacobs-Sera, D.; Hendrix, R.W.; Hatfull, G.F. Genomic characterization of mycobacteriophage Giles: Evidence for phage acquisition of host DNA by illegitimate recombination. *J. Bacteriol.* **2008**, *190*, 2172–2182. [CrossRef] [PubMed]
18. Tétart, F.; Repoila, F.; Monod, C.; Krisch, H. *Bacteriophage T4 Host Range Is Expanded by Duplications of a Small Domain of the Tail Fiber Adhesin*; Elsevier: Amsterdam, The Netherlands, 1996.
19. Shioda, T.; Levy, J.A.; Cheng-Mayer, C. Small amino acid changes in the V3 hypervariable region of gp120 can affect the T-cell-line and macrophage tropism of human immunodeficiency virus type 1. *Proc. Natl. Acad. Sci. USA* **1992**, *89*, 9434–9438. [CrossRef] [PubMed]
20. Morona, R.; Henning, U. Host range mutants of bacteriophage Ox2 can use two different outer membrane proteins of Escherichia coli K-12 as receptors. *J. Bacteriol.* **1984**, *159*, 579–582. [PubMed]
21. Werts, C.; Michel, V.; Hofnung, M.; Charbit, A. Adsorption of bacteriophage lambda on the LamB protein of Escherichia coli K-12: Point mutations in gene J of lambda responsible for extended host range. *J. Bacteriol.* **1994**, *176*, 941–947. [CrossRef]
22. Drexler, K.; Riede, I.; Montag, D.; Eschbach, M.-L.; Henning, U. Receptor specificity of the Escherichia coli T-even type phage Ox2: Mutational alterations in host range mutants. *J. Mol. Biol.* **1989**, *207*, 797–803. [CrossRef]
23. Drexler, K.; Dannull, J.; Hindennach, I.; Mutschler, B.; Henning, U. Single mutations in a gene for a tail fiber component of an Escherichia coli phage can cause an extension from a protein to a carbohydrate as a receptor. *J. Mol. Biol.* **1991**, *219*, 655–663. [PubMed]

24. Kwan, T.; Liu, J.; DuBow, M.; Gros, P.; Pelletier, J. Comparative genomic analysis of 18 Pseudomonas aeruginosa bacteriophages. *J. Bacteriol.* **2006**, *188*, 1184–1187. [PubMed]
25. Mesyanzhinov, V.; Robben, J.; Grymonprez, B.; Kostyuchenko, V.; Bourkaltseva, M.; Sykilinda, N.; Krylov, V.; Volckaert, G. The genome of bacteriophage phiKZ of Pseudomonas aeruginosa. *J. Mol. Biol.* **2002**, *317*, 1–19. [CrossRef] [PubMed]
26. McVay, C.S.; Velásquez, M.; Fralick, J.A. Phage therapy of Pseudomonas aeruginosa infection in a mouse burn wound model. *Antimicrob. Agents Chemother.* **2007**, *51*, 1934–1938. [CrossRef] [PubMed]
27. Rahme, L.G.; Stevens, E.J.; Wolfort, S.F.; Shao, J.; Tompkins, R.G.; Ausubel, F.M. Common virulence factors for bacterial pathogenicity in plants and animals. *Science* **1995**, *268*, 1899–1902. [PubMed]
28. Takeya, K.; Amako, K. A rod-shaped Pseudomonas phage. *Virology* **1966**, *28*, 163–165. [PubMed]
29. Carlson, K. *Appendix: Working with Bacteriophages: Common Techniques and Methodological Approaches. Bacteriophages: Biology and applications*; CRC Press: Boca Raton, FL, USA, 2005.
30. Adams, M. Enumeration of bacteriophage particles. In *Bacteriophages*; Interscience Publishers: New York, NY, USA, 1959; pp. 27–30.
31. Sambrook, J.; Russel, D.W. Extraction of Bacteriophage λ DNA from Large-scale Cultures Using Proteinase K and SDS. *Cold Spring Harb. Protoc.* **2006**. [CrossRef] [PubMed]
32. ParaBioSys. Available online: https://pga.mgh.harvard.edu/Parabiosys/projects/host-pathogen_interactions/sequencing.php (accessed on 1 February 2019).
33. Sastry, P.; Finlay, B.B.; Pasloske, B.; Paranchych, W.; Pearlstone, J.; Smillie, L. Comparative studies of the amino acid and nucleotide sequences of pilin derived from Pseudomonas aeruginosa PAK and PAO. *J. Bacteriol.* **1985**, *164*, 571–577. [PubMed]
34. Ceyssens, P.-J.; Brabban, A.; Rogge, L.; Lewis, M.S.; Pickard, D.; Goulding, D.; Dougan, G.; Noben, J.-P.; Kropinski, A.; Kutter, E. Molecular and physiological analysis of three Pseudomonas aeruginosa phages belonging to the "N4-like viruses". *Virology* **2010**, *405*, 26–30. [PubMed]
35. Ceyssens, P.J.; Noben, J.P.; Ackermann, H.W.; Verhaegen, J.; De Vos, D.; Pirnay, J.P.; Merabishvili, M.; Vaneechoutte, M.; Chibeu, A.; Volckaert, G. Survey of Pseudomonas aeruginosa and its phages: De novo peptide sequencing as a novel tool to assess the diversity of worldwide collected viruses. *Environ. Microbiol.* **2009**, *11*, 1303–1313. [PubMed]
36. Haggård-Ljungquist, E.; Halling, C.; Calendar, R. DNA sequences of the tail fiber genes of bacteriophage P2: Evidence for horizontal transfer of tail fiber genes among unrelated bacteriophages. *J. Bacteriol.* **1992**, *174*, 1462–1477. [CrossRef] [PubMed]
37. Labrie, S.J.; Samson, J.E.; Moineau, S. Bacteriophage resistance mechanisms. *Nat. Rev. Microbiol.* **2010**, *8*, 317–327. [CrossRef] [PubMed]
38. Labrie, S.J.; Moineau, S. Abortive infection mechanisms and prophage sequences significantly influence the genetic makeup of emerging lytic lactococcal phages. *J. Bacteriol.* **2007**, *189*, 1482–1487. [CrossRef] [PubMed]
39. Samson, J.E.; Magadán, A.H.; Sabri, M.; Moineau, S. Revenge of the phages: Defeating bacterial defences. *Nat. Rev. Microbiol.* **2013**, *11*, 675–687. [CrossRef] [PubMed]
40. Benzer, S. Fine structure of a genetic region in bacteriophage. *Proc. Natl. Acad. Sci. USA* **1955**, *41*, 344–354. [CrossRef] [PubMed]
41. Hershey, A.D. Mutation of bacteriophage with respect to type of plaque. *Genetics* **1946**, *31*, 620–640. [PubMed]
42. Doermann, A.H. Lysis and lysis inhibition with Escherichia coli bacteriophage. *J. Bacteriol.* **1948**, *55*, 257–276. [PubMed]
43. Burch, L.H.; Zhang, L.; Chao, F.G.; Xu, H.; Drake, J.W. The bacteriophage T4 rapid-lysis genes and their mutational proclivities. *J. Bacteriol.* **2011**, *193*, 3537–3545. [CrossRef] [PubMed]
44. Abedon, S.T.; Yin, J. Bacteriophage plaques: Theory and analysis. In *Bacteriophages*; Springer: New York, NY, USA, 2009; pp. 161–174.
45. Nguyen, A.H.; Molineux, I.J.; Springman, R.; Bull, J.J. Multiple genetic pathways to similar fitness limits during viral adaptation to a new host. *Evolution* **2012**, *66*, 363–374. [CrossRef] [PubMed]
46. Merril, C.R.; Biswas, B.; Carlton, R.; Jensen, N.C.; Creed, G.J.; Zullo, S.; Adhya, S. Long-circulating bacteriophage as antibacterial agents. *Proc. Natl. Acad. Sci. USA* **1996**, *93*, 3188–3192. [CrossRef] [PubMed]
47. Abe, M.; Izumoji, Y.; Tanji, Y. Phenotypic transformation including host-range transition through superinfection of T-even phages. *FEMS Microbiol. Lett.* **2007**, *269*, 145–152. [CrossRef] [PubMed]
48. Yoichi, M.; Abe, M.; Miyanaga, K.; Unno, H.; Tanji, Y. Alteration of tail fiber protein gp38 enables T2 phage to infect Escherichia coli O157: H7. *J. Biotechnol.* **2005**, *115*, 101–107. [CrossRef] [PubMed]

49. Meyer, J.R.; Dobias, D.T.; Weitz, J.S.; Barrick, J.E.; Quick, R.T.; Lenski, R.E. Repeatability and contingency in the evolution of a key innovation in phage lambda. *Science* **2012**, *335*, 428–432. [CrossRef] [PubMed]
50. Brüssow, H.; Hendrix, R.W. Phage genomics: Small is beautiful. *Cell* **2002**, *108*, 13–16. [CrossRef]
51. Juhala, R.J.; Ford, M.E.; Duda, R.L.; Youlton, A.; Hatfull, G.F.; Hendrix, R.W. Genomic Sequences of Bacteriophages HK97 and HK022: Pervasive Genetic Mosaicism in the Lambdoid Bacteriophages. *J. Mol. Biol.* **2000**, *299*, 27–51. [CrossRef] [PubMed]
52. Mmolawa, P.T.; Schmieger, H.; Heuzenroeder, M.W. Bacteriophage ST64B, a genetic mosaic of genes from diverse sources isolated from Salmonella enterica serovar typhimurium DT 64. *J. Bacteriol.* **2003**, *185*, 6481–6485. [CrossRef] [PubMed]
53. Pajunen, M.I.; Elizondo, M.R.; Skurnik, M.; Kieleczawa, J.; Molineux, I.J. Complete nucleotide sequence and likely recombinatorial origin of bacteriophage T3. *J. Mol. Biol.* **2002**, *319*, 1115–1132. [CrossRef]
54. Pedulla, M.L.; Ford, M.E.; Houtz, J.M.; Karthikeyan, T.; Wadsworth, C.; Lewis, J.A.; Jacobs-Sera, D.; Falbo, J.; Gross, J.; Pannunzio, N.R. Origins of Highly Mosaic Mycobacteriophage Genomes. *Cell* **2003**, *113*, 171–182. [CrossRef]
55. Kunisaki, H.; Tanji, Y. Intercrossing of phage genomes in a phage cocktail and stable coexistence with Escherichia coli O157: H7 in anaerobic continuous culture. *Appl. Microbiol. Biotechnol.* **2010**, *85*, 1533–1540. [CrossRef] [PubMed]
56. Abedon, S.T.; Thomas-Abedon, C. Phage therapy pharmacology. *Curr. Pharm. Biotechnol.* **2010**, *11*, 28–47. [CrossRef] [PubMed]
57. Payne, R.J.; Jansen, V.A. Phage therapy: The peculiar kinetics of self-replicating pharmaceuticals. *Clin. Pharmacol. Ther.* **2000**, *68*, 225–230. [CrossRef] [PubMed]
58. Payne, R.J.; Jansen, V.A. Pharmacokinetic principles of bacteriophage therapy. *Clin. Pharmacokinet.* **2003**, *42*, 315–325. [CrossRef] [PubMed]

13

Self-Assembled Nanoporous Biofilms from Functionalized Nanofibrous M13 Bacteriophage

Vasanthan Devaraj [1], Jiye Han [2,3], Chuntae Kim [2,3], Yong-Cheol Kang [4] and Jin-Woo Oh [1,2,3,5,*]

1 Research Center for Energy Convergence and Technology Division, Pusan National University, Busan 46241, Korea; devarajvasanthan@gmail.com
2 Department of Nano Fusion Technology, Pusan National University, Busan 46241, Korea; hanyksw20@naver.com (J.H.); chuntae1122@gmail.com (C.K.)
3 BK21 Plus Nanoconvergence Technology Division, Pusan National University, Busan 46241, Korea
4 Department of Chemistry, Pukyong National University, Busan 48513, Korea; yckang@pknu.ac.kr
5 Department of Nanoenergy Engineering, Pusan National University, Busan 46241, Korea
* Correspondence: ojw@pusan.ac.kr

Abstract: Highly periodic and uniform nanostructures, based on a genetically engineered M13 bacteriophage, displayed unique properties at the nanoscale that have the potential for a variety of applications. In this work, we report a multilayer biofilm with self-assembled nanoporous surfaces involving a nanofiber-like genetically engineered 4E-type M13 bacteriophage, which was fabricated using a simple pulling method. The nanoporous surfaces were effectively formed by using the networking-like structural layers of the M13 bacteriophage during self-assembly. Therefore, an external template was not required. The actual M13 bacteriophage-based fabricated multilayered biofilm with porous nanostructures agreed well with experimental and simulation results. Pores formed in the final layer had a diameter of about 150–500 nm and a depth of about 15–30 nm. We outline a filter application for this multilayered biofilm that enables selected ions to be extracted from a sodium chloride solution. Here, we describe a simple, environmentally friendly, and inexpensive fabrication approach with large-scale production potential. The technique and the multi-layered biofilms produced may be applied to sensor, filter, plasmonics, and bio-mimetic fields.

Keywords: M13 bacteriophage; biofilm; porous structure; filters; self-assembly

---

1. Introduction

Complex-free, cost-effective, non-lithographic, and highly ordered nanostructure fabrication approaches are necessary for diverse applications. Self-assembly methods have attracted interest because of their versatile functionalities resulting from the formations of highly ordered and well-defined nanostructures [1–5]. The self-assembly approach has major advantages, including large scale manufacturing capability, compatibility with experimental systems, device efficiency possibilities, and low-cost manufacturing support [6–10]. Highly ordered, self-assembly based nanostructures are generating interest in the filter, sensor, plasmonics, and photonics fields [11–13]. Despite the developmental progress, difficulties remain regarding building hierarchically ordered and highly complex nanostructures using self-assembled functional materials [14]. Understanding self-assembly and the related highly ordered complex structural arrangements can be derived from biological systems. Furthermore, biological materials and their self-assembled nanostructures can aid in the fabrication of a variety of functional nanomaterials [15–21]. More importantly, such fabrication methods can be eco-friendly and energy efficient [22]. Of these biological systems, viruses provide unique tools and examples of self-assembly from natural building blocks [23]. Advances in genetic engineering of viruses are providing solutions to ongoing self-assembly problems. As such, we have many opportunities to integrate viruses into various materials comprised of biologically inspired and highly self-ordered nanostructures [24–28].

With a well-defined geometry and flexibility resulting from modifications introduced using genetic engineering and chemical approaches, the M13 bacteriophage (M13 phage) is distinguished from other viruses because of its unique self-assembly nature that has led to a variety of novel nanostructures and devices [29–31]. Phages can be produced easily and cost-effectively by infecting a host bacterium, hence the name bacteriophage. After infecting a host bacterium, the host's metabolism is adapted by the phage to continuously synthesize and secrete new phage particles, which leads to the production of millions of copies after overnight culture. The M13 phage geometry is nanofiber-like, with a diameter of approximately 6.6 nm and a height of about 880 nm, and it is covered by about 2700 copies of a major coat protein (pVIII) and five copies of minor coat proteins (pIII and pIX) located at its ends. Furthermore, due to the high fidelity of its biological reproduction, the M13 phage exhibits monodispersity and highly anisotropic shape properties. These M13 phage properties enable it to exhibit liquid crystalline (LC) behavior, and thereby form highly ordered nanostructures in suspensions. Numerous applications of the M13 phage have been reported, such as in solar cells, phage displays, imaging, biomimetic structures, photonic nose, battery, scaffolds, photo- and piezo-devices, cancer therapy, gels, filters, sensors, gene delivery, catalysis, and tissue regeneration [32–46].

Here, we report the fabrication of a multilayered porous biofilm consisting of alternating M13 phages and polydiallyldimethylammonium chloride (PDDA) layers for filter applications. The negative charge (E, glutamic acid) carried by the M13 bacteriophage was deliberately increased via genetic engineering to enhance its properties for this type of application. Our fabrication approach differs from those previously used to fabricate porous structures, such as previously described M13 phage-based porous structures and the breath figure method [47–50]. The differences and advantages of our fabrication from other such methods are discussed in the results and discussion section. The fabrication method used in the present study was straightforward, non-lithographic, and can be scaled up to large scale, low cost production. The formation of porous surfaces in the multi-layered biofilms produced were confirmed experimentally and by simulation results.

2. Materials and Methods

2.1. Genetic Engineering

M13 phages were purchased from New England Bio-labs (Ipswich, MA, USA) and genetically engineered using recombinant DNA methods. Using an inverse polymerase chain reaction (PCR) cloning method, the peptide sequence positioned at the 25th of the N-terminus of the wild-type phage pVIII coat proteins was engineered using PCR methods. The respective template and PCR primers were as follows: M13KE (NEB, #N0316) vector with an engineered PstI site, and the inset sequence was 5′-ATATATCTGCAGGAAGAAGAGG AACCCGCAAAAGCGGCCTTTAACTCCC-3′ respectively. The reverse primer was designed as 5′-GCTGTCTTTCGCTGC-AGAGGGTG-3′ to ensure the vector was linear and complementary to the engineered pVIII 3′-5′ region. Genetically engineered M13 phages with four glutamic (E) acids, Alu-Glu-Glu-Glu-Glu-Asp (or AEEED), were verified by DNA sequencing analysis (Cosmo-Gentech, Seoul, Republic of Korea).

2.2. Phage Quantification

Concentrations of M13 phage solutions were measured using an ultraviolet-visible (UV-vis) spectrometer (Evolution 300, Thermo Fisher Scientific, Waltham, MA, USA). 4E-type phage concentrations were calculated from the absorbance spectra using the previously reported formula [29,36,37]:

$$\text{mg of phages/mL} = (A_{269} - A_{320})/3.84$$

where A_{269} and A_{320} represent absorbances of a phage suspension at 269 nm and 320 nm, respectively.

2.3. Fabrication of Multilayer Biofilm

The biofilms were fabricated using a commercial syringe pump (LEGATO 270, KD Scientific, Holliston, MA, USA). Micro-centrifuge tubes were used to load the prepared 4E-type M13 phage and polydiallyldimethylammonium chloride (PDDA) solutions. The concentrations of the 4E-type M13 phage and PDDA solutions were 5 mg/mL and 2 wt %, respectively. The glass substrate was attached to metal tweezers. The syringe pump's built-in software was used to control dipping and pulling speeds. Our biofilm consisted of 10 layers, 5 layers each of M13 phage and PDDA deposited in an alternating manner. Oxygen (O_2) plasma was used to clean the glass substrate and imbue hydrophilicity, which was then dipped in cysteamine to facilitate cross-linking with the M13 phage. To produce the multilayers, the glass substrate was dipped in the respective solutions for 3 s to form each layer. The glass substrate was pulled out from solution(s) after every layer deposition and left to dry naturally at room temperature for 30 min. The processes were repeated until all 10 layers had been deposited.

2.4. Structural Characterization

AFM images were collected using an NX10 unit (Park Systems, Suwon, Korea) equipped with the XEP 3.0.4 data acquisition program (Park Systems). Obtained AFM images were analyzed using the XEI 1.8.2 image processing program (Park Systems). All images were collected in true non-contact mode. A specialized probe for non-contact mode was used for measurements (PPP-NCHR, Nanosensors, Neuchatel, Switzerland). Optical images of M13 phage/PDDA multi-laminated films were obtained by scanning electron microscopy (SEM; S4800, Hitachi, Tokyo, Japan).

2.5. Analysis of Adsorbed Ions on Multilayer Biofilms

The thin film surface elements were analyzed by X-ray photoelectron spectrometry (XPS; ESCALAB 250, Thermo Fisher Scientific, Waltham, MA, USA) equipped with a hemispherical analyzer and a twin anode non-monochromatic Al-Kα source. For selective ion filter application, two types of films were prepared on glass slides: M13 phage only thin film and M13 phage/PDDA film. These films were immersed in NaCl solutions with different concentrations (0, 0.1, 0.2, 0.5, and 1 M) to bind ions, washed in distilled water to remove excess ions, and dried. Ions bound on surfaces in films were then analyzed by XPS.

2.6. Theoretical Method

Reflectance simulations of multilayer biofilms were performed using three-dimensional (3D) finite-difference time-domain (FDTD) simulations using a commercial package from Lumerical Solutions (Vancouver, BC, Canada). The multilayered biofilm structure was surrounded by a perfectly matched layer (PML) boundary conditions in the x, y, and z directions. To obtain accurate results, $\lambda/400$ elements of mesh size were applied. A broadband plane wave source was used for optical excitation, and a power monitor was placed at top of the multi-layered biofilm structure in the air region to record reflectance. At $\lambda = 600$ nm, refractive indices n of the slide glass, the M13 bacteriophage, and the PDDA were 1.514, 1.45, and 1.375, respectively [51].

2.7. Reflectance Experiment

Reflectance measurements were recorded using a Thermo Scientific Evolution 300 UV-Vis spectrophotometer, a xenon lamp, high resolution (1200 lines/mm) grating, and a silicon photodiode detector.

3. Results and Discussion

Information on the geometry of the M13 phage is provided in Figure 1a. The M13 phage has a nanofiber-like structure about 880 nm long with a diameter of about 6.6 nm. The nanofibrous M13 phage contains about 2700 copies of pVIII amino acid sequences with pIII, pVI, PVII, and pIX amino acid sequences at its ends. As described earlier, multi-layer nanoporous biofilms were fabricated by pulling method (PM) using commercial syringe setup. A PM fabrication schematic is shown in Figure 1b. A glass slide (the substrate), attached to a commercial syringe pump, was dipped in and then removed from M13 phage and PDDA solutions in an alternating manner. The mechanism used to form nanoporous surfaces in biofilms involves three steps: (1) deposition of initial layers with poor quality rough surface quality based on initial M13 phage networking layer(s), (2) improve surface quality by depositing additional layers leading to early stage pore formation, and (3) to form nanoporous surfaces in final layers (Figure 2a).

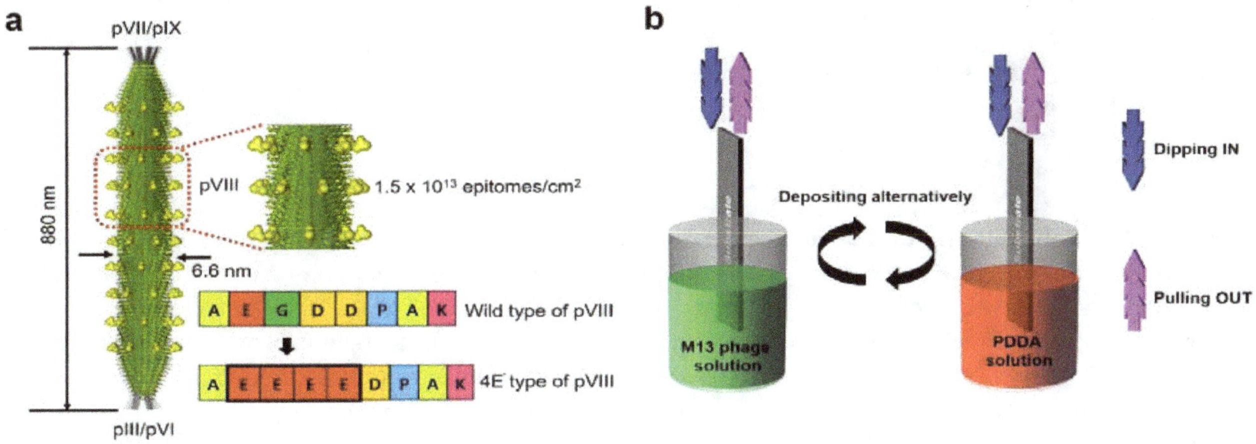

Figure 1. (**a**) Illustration of the structure of M13 phage and pVIII, pVII/pIX, and pIII/pVI amino acid sequences. pVIII genetic engineering site information shows conversion from wild type to 4E-type phage. (**b**) Schematic of the pulling method used to produce M13 phage/polydiallyldimethylammonium chloride (PDDA) multilayered films on glass.

To create the initial layers, the flexible nature of the geometry of the M13 phage plays a crucial role. During initial deposition, M13 phages are deposited randomly to form a nanofiber network, which results in a poor quality and uneven surface (Figure 2b,c). This network structure can be achieved by varying deposition time, speed, and M13 phage concentration. More importantly, these initial layers are made possible by the liquid crystalline (LC) behavior of M13 phages during self-assembly. When the substrate was subjected to PM, evaporation occurred more rapidly at the air-liquid-solid meniscus, which resulted in the local accumulation and deposition of M13 phages on the substrate. At this time, two crucial factors are thought to occur during self-assembly: local induction of chiral LC structure phase transitions occurring at the meniscus, and dominance of interfacial forces acting at the meniscus. By controlling the conditions mentioned above during biofilm growth, producing different self-assembled nanostructures is possible. LC phases are classified as thermotropic or lyotropic. Of these, the lyotropic phase plays a crucial role in the self-assembly of biological structures. By varying temperature and concentration of biological particles, producing lyotropic LCs with many phases is possible. Of the experiment conditions, the most critical factor is concentration (of M13 phage solution in this study), which dominantly affects the lyotropic LC phase. To fabricate a biofilm layer with a network structure with an uneven surface, we prepared M13 phage solution at a low concentration (five mg/mL). At this low concentration, M13 phages were distributed randomly on the substrate, that is, an isotropic phase. Such conditions satisfy the needs of the initial layers with a poor surface quality with a surface roughness of ~70 nm.

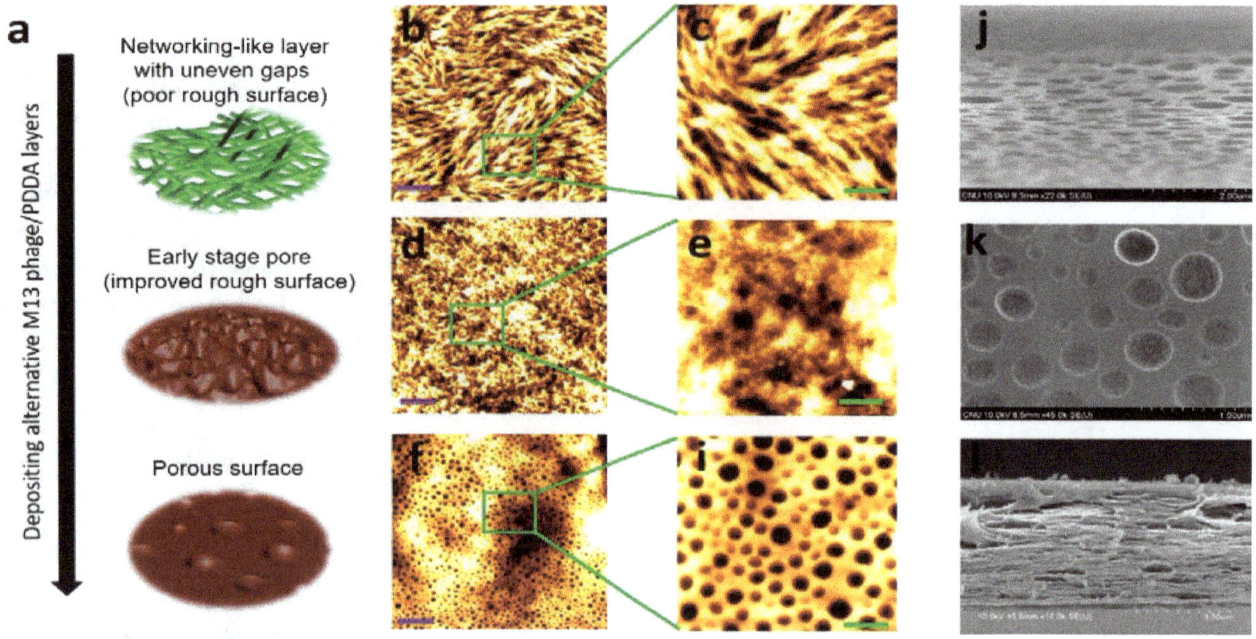

Figure 2. (**a**) Fabrication of porous surfaces by depositing M13 phage and PDDA layers in an alternating manner originating from a randomly distributed network-like structure. (**b**,**c**) AFM images of first layer, (**d**,**e**) sixth layer, and (**f**,**i**) 10th layer surfaces. Violet and green scale bar colors represent five and one µm, respectively. Scanning electron microscopy (SEM) images taken from a 10-layer biofilm: (**j**) perspective view, (**k**) top view, and (**l**) cross-sectional view. Scale bars for (**j**–**l**) are two, one, and three µm, respectively. The cracks shown in figure (**k**) were caused by platinum coating for SEM analysis.

PDDA films were then deposited in an alternating manner on the M13 phage layers. At this point, the role of PDDA layers was to reduce the surface roughness of previous M13 phage layers. Notably, uneven gaps created by poor surfaces in the network structures ranged from about 1 to 3 µm. These gaps in the layers were gradually filled by depositing PDDA layers, and when around six layers had been deposited, early stage pore formation was observed (Figure 2d,e) with better surface quality. The surface roughness was now ~37 nm, which is an improvement of about 50%. However, at this stage, a few surface nanopores with diameters of about 700–1200 nm (±20 nm) and depth of ~70–80 nm (±5 nm) were evident. Further depositions of M13 phage and PDDA layers improved surface quality and induced the formation of pores with decreasing diameters and depths. Clear porous surfaces were seen after depositing the eighth layer. The porous surfaces eventually formed by the 10th layer are shown in Figure 2f,i. Pore diameters ranged from ~150 to 500 nm (±20 nm) and depths from ~15 to 30 nm (±5 nm) in the 10th layer. The surface quality of the final layer improved by about 80% when compared with the first layer (surface roughness of ~14 nm). This porous surface information was well supported by the perspective (Figure 2j) and surface view (Figure 2k) of the SEM images. After 10 layers, clear porous structures were observed with diameters ranging from ~150 to 500 nm. The total thickness of the 10-layered bio-film was ~3.84 µm (SEM cross-sectional view, Figure 2l).

We then considered two porous structure fabrication concepts: breath figure (non-M13 phage-based structures) and M13 phage-based porous fabrication methods [47–50,52]. With the breath figure method, careful optimization of experimental parameters is required to achieve highly ordered porous structures. Furthermore, the open experimental process to fabricate porous structures is difficult. For example, in static breath figure methods, fabrication processes were sealed from external environmental lab conditions. Generally, in the breath figure process, water droplets formed from condensation, acting as a template to create a porous structure. Condensation of water vapours is vital in the breath figure process. Careful steps should be considered to prevent the water droplet template from effects such as a coalescence. Methods reported involving a combination of dip coating, similar to our pulling method, and the breath figure only showed porous structure formation in the presence of an external mesh template, such as nylon mesh [48,52].

When using fabrication methods involving M13 phages, porous structures were successfully formed either based on a supporting macroporous film, like anodic aluminum oxide (AAO), or by the freeze-drying method [49,50]. Nanoporous structures with diameters of about 200 nm formed using AAO template also used the etching process. In previously reported layer-by-layer biofilms involving M13 phages, ordered networking structures were formed but no nanoporous surfaces were reported [53]. Please note, when a wild type M13 phage was used instead of 4E type, networking structures were observed but no nanoporous surfaces were formed in the multilayer biofilm. Biofilms involving 4E type phage formed nanoporous surfaces successfully due to repulsive charge properties based on our fabrication method. Our fabrication method was simple, straightforward, and can be performed in an open, room temperature lab conditions. More importantly, our method does not require an external template to form a nanoporous structure. Our method is non-lithographic, and thus no etching process is required. Additionally, large scale fabrication is possible at comparable low costs. Given these advantages, our approach can facilitate the fabrication of nanoporous structures and might open interesting possibilities for a variety of applications, such as separation, bio-mimetic structures, plasmonics, sensors, etc. To justify our fabrication approach, we performed an optical investigation that confirms the structural transitions.

3.1. *Optical Results: Theoretical and Experimental*

We confirmed structural transitions of biofilms from poor surface quality to the porous surface using three-dimensional FDTD reflectance simulations and then performed experimental verification. Topographic information for each layer obtained from AFM data was used to create a layer geometry in FDTD. The biofilm structure was modeled as follows: no porous surfaces occurred during the first five layers, and after that, porous surfaces were created during the subsequent layers (6–10 layer). Surface roughness data obtained by AFM was found to be useful for creating uneven surfaces. We modeled the pore shape as parabolic one using the following equation [54]:

$$z = cx^2 \tag{1}$$

The term "c" is defined by:

$$c = 4 \times \frac{D}{W^2} \tag{2}$$

where D and W are pore diameter and depth, respectively (Figure 3a). At λ of 600 nm, no absorbance issues were observed for the glass substrate, M13 phage, or PDDA layer. This wavelength data will help analyze the reflectivity, as no other complex optical information is related to these three materials. Figure 3b shows the simulated and measured reflectivity data as function of layer number. Layer number 0 corresponds to the glass substrate. The reflectance of 4.2% is typical of glass substrates, and this well-matched the experiment and simulation results. From layer numbers 1 to 6, almost no difference was observed in reflectance data. At layer 1, a poor surface morphology increased the reflectance to ~6.5%, and up to layers 5 or 6, similar reflectance values were recorded. Furthermore, this reflectance concurred with experiment results, with average reflectance values of ~7% being measured for layers 1–6. Significant reflectance changes were observed from layer 7 to 10. At this point, the influence of the porous surface played crucial role. Reflectance values increased from ~7% to ~9%, due to the presence of porous nanostructures in the surface. For a 10-layered biofilm, ~8.7% reflectance was obtained via simulation. Experimental data showed a similar trend for reflectance values and a maximum value of ~9.5% was recorded for a 10-layer biofilm. This linear increase in reflectivity values can be explained using a solid immersion lens (SIL) structure. Our pores were shaped like an inverted SIL, which is parabolic shaped [54]. For pores with larger diameters and depths, light could be more easily transmitted, comparatively. With these dimensions, internal diffraction and reflection can be minimized resulting in a smaller reflectance value. At smaller pore dimensions (smaller diameter and depth), contributions from internal diffraction and reflections increase, resulting in an increase in reflectivity. These optical properties aligned well with our simulation and experimental results and support the proposed fabrication mechanism responsible for surface pore formation arising from initially poorly deposited layers and the presence of randomly distributed network-like structures. This approach opens an interesting path for the fabrication of self-assembled porous multilayer biofilms.

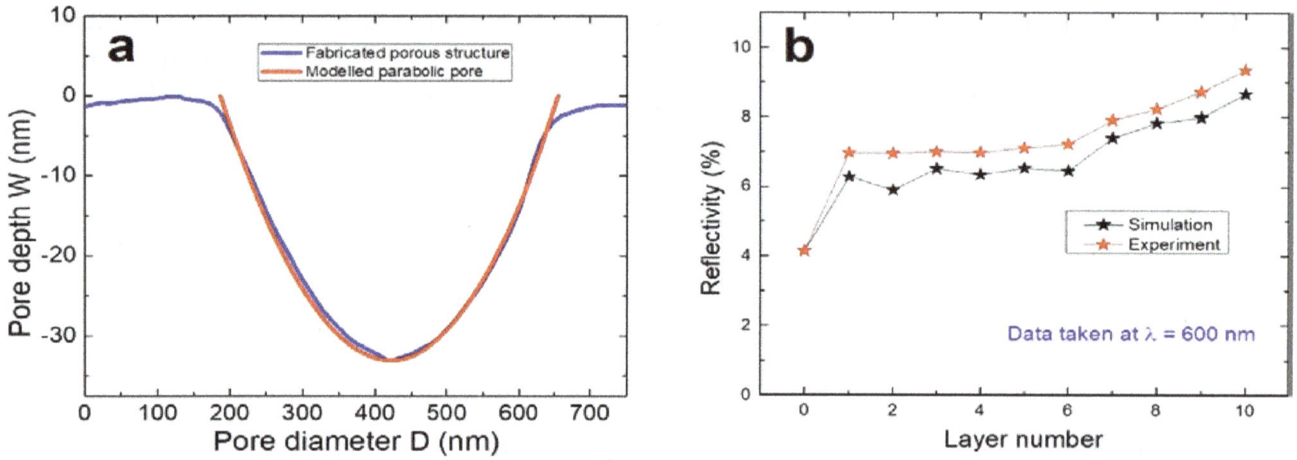

Figure 3. (**a**) Parabolic-shaped pore depth versus diameter plot obtained using Equation (1) ($z = cx^2$) aligns well with observed pore shapes as determined using the line profile AFM data for a 10-layer biofilm. (**b**) Experimental and modeled reflectance values obtained at λ = 600 nm for a 10-layer biofilm.

3.2. Filter Experiment

To investigate potential applications of the M13 phage and PDDA laminated nanoporous structure as a functional filter, we examined its ability to extract ions from sodium chloride (NaCl) solution. For this purpose, two types of films were tested: a M13 phage only film and a M13 phage-PDDA multilayered nanoporous film. Figure 4a,b show the relative atomic ratios of the captured ions at different NaCl solution concentrations for the two films. In the absence of a PDDA layer, Na^+ ions were captured more effectively by the negative charges of M13 phages.

However, in the M13 phage-PDDA multilayer biofilms, a larger number of Cl^- ions were captured by the positively charged PDDA. Although the laminated structure consists of the M13 phage, Na^+ ions that permeated into pores and bound to virus layers were not detected because of the limited skin depth of the XPS system. This result indicates that Cl^- ions bound effectively to the positive charges of PDDA in the multi-layered biofilm.

Eventhough the capture ratio of the Cl^- ions was smaller at this time, further improvements will be performed in the near future. Problems arise when performing selective ion extraction experiments on a single-layered PDDA film. The success rate of PDDA film attached to the substrate when dipped into NaCl solution was ~< 1%. Thus, we believe the ten layered biofilm is far better way to proceed. Finally, our results confirm that multi-layer biofilms with a nanoporous structure can act as membranes for selective ion extraction.

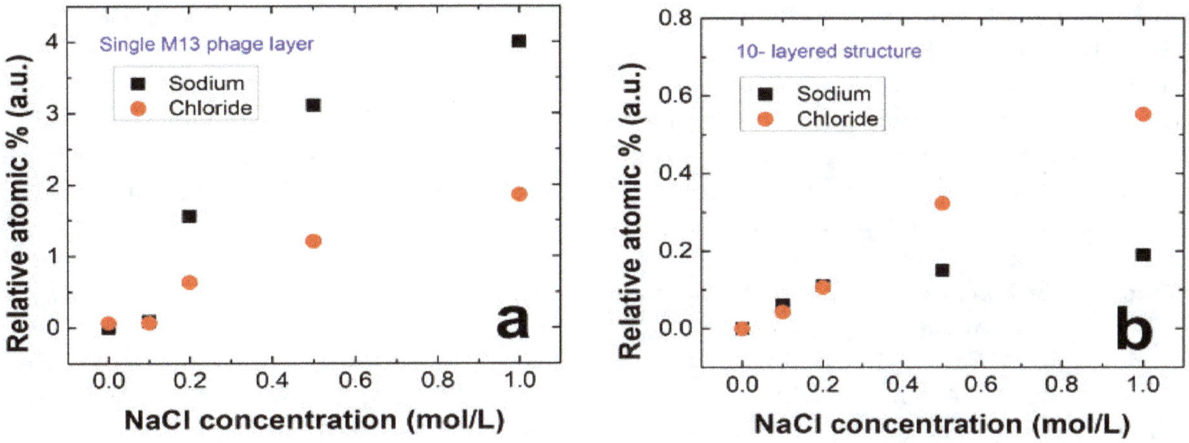

Figure 4. Calculated relative atomic percentages obtained from X-ray photoelectron spectroscopy (XPS) measurements for (**a**) M13 phage only and (**b**) 10-layer M13 phage-PDDA nanoporous biofilms, which confirm selective ion filtration.

4. Summary

In summary, we reported a straightforward fabrication approach to produce nanoporous multilayered M13 phage-PDDA films using a pulling method. Our fabrication approach was non-lithographic, and nanoporous structures with diameters of about 150 nm were formed without the support of an external template. Good agreement between experiment and simulation results support our suggested fabrication mechanism. M13 phage nanofiber geometry played a significant role in pore formation during our fabrication process. The described process has the advantages of straightforward fabrication, low-cost manufacturing, and compatibility with large-scale production. The devised nanoporous multi-layer biofilm provides a novel option for selective ion filtration, and may have interesting applications in the fields of plasmonics, photonics, and sensors.

Author Contributions: V.D. and J.-W.O. designed the structure and wrote the manuscript. J.H. carried out fabrication, and filter experiments. V.D. characterized the AFM data. Genetic engineering was carried out by C.K. Optical simulations were done by V.D. Y.-C.K. performed XPS measurements. J.-W.O. supervised the project. All the authors contributed to the work and participated in discussions.

Acknowledgments: This research work was supported by following funding sources: Creative Materials Discovery Program by the National Research Foundation of Korea (NRF) funded by Ministry of Science and ICT (NRF-2017M3D1A1039287), National Research Foundation of Korea (NRF) Grant funded by the Korean Government (MSIP, MOE) (NRF-2016R1C1B2014423), and by the Pioneer Research Center Program through the National Research Foundation of Korea funded by the Ministry of Science, ICT & Future Planning (NRF-2013M3C1A3065522).

References

1. Capito, R.M.; Azevedo, H.S.; Velichko, Y.S.; Mata, A.; Stupp, S.I. Self-Assembly of Large and Small Molecules into Hierarchically Ordered Sacs and Membranes. *Science* **2008**, *319*, 1812–1816. [CrossRef] [PubMed]
2. Zhang, S. Fabrication of novel biomaterials through molecular self-assembly. *Nat. Biotechnol.* **2003**, *21*, 1171–1178. [CrossRef] [PubMed]
3. Whiteslides, G.M.; Grzybowski, B. Self-Assembly at All Scales. *Science* **2002**, *295*, 2418–2421. [CrossRef] [PubMed]
4. Sandhage, K.H.; Dickerson, M.B.; Huseman, P.M.; Caranna, M.A.; Clifton, J.D.; Bull, T.A.; Heibel, T.J.; Overton, W.R.; Schoenwaelder, M.E.A. Novel, Bioclastic Route to Self-Assembled, 3D, Chemically Tailored Meso/Nanostructures: Shape-Preserving Reactive Conversion of Biosilica (Diatom) Microshells. *Adv. Mater.* **2002**, *6*, 429–433. [CrossRef]

5. Wang, D.; Choi, D.; Li, J.; Yang, Z.; Nie, Z.; Kou, R.; Hu, D.; Wang, C.; Saraf, L.V.; Zhang, J.; et al. Self-Assembled TiO_2-Graphene Hybrid Nanostructures for Enhanced Li-Ion Insertion. *ACS Nano* **2009**, *3*, 907–914. [CrossRef] [PubMed]
6. Nie, Z.; Petukhova, A.; Kumacheva, E. Properties and emerging applications of self-assembled structures made from inorganic nanoparticles. *Nat. Nanotechnol.* **2009**, *5*, 15–25. [CrossRef] [PubMed]
7. Yu, X.; Xiao, K.; Chen, J.; Lavrik, N.V.; Hong, K.; Sumpter, B.G.; Geohegan, D.B. High-Performance Field-Effect Transistors Based on Polystyrene-*b*-Poly(3-hexylthiophene) Diblock Copolymers. *ACS Nano* **2011**, *5*, 3559–3567. [CrossRef] [PubMed]
8. Zhao, B.; Wang, Q.; Zhnag, S.; Deng, C. Self-assembled wafer-like porous $NaTi_2(PO_4)_3$ decorated with hierarchical carbon as a high-rate anode for aqueous rechargeable sodium batteries. *J. Mater. Chem. A* **2015**, *3*, 12089–12096. [CrossRef]
9. Liu, J.; Lan, Y.; Yu, Z.; Tan, C.S.Y.; Parker, R.M.; Abell, C.; Scherman, O.A. Cucurbit[*n*]uril-Based Microcapsules Self-Assembled within Microfluidic Droplets: A Versatile Approach for Supramolecular Architectures and Materials. *Acc. Chem. Res.* **2017**, *50*, 208–217. [CrossRef] [PubMed]
10. Love, J.C.; Estroff, L.A.; Kriebel, J.K.; Nuzzo, R.G.; Whiteslides, G.M. Self-Assembled Monolayers of Thiolates on Metals as a Form of Nanotechnology. *Chem. Rev.* **2005**, *105*, 1103–1169. [CrossRef] [PubMed]
11. Kim, S.H.; Lee, S.Y.; Yang, S.M.; Yi, G.R. Self-assembled colloidal structures for photonics. *NPG Asia Mater.* **2011**, *3*, 25–33. [CrossRef]
12. Barnes, W.L.; Dereux, A.; Ebbesen, T.W. Surface plasmon subwavelength optics. *Nature* **2003**, *424*, 824–830. [CrossRef] [PubMed]
13. Medintz, I.L.; Clapp, A.R.; Mattoussi, H.; Goldman, E.R.; Fisher, B.; Mauro, J.M. Self-assembled nanoscale biosensors based on quantum dot FRET donors. *Nat. Mater.* **2003**, *2*, 630–638. [CrossRef] [PubMed]
14. Chen, S.; Slattum, P.; Wang, C.; Zang, L. Self-Assembly of Perylene Imide Molecules into 1D Nanostructures: Methods, Morphologies, and Applications. *Chem. Rev.* **2015**, *115*, 11967–11998. [CrossRef] [PubMed]
15. Yan, H.; Park, S.H.; Finkelstein, G.; Reif, J.H.; LaBean, T.H. DNA-Templated Self-Assembly of Protein Arrays and Highly Conductive Nanowires. *Science* **2003**, *301*, 1882–1884. [CrossRef] [PubMed]
16. Malvankar, N.S.; Vargas, M.; Nevin, K.P.; Franks, A.E.; Leang, C.; Kim, B.C.; Inoue, K.; Mester, T.; Covalla, S.F.; Johnson, J.P.; et al. Tunable metallic-like conductivity in microbial nanowire networks. *Nat. Nanotechnol.* **2011**, *6*, 573–579. [CrossRef] [PubMed]
17. Song, F.; Su, H.; Che, J.; Moon, W.J.; Lau, W.M.; Zhang, D. 3D hierarchical porous SnO_2 derived from self-assembled biological systems for superior gas sensing application. *J. Mater. Chem.* **2012**, *22*, 1121–1126. [CrossRef]
18. Wang, J.; Yang, M.; Zhu, Y.; Wang, L.; Tomsia, A.P.; Mao, C. Phage Nanofibers Induce Vascularized Osteogenesis in 3D Bone Scaffolds. *Adv. Mater.* **2014**, *26*, 4961–4966. [CrossRef] [PubMed]
19. Palmer, L.C.; Newcomb, C.J.; Kaltz, S.R.; Spoerke, E.D.; Stupp, S.I. Biomimetic Systems for Hydroxyapatite Mineralization Inspired by Bone and enamel. *Chem. Rev.* **2008**, *108*, 4754–4783. [CrossRef] [PubMed]
20. Whaley, S.R.; English, D.S.; Hu, E.L.; Barbara, P.F.; Belcher, A.M. Selection of peptides with semiconductor binding specificity for directed nanocrystal assembly. *Nature* **2000**, *205*, 665–668. [CrossRef] [PubMed]
21. Tahara, Y.; Akiyoshi, K. Current advances in self-assembled nanogel delivery systems for immunotherapy. *Adv. Drug Deliv. Rev.* **2015**, *95*, 65–76. [CrossRef] [PubMed]
22. Liu, M.; Zhang, L.; Wang, T. Supramolecular Chirality in Self-Assembled Systems. *Chem. Rev.* **2015**, *115*, 7304–7397. [CrossRef] [PubMed]
23. Liu, Z.; Qiao, J.; Niu, Z.; Wang, Q. Natural supramolecular building blocks: From virus coat proteins to viral nanoparticles. *Chem. Soc. Rev.* **2012**, *41*, 6178–6194. [CrossRef] [PubMed]
24. Sawada, T. Filamentous virus-based soft materials based on controlled assembly through liquid crystalline formation. *Polym. J.* **2017**, *49*, 639–647. [CrossRef]
25. Lee, S.W.; Mao, C.; Flynn, C.E.; Belcher, A.M. Ordering of Quantum Dots Using Genetically Engineered Viruses. *Science* **2002**, *296*, 892–895. [CrossRef] [PubMed]
26. Mao, C.; Solis, D.J.; Reiss, B.D.; Kottmann, S.T.; Sweeney, R.Y.; Hayhurst, A.; Georgiou, G.; Iverson, B.; Belcher, A.M. Virus-Based Toolkit for the Directed Synthesis of Magnetic and Semiconducting Nanowires. *Science* **2004**, *303*, 213–217. [CrossRef] [PubMed]
27. He, T.; Abbineni, G.; Cao, B.; Mao, C. Nanofibrous Bio-inorganic Hybrid Structures Formed through Self-Assembly and Oriented Mineralization of Genetically Engineered Phage Nanofibers. *Small* **2010**, *6*, 2230–2235. [CrossRef] [PubMed]

28. Li, K.; Chen, Y.; Li, S.; Nguyen, H.G.; Niu, Z.; You, S.; Mello, M.C.; Lu, X.; Wang, Q. Chemical Modification of M13 Bacteriophage and Its Application in Cancer Cell Imaging. *Bioconjug. Chem.* **2010**, *21*, 1369–1377. [CrossRef] [PubMed]
29. Chung, W.J.; Oh, J.W.; Kwak, K.; Lee, B.Y.; Meyer, J.; Wang, E.; Hexemer, A.; Lee, S.W. Biomimetic self-templating supramolecular structures. *Nature* **2011**, *478*, 364–368. [CrossRef] [PubMed]
30. Yang, S.H.; Chung, W.J.; Mcfarland, S.; Lee, S.W. Assembly of Bacteriophage into Functional Materials. *Chem. Rec.* **2013**, *13*, 43–59. [CrossRef]
31. Moon, J.S.; Kim, W.G.; Kim, C.; Park, G.T.; Heo, J.; Yoo, S.Y.; Oh, J.W. M13 Bacteriophage-Based Self-Assembly structures and Their Functional Capabilities. *Mini Rev. Org. Chem.* **2015**, *12*, 271–281. [CrossRef] [PubMed]
32. Che, P.Y.; Dang, X.; Klug, M.T.; Qi, J.; Courchesne, N.M.D.; Burpo, F.J.; Fang, N.; Hammond, P.T.; Belcher, A.M. Versatile Three-Dimensional Virus based Template for Dye-Sensitized Solar Cells with Improved Electron Transport and Light Harvesting. *ACS Nano* **2013**, *8*, 6563–6574.
33. Kim, W.G.; Kim, K.; Ha, S.H.; Song, H.; Yu, H.W.; Kim, C.; Kim, J.M.; Oh, J.W. Virus based Full Colour Pixels using a Microheater. *Sci. Rep.* **2015**, *5*, 13757, doi:10.1038/srep13757. [CrossRef] [PubMed]
34. Wang, Y.; Ju, Z.; Cao, B.; Gao, X.; Zhu, Y.; Qiu, P.; Xu, H.; Pan, P.; Bao, H.; Wang, L.; et al. Ultrasensitive rapid Detection of Human Serum Antibody Biomarkers by Biomarker-Capturing Viral Nanofibers. *ACS Nano* **2015**, *9*, 4475–4483. [CrossRef] [PubMed]
35. Lee, Y.J.; Lee, Y.; Oh, D.; Chen, T.; Ceder, G.; Belcher, A.M. Biologically Activated Noble Metal Alloys at the Nanoscale: For Lithium Ion Battery Anodes. *Nano Lett.* **2010**, *10*, 2433–2440. [CrossRef] [PubMed]
36. Lee, B.Y.; Zhnag, J.; Zueger, C.; Chung, W.J.; Yoo, S.Y.; Wang, E.; Meyer, J.; Ramesh, R.; Lee, S.W. Virus-based piezoelectric energy generation. *Nat. Nanotechnol.* **2012**, *7*, 351–356. [CrossRef] [PubMed]
37. Shin, D.M.; Han, H.J.; Kim, W.G.; Kim, E.; Kim, C.; Hong, S.W.; Kim, H.K.; Oh, J.W.; Hwang, Y.H. Bioinspired piezoelectric nanogenerators based on vertically aligned phage nanopillars. *Energy Environ. Sci.* **2015**, *8*, 3198–3203. [CrossRef]
38. Moon, J.S.; Kim, W.G.; Shin, D.M.; Lee, S.Y.; Kim, C.; Lee, Y.; Han, J.; Kim, K.; Yoo, S.Y.; Oh, J.W. Bioinspired M-13 bacteriophage-based photonic nose for differential cell recognition. *Chem. Sci.* **2017**, *8*, 921–927.
39. Yang, M.; Li, Y.; Huai, Y.; Wang, C.; Yi, W.; Mao, C. Evolutionary selection of personalized melanoma cell/tissue dual-homing peptides for guiding bionanofibers to malignant tumors. *Chem. Commun.* **2018**, *54*, 1631–1634. [CrossRef] [PubMed]
40. Ghosh, D.; Lee, Y.; Thomas, S.; Kohli, A.G.; Yun, D.S.; Belcher, A.M.; Kelly, K.A. M13-templated magnetic nanoparticles for targeted in vivo imaging of prostrate cancer. *Nat. Nanotechnol.* **2012**, *7*, 677–682. [CrossRef] [PubMed]
41. Wang, Y.; Wang, J.; Hao, H.; Cai, M.; Wang, S.; Ma, J.; Li, Y.; Mao, C.; Zhang, S. In Vitro and in Vivo Mechanism of Bone Tumors Inhibition by Selenium-Doped Bone Mineral Nanoparticles. *ACS Nano* **2016**, *10*, 9927–9937. [CrossRef] [PubMed]
42. Ma, K.; Wang, D.D.; Lin, Y.; Wang, J.; Valery, P.; Mao, C. Synergetic Targeted Delivery of Sleeping-Beauty Transposon System to Mesenchymal Stem Cells Using LPD Nanoparticles Modified with a Phage-Displayed Targeting Peptide. *Adv. Funct. Mater.* **2013**, *23*, 1172–1181. [CrossRef] [PubMed]
43. Sunderland, K.S.; Yang, M.; Mao, C. Phage-Enabled Nanomedicine: From Probes to Therapeutics in Precision Medicine. *Angew. Chem. Int. Ed.* **2017**, *56*, 1964–1992. [CrossRef] [PubMed]
44. Kim, I.; Kang, K.; Oh, M.H.; Yang, M.Y.; Park, I.; Nam, Y.S. Virus-Templated Self-Mineralization of Ligand-Free Colloidal Palladium Nanostructures for High Surface Activity and Stability. *Adv. Funct. Mater.* **2017**, *27*, 1703262. [CrossRef]
45. Nam, Y.S.; Magyar, A.P.; Lee, D.; Kim, J.W.; Yun, D.S.; Park, H.; Pollar, T.S., Jr.; Weitz, D.A.; Belcher, A.M. Biologically templated photocatalytic nanostructures for sustained light-driven water oxidation. *Nat. Nanotechnol.* **2010**, *5*, 340–344. [CrossRef] [PubMed]
46. Maeda, Y.; Javid, N.; Duncan, K.; Birchall, L.; Gibson, K.F.; Cannon, D.; Kanetsuki, Y.; Knapp, C.; Tuttle, T.; Ulijin, R.V. Discovery of Catalytic Phages by Biocatalytic Self-Assembly. *J. Am. Chem. Soc.* **2014**, *136*, 15893–15896. [CrossRef] [PubMed]
47. Boker, A.; Lin, Y.; Chiapperini, K.; Horowitz, R.; Thompson, M.; Carreon, V.; Xu, T.; Abetz, C.; Skaff, H.; Dinsmore, A.D.; et al. Hierarchical nanoparticle assemblies formed by decorating breath figures. *Nat. Mater.* **2004**, *3*, 302–306. [CrossRef] [PubMed]
48. Zhang, A.; Bai, H.; Li, L. Breath Figure: A Nature-Inspired Preparation Method for Ordered Porous Films. *Chem. Rev.* **2015**, *115*, 9801–9868. [CrossRef] [PubMed]

49. Jung, S.M.; Qi, J.; Oh, D.; Belcher, A.M.; Kong, J. M13 Virus Aerogels as a Scaffold for Functional Inorganic Materials. *Adv. Funct. Mater.* **2017**, *27*, 1603203, doi:10.1002/adfm.201603203. [CrossRef]
50. Lee, Y.M.; Jung, B.; Kim, Y.H.; Park, A.R.; Han, S.; Choe, W.S.; Yoo, P.J. Nanomesh-Structured Ultrathin Membranes Harnessing the Unidirectional Alignment of Viruses on a Graphene-Oxide Film. *Adv. Mater.* **2014**, *26*, 3899–3904. [CrossRef] [PubMed]
51. Kim, W.G.; Song, H.; Kim, C.; Moon, J.S.; Kim, K.; Lee, S.W.; Oh, J.W. Biomimetic self-templating optical structures fabricated by genetically engineered M13 bacteriophage. *Biosens. Bioelectron.* **2016**, *85*, 853–859. [CrossRef] [PubMed]
52. Mansouri, J.; Yapit, E.; Chen, V. Polysulfone filtration membranes with isoporous structures prepared by a combination of dip-coatting and breath figure approach. *J. Membr. Sci.* **2013**, *444*, 237–251. [CrossRef]
53. Yoo, P.J.; Nam, K.T.; Qi, J.; Lee, S.K.; Park, J.; Belcher, A.M.; Hammond, P.T. Spontaneous assembly of viruses on multilayered polymer surfaces. *Nat. Mater.* **2006**, *5*, 234–240. [CrossRef] [PubMed]
54. Devaraj, V.; Baek, J.; Jang, Y.; Jeong, H.; Lee, D. Design for an efficient single photon source based on a single quantum dot embedded in a parabolic solid immersion lens. *Opt. Express* **2016**, *24*, 8045–8053. [CrossRef] [PubMed]

14

Safety Studies of Pneumococcal Endolysins Cpl-1 and Pal

Marek Harhala [1], Daniel C. Nelson [2], Paulina Miernikiewicz [1], Ryan D. Heselpoth [2,†], Beata Brzezicka [1], Joanna Majewska [1], Sara B. Linden [2], Xiaoran Shang [2], Aleksander Szymczak [1], Dorota Lecion [1], Karolina Marek-Bukowiec [3], Marlena Kłak [3], Bartosz Wojciechowicz [4], Karolina Lahutta [1], Andrzej Konieczny [3] and Krystyna Dąbrowska [1,3,*]

1 Bacteriophage Laboratory, Institute of Immunology and Experimental Therapy, Polish Academy of Sciences, Rudolfa Weigla Street 12, 53-114 Wroclaw, Poland; marek.harhala@iitd.pan.wroc.pl (M.H.); pola@iitd.pan.wroc.pl (P.M.); beatabrzezicka94@gmail.com (B.B.); joanna.majewska@iitd.pan.wroc.pl (J.M.); szymczakaleksander@gmail.com (A.S.); dorota.lecion@wp.pl (D.L.); karolina.wojtyna@iitd.pan.wroc.pl (K.L.)

2 Institute for Bioscience and Biotechnology Research, University of Maryland, Rockville, MD 20850, USA; nelsond@umd.edu (D.C.N.); rheselpoth@rockefeller.edu (R.D.H.); slinden1@umd.edu (S.B.L.); sxr520@umd.edu (X.S.)

3 Research and Development Center, Regional Specialized Hospital, 51-124 Wrocław, Poland; marek-bukowiec@wssk.wroc.pl (K.M.-B.); klak@wssk.wroc.pl (M.K.); andrzej_konieczny@yahoo.com (A.K.)

4 Perlan Technologies Sp. z o. o., 02-785 Warsaw, Poland; bwojciechowicz@perlan.com.pl

* Correspondence: dabrok@iitd.pan.wroc.pl

† Current address: Laboratory of Bacterial Pathogenesis and Immunology, The Rockefeller University, New York, NY 10065, USA.

Abstract: Bacteriophage-derived endolysins have gained increasing attention as potent antimicrobial agents and numerous publications document the in vivo efficacy of these enzymes in various rodent models. However, little has been documented about their safety and toxicity profiles. Here, we present preclinical safety and toxicity data for two pneumococcal endolysins, Pal and Cpl-1. Microarray, and gene profiling was performed on human macrophages and pharyngeal cells exposed to 0.5 μM of each endolysin for six hours and no change in gene expression was noted. Likewise, in mice injected with 15 mg/kg of each endolysin, no physical or behavioral changes were noted, pro-inflammatory cytokine levels remained constant, and there were no significant changes in the fecal microbiome. Neither endolysin caused complement activation via the classic pathway, the alternative pathway, or the mannose-binding lectin pathway. In cellular response assays, IgG levels in mice exposed to Pal or Cpl-1 gradually increased for the first 30 days post exposure, but IgE levels never rose above baseline, suggesting that hypersensitivity or allergic reaction is unlikely. Collectively, the safety and toxicity profiles of Pal and Cpl-1 support further preclinical studies.

Keywords: endolysin; Pal; Cpl-1; safety; toxicity; immune response; *Streptococcus pneumoniae*

---

1. Introduction

Streptococcus pneumoniae is the most common cause of bacteremia, pneumonia, meningitis, and otitis media in children. Despite a successful vaccine campaign against pneumococcal disease over the past two decades, a Global Burden of Disease Study suggests that over 500,000 deaths still occur annually due to *S. pneumoniae* infection [1] and the World Health Organization estimated that 5% of all-cause mortality in children under five years of age is related to pneumococcal infection [2]. Furthermore, there has been a worldwide increase in antibiotic-resistant strains of *S. pneumoniae* [3,4], with high rates of resistance reported for penicillin (34.2%), trimethoprim-sulfamethoxazole (31.9%), and erythromycin (29.5%) [5]. The remarkable variability among *S. pneumoniae* strains further complicates vaccine approaches, with at least 92 capsular serotypes having been identified to date [6] and current conjugate vaccines only covering a small subset (i.e., 7–13 serovars) of the most common capsule serotypes. Therefore, alternative antimicrobial approaches to pneumococcal disease are highly desirable.

Bacteriophages, as viruses that infect and kill bacteria, use lytic proteins to destroy the bacterial membrane and peptidoglycan, resulting in lysis of the cell and the release of progeny phage. Among these proteins, endolysins are enzymes that function to break chemical bonds in the peptidoglycan. Appreciably, these enzymes can cause "lysis from without" on susceptible Gram-positive bacteria in the absence of phage due to their actions on the external cell wall and subsequent osmotic lysis of the unprotected bacterial membrane. Because of this unique property, endolysins have been shown to be highly effective antibacterial agents that represent an alternative to antibiotics [7–9].

Several pneumococcal-specific endolysins have been described in the literature. The two most notable of these endolysins are Pal, which is derived from the streptococcal Dp-1 phage [10], and Cpl-1, which is derived from the streptococcal Cp-1 phage [11]. These enzymes have been validated for efficacy in mouse pneumococcal bacteremia models [12,13]. Additionally, Pal and Cpl-1 have been validated in a mouse nasopharyngeal colonization model [14,15], and Cpl-1 has shown efficacy against *S. pneumoniae*-induced endocarditis in rats [16], meningitis in rats [17], and by aerosolized delivery in a mouse model of fatal pneumococcal pneumonia [18]. Lastly, the combinational use of Pal, which possesses an *N*-acetylmuramoyl-L-alanine amidase activity, and Cpl-1, which has an N-acetylmuramidase activity, displayed in vitro and in vivo synergistic efficacy [13,19].

Preclinical safety and toxicity profiles on mammalian cells and tissues are critically important for future translational development of endolysins. Here, we present a safety assessment of Pal and Cpl-1, including an in vivo assessment of safety in a mouse model and gene expression profiling to identify potential effects of these enzymes on human cell functions as well as non-cellular immunological complement cascades in vitro and ex vivo. Immune response assessment further includes inflammatory maker testing (i.e., IL-6) and endolysin-specific IgG and IgE antibody induction, with the latter providing an overall assessment of potential allergic reactions (type I hypersensitivity) [20,21].

2. Materials and Methods

2.1. Protein Expression

Two endolysins—Pal (Acc. no. YP_004306947) from *Streptococcus* phage Dp-1, and Cpl-1 (Acc. no. CAA87744) from *Streptococcus* phage Cp-1—were investigated. The genes for Cpl-1 and Pal were originally amplified directly from the Cp-1 phage and the Dp-1 phage, respectively, into the Gateway cloning system. Subsequently, primers incorporating a C-terminal 6×-His tag were used to amplify from template plasmids containing the native genes. The resulting amplicons were cloned into a pBAD24 expression plasmid. Vectors were expressed in *E. coli* B834(DE3) cells (EMD), and grown at 37 °C with shaking in Luria-Bertani (LB) broth, supplemented with 10 g/L NaCl and ampicillin (100 µg/mL) (all from Sigma-Aldrich, Europe), until OD_{600} reached 1.0. Then, the bacterial culture was cooled to 20 °C and protein expression was induced by the addition of arabinose at a final concentration of 2 g/L. The culture was subsequently incubated overnight at 20 °C with aeration by shaking.

2.2. Protein Purification

Bacteria were harvested using centrifugation and suspended in phosphate buffer (50 mM Na_2HPO_4, 300 mM NaCl, pH 8), which was supplemented with PMSF (1 mM) and lysozyme (0.5 mg/mL). The slurry was incubated for 6–7 h on ice and lysed using the freeze-thaw method. Mg^{2+} (up to 0.25 mM), DNAse (10 µg/mL), and RNAse (20 µg/mL) were then added to the extract and allowed to incubate on ice for an additional 3 h. The fractions were separated using centrifugation

(12,000× g, 45 min, 4 °C) and the soluble fraction (supernatant) was collected. The samples were then incubated with NiNTA agarose (Qiagen, Hilden, Germany) at room temperature for 10 min and washed with PBS (5× volume of the agarose) and an increasing concentration of imidazole (0 mM, 20 mM, 100 mM, 250 mM, and 500 mM). The 100 mM and 250 mM fractions containing the eluted endolysins were dialyzed against PBS at 4 °C. Next, proteins were separated using gel filtration (fast protein liquid chromatography) on a Superdex 75 10/300 GL column (GE Healthcare Life Sciences, Chicago, IL, USA). The final step was LPS removal, which was performed with an EndoTrap Blue column (Hyglos GmbH, Munich, Germany). Purified, endotoxin-free protein samples were dialyzed against PBS and filtered through sterile 0.22-μm polyvinylidene difluoride filters (Millipore, Burlington, MA, USA). All the purification steps were monitored using SDS-PAGE electrophoresis and the LPS (endotoxin) content was determined using the EndoLISA assay (Hyglos GmbH, Munich, Germany) and confirmed to be 0.6–1.6 endotoxin units (EU) per mL for all of the microarray gene expression profiling and less than 10 EU per animal in animal experiments.

2.3. *Microarray Gene Expression Profiling*

The cell lines FaDu (human pharynx squamous cell carcinoma, ATCC HTB-43) and SC (human peripheral blood macrophages, ATCC CRL-9855) were cultured in media recommended by the manufacturer with and without protein solutions containing Pal or Cpl-1 (0.5 μM, i.e., 17.5 and 20 μg/mL, respectively) for 6 h. This time was chosen to allow for the development of cellular responses to external factors at the cellular gene expression level. Control cells were cultured with an equivalent amount of albumin (Sigma, Poznan, Poland). After incubation, cells were harvested and the total RNA was immediately isolated with RNeasy Mini Kit (Qiagen). For this study, we used SurePrint G3 Human Gene Expression v3 8 × 60 K Microarrays (Agilent Technologies, Santa Clara, CA, USA). The One-Color Microarray-Based Gene Expression Analysis Protocol (version 6.9.1) was used to process the arrays.

After amplification, the total RNA was labeled with Cy3 using the Low Input Quick Amp Labeling Kit (Agilent Technologies, USA). The labeled RNA was purified (RNeasy Kit, Qiagen, USA), and the RNA yield (nanograms of complementary RNA (cRNA)) as well as specific activity (picomoles of Cy3 per microgram of cRNA) were measured using a NanoQuant plate (Tecan Group, Germany) in an Infinite 200 PRO reader. Next, the labeled cRNA was fragmented and placed on the microarray slide after mixing with the hybridization buffer. Microarrays were incubated for 17 h at 65 °C and then washed twice in GE wash buffer (Agilent). Agilent's High-Resolution C Microarray Scanner was used to scan the slides according to the 8 × 60 K array format. The scanned images were analyzed with the Agilent Feature Extraction software v. 12.1 (Agilent Technologies, Santa Clara, CA, USA). The final analysis included dye normalization (linear and LOWESS), background subtractions, and filtering of outlier spots.

2.3.1. Differentially Expressed Genes

GeneSpring GX 13 (Agilent, USA) was used to further analyze the data after extraction. The cut-off was set to 1.5-fold for the determination of significant differential expression (up or down regulation). A moderated Student's *t* test was used to determine significant differences, defined as $p \leq 0.05$, for gene expression.

2.3.2. Enriched Gene Ontology Terms & Pathway Analysis

Lists of differentially expressed genes were uploaded to the DAVID 6.7—Database for Annotation, Visualization and Integrated Discovery Classification System to analyze their ontologies and participation in curated pathways, and to perform a Gene Set Enrichment Analysis (GSEA). Ontologies and pathways were assigned independently to upregulated and downregulated gene lists. Annotations (official gene symbols) were limited to *Homo sapiens* and the genome of this species was used as a background for analyses. For the pathway analysis, a Kyoto Encyclopedia of Genes and Genomes (KEGG) was used as a reference database.

2.4. Animal Experiments

Six- to twelve-week-old male mice, either C57BL6/J ($N = 6$) or BALB/c ($N = 5$ or 7), were bred at the Animal Breeding Centre of the Institute of Immunology and Experimental Therapy (IIET). All the animal experiments were approved by the 1st Local Committee for Experiments with the Use of Laboratory Animals, Wrocław, Poland (projects no. 76/2011) and performed according to EU directive 2010/63/EU. All the animal experiments adhered to the ARRIVE (Animal Research: Reporting of In Vivo Experiments) guidelines [22].

For the general health condition (Section 2.5), the microbiome assessment (Section 2.6), and the inflammatory/cytokine assessment (Section 2.7), BALB/c mice were injected with Pal or Cpl-1 intraperitoneally in one dose at 0.3 mg per mouse (15 mg/kg) for each protein. The negative control was inoculated with PBS and a positive control of inflammation was inoculated with LPS (2000 EU/mouse). Murine blood was collected into clotting tubes under anesthesia from the tail vein. Serum was separated from the blood using double centrifugation (2250× g for 5 min and 10,000× g for 10 min). The sera were stored at −20 °C.

2.5. Overall Health Scoring Matrix of Mice

A composite scoring matrix indicating the general health of mice was calculated on a 15-point scale by adding the scores of five aspects, each rated from 0 to 3 (with mid-values possible), where the highest score represented the worst condition. The specific criteria used included: Activity (0 = alert and active, or calm and resting; 3 = slow to move or non-responsive when coaxed or violent reaction to stimuli); Fur (0 = normal, well groomed; 3 = very rough hair coat); Eyes (0 = normal, open, clean, no exudate; 3 = closed, sunken or covered with suppurate exudates); Abdomen (0 = normal; 3 = large abdominal mass and/or edema); and Skin (0 = normal, healthy skin; 3 = wounds, dermatitis, lesions).

2.6. Microbiome Assessment of Mice

Mice were injected *i.p.* with Pal, Cpl-1, or PBS (control), as described in Section 2.4, and fresh fecal samples were analyzed prior to treatment or 24 h after treatment. DNA isolation from mice feces was performed with a QIAamp DNA Stool Mini Kit preceded by physical homogenization with 0.1 mm zirconia beads. A total of 1 mL of InhibitEX buffer was added to the tubes containing the beads and stool samples. The samples were vortexed for 1 min and then incubated at 95 °C for 5 min. Next, each sample was vigorously vortexed for 3 min and centrifuged for 1 min at 14,500 rpm. Further isolation was performed according to the manufacturer's instructions with additional cleaning utilizing the Zymo Clean & Concentrator. The DNA was eluted with deionized water and the samples were stored at –20 °C for further use.

A preliminary measurement of the isolated DNA was performed with the Qubit 2.0 fluorometer using the Qubit™ dsDNA HS assay kit (Life Technologies Corp., Eugene, OR, USA). Three samples of the highest quality from each group were selected for further processing. The Ion 16S Metagenomics Kit (ThermoFisher, Waltham, MA, USA) was used to amplify DNA coding for the 16S rRNA V2, V3, V4, V5, V6–7, V8, and V9 hypervariable regions, according to the manufacturer's instructions using 5 ng of DNA for each sample. Barcoded libraries were created using the Ion Xpress™ Plus Fragment Library Kit with the Ion Xpress Barcodes. The final library concentration was quantified by RT-qPCR with the Ion Library TaqMan Quantitation Kit according to manufacturer's protocol. Emulsion PCR and a bead enrichment step were performed on an Ion OneTouch™ 2 System with the Ion PGM Hi-Q

View OT2 kit. Enriched template beads were mixed with reagents from the Hi-Q View 400 Sequencing kit and loaded onto Ion Torrent 314 V2 chips. Sequencing parameters standard for 16S rRNA Targeted Sequencing were used based on the manufacturer's protocol.

Unaligned binary data files (Binary Alignment Map (.BAM)) generated by the Ion Torrent PGM were uploaded to IonReporter version 5.6. An analysis was performed with the base pair cut-off number set at 150, minimum alignment coverage at 90%, and minimum abundance at 10 copies. Curated MicroSEQ 16S Reference Library v2013.1 was used as a reference database to identify the reads obtained. The results received by the workflow described were visualized using KRONA software integrated into IonReporter 5.6.

2.7. Cytokine Assay

The progression of the inflammatory reaction in the murine blood was monitored by measuring interleukin-6 (IL-6) serum levels using a commercially available Human/Mouse IL-6 Mini ABTS ELISA Development Kit (PeproTech Inc., Rocky Hill, NJ, USA) following the recommendations of the manufacturer.

2.8. Specific Sera Induction

For specific IgG induction, C57BL6/J mice were challenged *s.c.* with Pal or Cpl-1 (0.05 mg per mouse) on day 0. Murine blood was collected into clotting tubes under anesthesia from the tail vein every 3–5 days. Serum was separated from the blood using double centrifugation (2250× *g* for 5 min and 10,000× *g* for 10 min). The sera were stored at −20 °C.

2.9. Measurement of Specific IgG and IgE Antibody Levels

MaxiSorp flat-bottom 96-well plates (Nunc, Thermo Scientific, Waltham, MA, USA) were sterilely coated overnight at 4 °C with endolysins or PBS as control using 100 μL per well at a concentration of 10 μg/mL. Subsequently, wells were washed 5 times with PBS and blocked for 1 h with 1% albumin or SuperBlock Blocking buffer (Life Technologies Europe BV, Bleiswijk, The Netherlands) at 150 μL per well at room temperature. Solution was removed and the plates were washed 5 times with 0.05% Tween 20 (AppliChem GmbH, Darmstadt, Germany) in PBS at 100 μL per well. One hundred microliters per well of diluted serum (1:100 in PBS) was applied to the wells coated with endolysins as well as control wells. Each sample was investigated in duplicate. The plates were incubated at 37 °C for 2 h after which serum was removed and the plates were washed 5 times with 0.05% Tween 20 in PBS at 100 μL per well. One hundred microliters per well of diluted detection antibody (peroxidase-conjugated goat anti-mouse IgG (Jackson ImmunoResearch Laboratories, Cambridgeshire, UK) or IgE (ThermoFisher, Waltham, MA, USA)) was applied to the plates and incubated for 1 h at room temperature in the dark. The antibody solution was removed and the plates were washed with PBS with 0.05% Tween 20 5 times at 100 μL per well. TMB (50 μL) was used as a substrate reagent for peroxidase according to the manufacturer's instructions (R&D Systems, Minneapolis, MN, USA) and incubated for 30 min. Twenty-five microliters of 2N H_2SO_4 was added to stop the reaction and the absorbance was measured at 450 nm (main reading) and normalized by subtracting the background absorbance at 570 nm.

As a reference level of specific IgE antibody induction, an oral mouse allergy model to ovalbumin (OVA) was utilized. Allergy to OVA was induced in mice using a dedicated adjuvant as described [23]. Briefly, the mice were injected subcutaneously with OVA (50 μg/mouse) with $Al(OH)_3$ as an adjuvant promoting the hypersensitivity reaction. The injection was repeated after 14 days. Seven days after the second sensitization, the mice were given 20% Egg White Solution (EWS) (Sigma, Poznan, Poland) in the drinking water. IgE levels specific for OVA were evaluated using ELISA as described above and compared to control mice injected with PBS instead of OVA.

2.10. Complement System Activity Test

Complement assays were performed with the WIESLAB® Complement System kits (Euro Diagnostica, Lundavagen, Sweden) to test the activation of the three complement pathways in the presence of Pal or Cpl-1. Briefly, the kits contained pre-coated plates with specific activators for the classic pathway (CP), alternative pathway (AP), or mannose-binding lectin (MBL) pathway (LP). Blood samples were collected from six healthy human donors. Blood was collected in clotting tubes and incubated for 60 min at room temperature, followed by 10 min centrifugation at 2000× *g*, and isolated serum was immediately used for the tests. First, human serum was incubated at 37 °C for 10 min with 2 µg/mL of Pal, Cpl-1, or PBS in equal volumes. Then, sera were diluted with the provided buffers at 1: 200, 1:18, 1:100 for the CP, AP and LP pathways, respectively, and transferred to the WIESLAB plate and processed according the manufacturer's guidelines. Positive and negative controls included in the test kit served as quality control of the assay.

2.11. Research Ethics Involving Human Subjects

All the subjects gave their informed consent for inclusion before they participated in the study. The study was conducted in accordance with the Declaration of Helsinki, and the protocol was approved by the Bioethics Committee of Regional Specialist Hospital in Wrocław (Project identification code: KB/nr 2/2017).

3. Results

3.1. Microarray Gene Expression Profiling in Human Cells Exposed to Pal or Cpl-1

Analysis of gene expression profiles in eukaryotic cells and their potential changes after exposure to investigated agents provides a sensitive tool for the detection of effects that those agents exert in eukaryotic cells, including potential toxicity and harmful effects. Therefore, gene expression patterns were tested in two human cell lines, normal SC macrophages and a pharyngeal carcinoma cell line, FaDu. Cell lines were exposed to 0.5 µM Pal or Cpl-1 (17.5 and 20 µg/mL, respectively), or albumin (BSA) as a control, for 6 h. DNA microarray analysis (SurePrint G3 Human Gene Expression v3 × 60 K Microarrays; Agilent Technologies, USA) showed no statistically significant ($p < 0.05$) changes in gene expression in either cell line when exposed to endolysins using two different types of statistical analysis, the Bonferroni Holm method for the family-wise error rate (FWER) and the Benjamini Hochberg method for the false discovery rate (FDR) (Supplementary Tables S1 and S2). Taken together, these results demonstrate that Pal and Cpl-1 had no negative effects on the human cells, including provocation of a toxicity or inflammatory response.

3.2. Effect of Pal and Cpl-1 on Complement System Activity in Human Blood Ex Vivo

The complement system is a non-cellular component of blood that plays an important role in immune responses. Therefore, in addition to the cellular assays, potential effects of endolysins on the complement system were also measured ex vivo in human blood with or without Cpl-1 and Pal. Diagnostic complement activity tests included those for the classic, alternative, and MBL-dependent pathways. As can be seen in Figure 1, neither Cpl-1 nor Pal had any effect on the activation of the complement system compared to the control. These results demonstrate that Cpl-1 and Pal do not activate the first line of non-cellular immune response in humans.

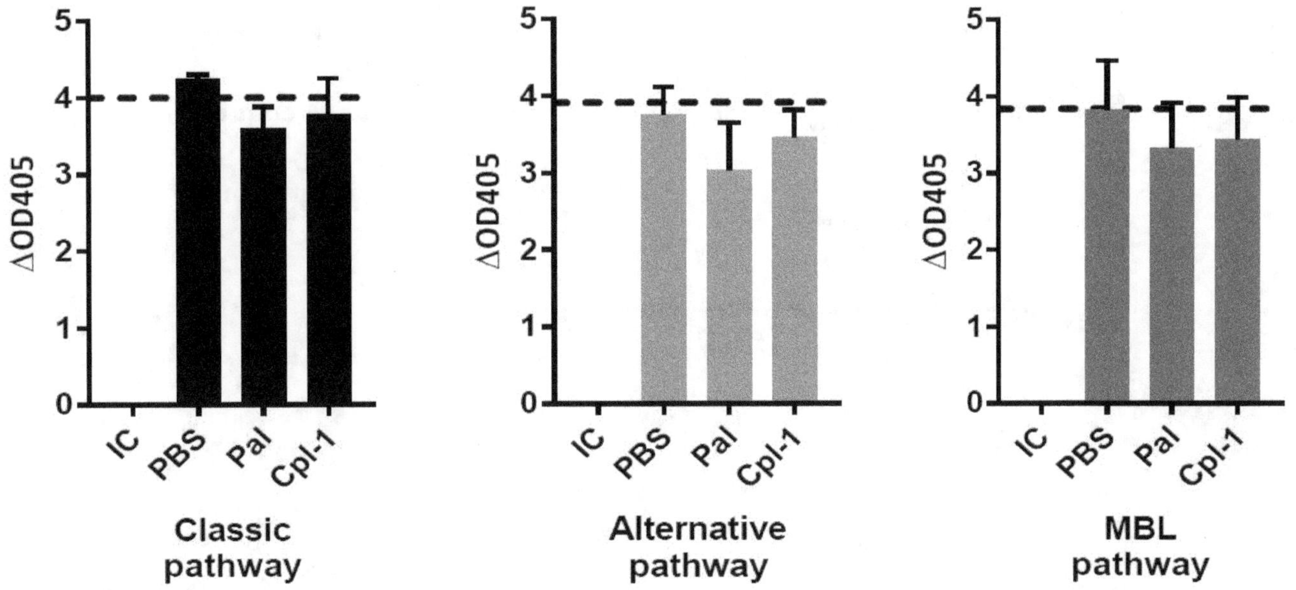

Figure 1. Ex vivo activity of the complement system in human serum incubated with Cpl-1, Pal or PBS. Pal and Cpl-1 tested at a concentration of 2 μg/mL. IC: control of inactivated complement system (as supplied by the manufacturer), dash line—control of normal complement activity (as supplied by the manufacturer).

3.3. Overall Health Condition, Inflammation, and Microbiome Assessment in Mice Challenged with Pal or Cpl-1

Side effects of potential therapeutics may be demonstrated in vivo even though the in vitro and ex vivo assays suggest relative safety of the investigated therapeutics. To this end, we assessed the overall health condition of mice treated with Pal or Cpl-1, as well as the levels of the inflammatory marker, interleukin 6 (IL-6). The composite health status of the mice was assessed and scored 2 h and 5 h after injection with Pal and Cpl-1. Negative controls were treated with the same level of LPS as contained in the purified endolysin solutions and a toxic dose of LPS (2000 EU per mouse) served as a positive control of toxicity. Neither Pal nor Cpl-1 displayed any negative effects on the composite health status of mice compared to the negative controls (Figure 2). In contrast, the mice that received the high LPS dose had a marked, negative effect on their scoring matrix at both time points. A potential inflammatory response to Pal or Cpl-1 was further assessed using measurements of IL-6 levels in murine blood 5 hours after treatment with the endolysins. No significant differences were noted between Pal- or Cpl-1-treated mice compared to control mice for IL-6 production, yet a significant response was seen with mice given a high LPS dose (Figure 3), which recapitulates the health scoring results.

Because endolysins specifically destroy bacterial cells, their therapeutic use can potentially affect the microbiome of living individuals similar to the actions of traditional antibiotics. Abnormalities of microbiome composition can, in turn, affect the overall health status of an individual. Therefore, in mice treated *i.p.* with Pal or Cpl-1, the fecal bacterial microbiome composition was assessed using 16S rRNA targeted sequencing. Analysis of microbiome before treatment and 24 h after treatment revealed no meaningful changes in the general composition of bacterial groups (Supplementary Table S3, Figures S1 and S2). Specifically, the major groups remained unchanged (and no loss was observed in the Firmicutes group, which contains bacteria sensitive to Pal and Cpl-1).

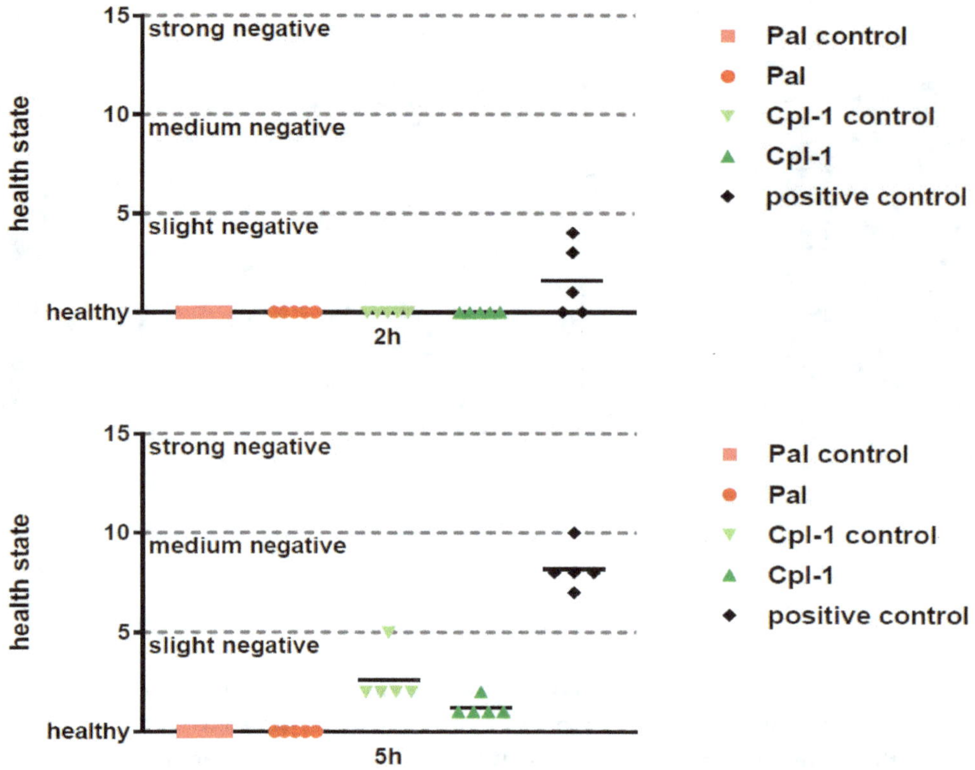

Figure 2. Composite health matrix of mice treated intraperitoneally with Pal and Cpl-1. Pal: mice treated with 0.3 mg of Pal; Pal control: mice treated with PBS containing the same residual LPS content as Pal (0.6 EU); Cpl-1: mice treated with 0.3 mg of Cpl-1; Cpl-1 control: mice treated with PBS containing the same residual LPS content as Cpl-1 (8 EU); positive control: mice treated with a toxic LPS dose (2000 EU). Upper panel: 2 h after treatment; Lower panel: 5 h after treatment.

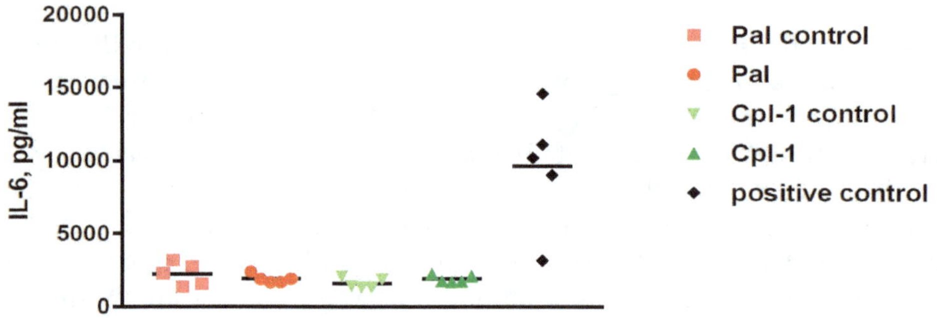

Figure 3. IL-6 cytokine levels in murine blood 5 h after intraperitoneal treatment with Pal or Cpl-1. Pal: mice treated with 0.3 mg of Pal; Pal control: mice treated with PBS containing the same residual LPS content as Pal (0.6 EU); Cpl-1: mice treated with 0.3 mg of Cpl-1; Cpl-1 control: mice treated with PBS containing the same residual LPS content as Cpl-1 (8 EU); positive control: mice treated with a toxic LPS dose (2000 EU).

3.4. Antibody Induction in Mice Challenged with Pal and Cpl-1

Specific immune responses to Pal and Cpl-1 were assessed in a mouse model through the analysis of specific antibody induction. The serum levels of IgG were measured to evaluate the normal antibody response and serum IgE levels were determined for a potential hypersensitivity response to the endolysins. Murine sera were collected at days 1, 5, 10, 15, 20, 25, 30 and 50 after a single-dose application of each enzyme. As expected, a typical pattern of IgG induction was observed, with a slow increase in the specific antibody titer until ~day 30, then the titers leveled off (Figure 4). Importantly, during the entire 50-day evaluation period, no increase was observed for

Pal- or Cpl-1-specific IgE relative to control mice ($p < 0.05$) (Figure 4). As a control to demonstrate the induction of IgE hypersensitivity, the same assays were performed on mice sensitized to OVA, which showed a large IgE response after 21 days (Supplemental Figure S3). Thus, we detected no propensity for the development of an IgE-mediated allergic reaction to Pal or Cpl-1.

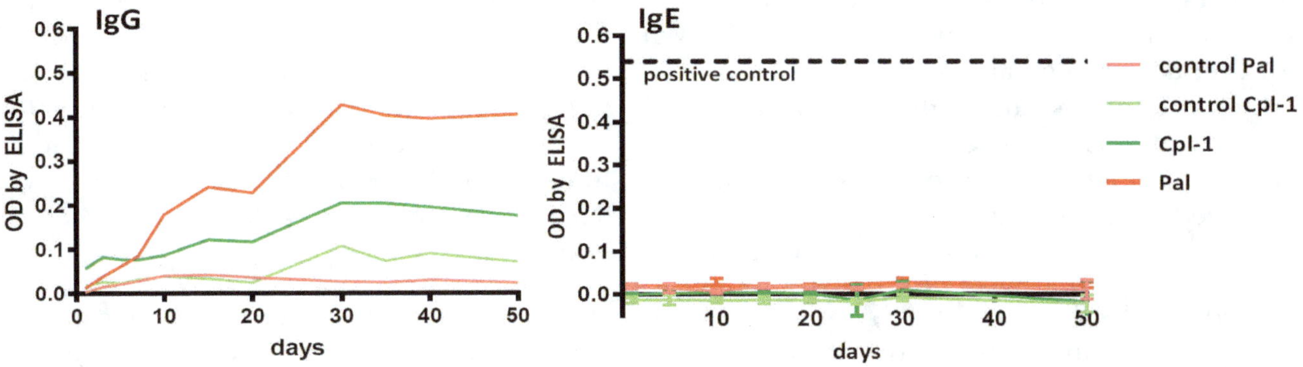

Figure 4. Specific IgG and IgE antibody levels in mice challenged with Pal or Cpl-1. Mice ($N = 6$) were injected intraperitoneally with enzymes 50 µg per mouse and control animals were injected with the relevant amount of LPS. Serum samples were collected on days 1, 5, 10, 15, 20, 25, 30, and 50 and tested using ELISA with Pal or Cpl-1 as the antigens. Pal: Pal-specific antibody level in mice treated with Pal; control Pal: Pal-specific antibody level in control mice; Cpl-1: Cpl-1-specific antibody level in mice treated with Cpl-1; control Cpl-1: Cpl-1-specific antibody level in control mice. Positive control reference level in IgE: antibodies specific to a model antigen (OVA) in allergic mice (Supplementary Figure S3).

4. Discussion

Bacteriophage endolysins have received increasing attention as potent alternatives to traditional antibiotics in the past decade and at least three companies are enrolling patients for Phase 2 clinical trials with endolysin-based therapeutics, according to clinicaltrials.gov. Nonetheless, there are few published articles describing the safety or toxicity of endolysins. The exception is SAL200, a pharmaceutical composition containing the SAL-1 endolysin, which targets *Staphylococcus aureus*. SAL200 showed no toxicity in rodent and dog single- and repeated-dose studies and pharmacology studies demonstrated no signs of toxicity in central nervous and respiratory system function tests [24]. Next, SAL200 was tested in monkeys to obtain pharmacokinetic and safety information. No laboratory abnormalities or adverse events were detected after injection of a single dose (up to 80 mg/kg per day) or injection of multiple doses (up to 40 mg/kg per day) [25]. Finally, the results for the human Phase 1 study of SAL200, the first in-human study for a bacteriophage endolysin-based drug, were recently published [26]. SAL200 was well-tolerated, and no serious adverse events were observed in this clinical study, with most adverse events being mild, self-limiting, and transient. Furthermore, no clinically significant values with respect to clinical chemistry, hematology, coagulation, urinalysis, vital signs, or physical examinations were observed.

In the present study, two pneumococcal endolysins, Pal and Cpl-1, were investigated for their potential effect on mammalian cells, cellular and non-cellular immune responses, microbiome changes, inflammatory response, and overall safety in a single-dose study in mice. We found that in most cases, Pal and Cpl-1 were neutral and exerted minor or no effect on cell or systemic functions.

Microarray and gene profiling analyses of human SC macrophages and human pharyngeal FaDu cells were analyzed after exposure to Pal and Cpl-1. Macrophages were chosen as they allow for the identification of potential effects of the endolysins on immune responses, whereas the pharyngeal cells represent non-immunological cells from a body site commonly colonized by *S. pneumoniae* [27,28]. In both types of cells, no statistically significant changes in the expression levels of over 16,000 genes were noted compared to albumin-treated cells (Tables S1 and S2), indicating that no specific pathways, like apoptosis or inflammatory responses, were activated by the enzymes, thus strongly supporting the safety of Pal and Cpl-1. It should be noted that Entenza et al. found increased serum pro-inflammatory cytokine (i.e., IL-1β, IL-6, TNF-α, and INF-γ) levels in a rat endocarditis model of *S. pneumoniae* when treated with high, continuous levels of Cpl-1 [16]. However, both our gene profiling results as well as the direct serum measurements of IL-6 (Figure 3) showed no increase due to endolysins, suggesting that the results of Entenza et al. reflect responses to the lysed bacterial cells rather than endolysins themselves. This is supported by Witzenrath et al., who showed that lower levels of Cpl-1 reduced pro-inflammatory cytokines in infected mice relative to untreated, infected mice [29]. It was hypothesized that higher concentrations of Cpl-1 generate a more complete digestion of the peptidoglycan, thereby generating more pro-inflammatory fragments to induce a response. Due to the lytic nature of these enzymes, detailed dosing studies in conjunction with safety profiles will be needed to achieve an optimal safe, effective dose. Nonetheless, our results demonstrate that the purified enzymes themselves cause no inflammatory responses.

Our results show that an IgG response is generated toward endolysins, which is not unexpected due to the proteinaceous nature of these enzymes. However, it has been reported that antibodies do not neutralize the catalytic activity of endolysins. Specifically, hyperimmune serum raised against Cpl-1 only slowed the killing kinetics of Cpl-1 against *S. pneumoniae* in vitro compared to identical experiments using pre-immune serum [12] (Figure S4). Likewise, Cpl-1 equally reduced the pneumococcal titer in the blood of both naïve mice and mice previously vaccinated with Cpl-1 in a bacteremia model. Similar results demonstrating high titer, but non-neutralizing effects of antibodies directed against endolysins have been reported for the MV-L and ClyS staphylococcal endolysins [30,31]. At present, it is not known why antibodies do not neutralize the activity of endolysins, but it is known that endolysins bind epitopes on the bacterial surface with nanomolar to picomolar affinities [32,33], which could out-compete the binding of the antibodies. The accumulated results suggest that endolysins, including Pal and Cpl-1, may be used as therapeutics repeatedly without diminished activity due to antibody response. Moreover, our results demonstrate for the first time in any pneumococcal endolysin that Pal and Cpl-1 do not generate an IgE response associated with hypersensitivity and allergic reaction.

One advantage of endolysin therapy is the exquisite specificity, often confined to a single species, displayed by the endolysins [8,9]. The targeted killing by endolysins prevents collateral bacteriolytic effects on commensal organisms and dysbiosis often associated with conventional antibiotic therapy. In the case of Pal and Cpl-1, the pneumococcal specificity is defined by C-terminal choline binding domains in each enzyme that bind with high affinity to the choline-containing wall teichoic acids of *S. pneumoniae* [34]. This is an advantage of Cpl-1 and Pal over other endolysins; only a few related oral bacteria from the mitis group are poorly susceptible compared to the rapid bactericidal activity on pneumococci. In this line of evidence, it is unlikely that Cpl-1 or Pal would have substantial bactericidal activity on the bacteria present in the fecal microbiome. As such, changes in the microbiome were not anticipated, and indeed were not observed (Supplementary Table S2, Figure S2). Cheng et al. showed that low doses (5 μg) of LysEF-P10 did not affect the gut microbiome, but that high doses (100 μg) did affect the microbiome, specifically lowering members of the geneus *Enterococcus* [35]. In contrast, our results showed that 300 μg of either Cpl-1 or Pal had no effect on the fecal microbiome. This is likely attributable to either the broad specificity of LysEF-P10 or the natural abundance of the enterococci in the fecal microbiome compared to the very narrow specificity of Cpl-1/Pal and the low natural abundance of pneumococci in the fecal microbiome.

We also tested the safety of both endolysins in vivo, at concentrations (0.3 mg/mouse; ~15 mg/kg based on a 20 g mouse) higher than previously used to show the efficacy of Cpl-1 in a pneumococcal sepsis model [13]. The overall composite health score indicated no adverse effects of endolysin treatment (Figure 2). Only the Cpl-1 and the Cpl-1 control showed any signs that deviated from perfect health, although the signs did not even reach the level of a slight effect. Additionally, the endotoxin levels in the Cpl-1 preparations was >10 times higher than the levels in the Pal preparations (8 EU vs.

0.6 EU), underscoring the importance of endotoxin removal and testing. Taken together, our preclinical safety results of Pal and Cpl-1 support the implementation of future escalating-dose safety studies, as well as pharmacokinetic and pharmacodynamic studies in larger mammals and primates.

Supplementary Materials
Table S1: Statistical data from the analysis of gene expression in SC cells exposed to Pal or Cpl-1 compared to albumin (Alb), Table S2: Statistical data from the analysis of gene expression in FaDu cells exposed to Pal or Cpl-1 compared to albumin (Alb), Table S3: Composition of bacterial microbiome component in mice after treatment with Pal or Cpl-1, Figure S1: Shannon and Simpson indexes for bacterial microbiome in the gut of mice challenged with Pal or Cpl-1, Figure S2: Composition of bacterial microbiome component in mice before and after treatment with Pal or Cpl-1, Figure S3: Serum levels of IgE specific for OVA, positive reference of allergy in mice. Figure S4: Bacterial culture density 15 min after treatment with Cpl-1 or Pal lysins (final concentration: 25 μg/mL).

Author Contributions: K.D., M.H., and D.C.N. conceived and designed the experiments; M.H., R.D.H, S.B.L. and X.S. cloned, expressed, and purified the endolysins; M.H., K.D., P.M., B.B, J.M., A.S., M.K., K.L. and D.L performed the experiments; K.D., B.W., M.H., K.M.-B. and A.K. analyzed the data; the manuscript was written by K.D., M.H., and D.C.N. The overall editing was performed by D.C.N.

References

1. Abubakar, I.I.; Tillmann, T.; Banerjee, A. Global, regional, and national age-sex specific all-cause and cause-specific mortality for 240 causes of death, 1990–2013: A systematic analysis for the global burden of disease study 2013. *Lancet* **2015**, *385*, 117–171.
2. Estimated Hib and Pneumococcal Deaths for Children under 5 Years of Age. 2008. Available online: http://www.who.int/immunization/monitoring_surveillance/burden/estimates/Pneumo_hib/en/ (accessed on 15 March 2017).
3. Butler, J.C.; Hofmann, J.; Cetron, M.S.; Elliott, J.A.; Facklam, R.R.; Breiman, R.F. The continued emergence of drug-resistant *Streptococcus pneumoniae* in the United States: An update from the centers for disease control and prevention's pneumococcal sentinel surveillance system. *J. Infect. Dis.* **1996**, *174*, 986–993. [CrossRef] [PubMed]
4. Song, J.Y.; Nahm, M.H.; Moseley, M.A. Clinical implications of pneumococcal serotypes: Invasive disease potential, clinical presentations, and antibiotic resistance. *J. Korean Med. Sci.* **2013**, *28*, 4–15. [CrossRef] [PubMed]
5. Doern, G.V.; Richter, S.S.; Miller, A.; Miller, N.; Rice, C.; Heilmann, K.; Beekmann, S. Antimicrobial resistance among *Streptococcus pneumoniae* in the United States: Have we begun to turn the corner on resistance to certain antimicrobial classes? *Clin. Infect. Dis.* **2005**, *41*, 139–148. [CrossRef] [PubMed]
6. Steel, H.C.; Cockeran, R.; Anderson, R.; Feldman, C. Overview of community-acquired pneumonia and the role of inflammatory mechanisms in the immunopathogenesis of severe pneumococcal disease. *Mediat. Inflamm.* **2013**, *2013*, 490346. [CrossRef] [PubMed]
7. Gutierrez, D.; Fernandez, L.; Rodriguez, A.; Garcia, P. Are phage lytic proteins the secret weapon to kill *Staphylococcus aureus*? *mBio* **2018**, *9*, e01923-17. [CrossRef] [PubMed]
8. Nelson, D.C.; Schmelcher, M.; Rodriguez-Rubio, L.; Klumpp, J.; Pritchard, D.G.; Dong, S.; Donovan, D.M. Endolysins as antimicrobials. *Adv. Virus Res.* **2012**, *83*, 299–365. [PubMed]
9. Oliveira, H.; Sao-Jose, C.; Azeredo, J. Phage-derived peptidoglycan degrading enzymes: Challenges and future prospects for in vivo therapy. *Viruses* **2018**, *10*, 292. [CrossRef] [PubMed]
10. Garcia, P.; Mendez, E.; Garcia, E.; Ronda, C.; Lopez, R. Biochemical characterization of a murein hydrolase induced by bacteriophage dp-1 in *Streptococcus pneumoniae*: Comparative study between bacteriophage-associated lysin and the host amidase. *J. Bacteriol.* **1984**, *159*, 793–796. [PubMed]
11. Garcia, J.L.; Garcia, E.; Arraras, A.; Garcia, P.; Ronda, C.; Lopez, R. Cloning, purification, and biochemical characterization of the pneumococcal bacteriophage cp-1 lysin. *J. Virol.* **1987**, *61*, 2573–2580. [PubMed]
12. Loeffler, J.M.; Djurkovic, S.; Fischetti, V.A. Phage lytic enzyme cpl-1 as a novel antimicrobial for pneumococcal bacteremia. *Infect. Immun.* **2003**, *71*, 6199–6204. [CrossRef] [PubMed]
13. Jado, I.; Lopez, R.; Garcia, E.; Fenoll, A.; Casal, J.; Garcia, P. Phage lytic enzymes as therapy for antibiotic-resistant *Streptococcus pneumoniae* infection in a murine sepsis model. *J. Antimicrob. Chemother.* **2003**, *52*, 967–973. [CrossRef] [PubMed]

14. Loeffler, J.M.; Nelson, D.; Fischetti, V.A. Rapid killing of *Streptococcus pneumoniae* with a bacteriophage cell wall hydrolase. *Science* **2001**, *294*, 2170–2172. [CrossRef] [PubMed]
15. Corsini, B.; Diez-Martinez, R.; Aguinagalde, L.; Gonzalez-Camacho, F.; Garcia-Fernandez, E.; Letrado, P.; Garcia, P.; Yuste, J. Chemotherapy with phage lysins reduces pneumococcal colonization of the respiratory tract. *Antimicrob. Agents Chemother.* **2018**, *62*, AAC-02212. [CrossRef] [PubMed]
16. Entenza, J.M.; Loeffler, J.M.; Grandgirard, D.; Fischetti, V.A.; Moreillon, P. Therapeutic effects of bacteriophage Cpl-1 lysin against *Streptococcus pneumoniae* endocarditis in rats. *Antimicrob. Agents Chemother.* **2005**, *49*, 4789–4792. [CrossRef] [PubMed]
17. Grandgirard, D.; Loeffler, J.M.; Fischetti, V.A.; Leib, S.L. Phage lytic enzyme Cpl-1 for antibacterial therapy in experimental pneumococcal meningitis. *J. Infect. Dis.* **2008**, *197*, 1519–1522. [CrossRef] [PubMed]
18. Doehn, J.M.; Fischer, K.; Reppe, K.; Gutbier, B.; Tschernig, T.; Hocke, A.C.; Fischetti, V.A.; Loffler, J.; Suttorp, N.; Hippenstiel, S.; et al. Delivery of the endolysin Cpl-1 by inhalation rescues mice with fatal pneumococcal pneumonia. *J. Antimicrob. Chemother.* **2013**, *68*, 2111–2117. [CrossRef] [PubMed]
19. Loeffler, J.M.; Fischetti, V.A. Synergistic lethal effect of a combination of phage lytic enzymes with different activities on penicillin-sensitive and -resistant *Streptococcus pneumoniae* strains. *Antimicrob. Agents Chemother.* **2003**, *47*, 375–377. [CrossRef] [PubMed]
20. Gould, H.J.; Sutton, B.J.; Beavil, A.J.; Beavil, R.L.; McCloskey, N.; Coker, H.A.; Fear, D.; Smurthwaite, L. The biology of IgE and the basis of allergic disease. *Annu. Rev. Immunol.* **2003**, *21*, 579–628. [CrossRef] [PubMed]
21. Sampson, H.A.; Ho, D.G. Relationship between food-specific IgE concentrations and the risk of positive food challenges in children and adolescents. *J. Allergy Clin. Immunol.* **1997**, *100*, 444–451. [CrossRef]
22. Kilkenny, C.; Browne, W.J.; Cuthill, I.C.; Emerson, M.; Altman, D.G. Improving bioscience research reporting: The arrive guidelines for reporting animal research. *Osteoarthr. Cartil.* **2012**, *20*, 256–260. [CrossRef] [PubMed]
23. Saldanha, J.C.; Gargiulo, D.L.; Silva, S.S.; Carmo-Pinto, F.H.; Andrade, M.C.; Alvarez-Leite, J.I.; Teixeira, M.M.; Cara, D.C. A model of chronic IgE -mediated food allergy in ovalbumin-sensitized mice. *Braz. J. Med. Biol. Res.* **2004**, *37*, 809–816. [CrossRef] [PubMed]
24. Jun, S.Y.; Jung, G.M.; Yoon, S.J.; Choi, Y.J.; Koh, W.S.; Moon, K.S.; Kang, S.H. Preclinical safety evaluation of intravenously administered sal200 containing the recombinant phage endolysin sal-1 as a pharmaceutical ingredient. *Antimicrob. Agents Chemother.* **2014**, *58*, 2084–2088. [CrossRef] [PubMed]
25. Jun, S.Y.; Jung, G.M.; Yoon, S.J.; Youm, S.Y.; Han, H.Y.; Lee, J.H.; Kang, S.H. Pharmacokinetics of the phage endolysin-based candidate drug sal200 in monkeys and its appropriate intravenous dosing period. *Clin. Exp. Pharmacol. Physiol.* **2016**, *43*, 1013–1016. [CrossRef] [PubMed]
26. Jun, S.Y.; Jang, I.J.; Yoon, S.; Jang, K.; Yu, K.S.; Cho, J.Y.; Seong, M.W.; Jung, G.M.; Yoon, S.J.; Kang, S.H. Pharmacokinetics and tolerance of the phage endolysin-based candidate drug sal200 after a single intravenous administration among healthy volunteers. *Antimicrob. Agents Chemother.* **2017**, *61*, AAC-02629. [CrossRef] [PubMed]
27. Principi, N.; Preti, V.; Gaspari, S.; Colombini, A.; Zecca, M.; Terranova, L.; Cefalo, M.G.; Ierardi, V.; Pelucchi, C.; Esposito, S. *Streptococcus pneumoniae* pharyngeal colonization in school-age children and adolescents with cancer. *Hum. Vaccin. Immunother.* **2016**, *12*, 301–307. [CrossRef] [PubMed]
28. Principi, N.; Terranova, L.; Zampiero, A.; Manzoni, F.; Senatore, L.; Rios, W.P.; Esposito, S. Oropharyngeal and nasopharyngeal sampling for the detection of adolescent *streptococcus pneumoniae* carriers. *J. Med. Microbiol.* **2014**, *63*, 393–398. [CrossRef] [PubMed]
29. Witzenrath, M.; Schmeck, B.; Doehn, J.M.; Tschernig, T.; Zahlten, J.; Loeffler, J.M.; Zemlin, M.; Muller, H.; Gutbier, B.; Schutte, H.; et al. Systemic use of the endolysin Cpl-1 rescues mice with fatal pneumococcal pneumonia. *Crit. Care Med.* **2009**, *37*, 642–649. [CrossRef] [PubMed]
30. Daniel, A.; Euler, C.; Collin, M.; Chahales, P.; Gorelick, K.J.; Fischetti, V.A. Synergism between a novel chimeric lysin and oxacillin protects against infection by methicillin-resistant *Staphylococcus aureus*. *Antimicrob. Agents Chemother.* **2010**, *54*, 1603–1612. [CrossRef] [PubMed]
31. Rashel, M.; Uchiyama, J.; Ujihara, T.; Uehara, Y.; Kuramoto, S.; Sugihara, S.; Yagyu, K.; Muraoka, A.; Sugai, M.; Hiramatsu, K.; et al. Efficient elimination of multidrug-resistant *Staphylococcus aureus* by cloned lysin derived from bacteriophage phi MR11. *J. Infect. Dis.* **2007**, *196*, 1237–1247. [CrossRef] [PubMed]

32. Loessner, M.J.; Kramer, K.; Ebel, F.; Scherer, S. C-terminal domains of *Listeria monocytogenes* bacteriophage murein hydrolases determine specific recognition and high-affinity binding to bacterial cell wall carbohydrates. *Mol. Microbiol.* **2002**, *44*, 335–349. [CrossRef] [PubMed]
33. Schmelcher, M.; Shabarova, T.; Eugster, M.R.; Eichenseher, F.; Tchang, V.S.; Banz, M.; Loessner, M.J. Rapid multiplex detection and differentiation of *listeria* cells by use of fluorescent phage endolysin cell wall binding domains. *Appl. Environ. Microbiol.* **2010**, *76*, 5745–5756. [CrossRef] [PubMed]
34. Maestro, B.; Sanz, J.M. Choline binding proteins from *Streptococcus pneumoniae*: A dual role as enzybiotics and targets for the design of new antimicrobials. *Antibiotics* **2016**, *5*, 21. [CrossRef] [PubMed]
35. Cheng, M.; Zhang, Y.; Li, X.; Liang, J.; Hu, L.; Gong, P.; Zhang, L.; Cai, R.; Zhang, H.; Ge, J.; et al. Endolysin lysef-p10 shows potential as an alternative treatment strategy for multidrug-resistant *Enterococcus faecalis* infections. *Sci. Rep.* **2017**, *7*, 10164. [CrossRef] [PubMed]

15

Nanomedicine and Phage Capsids

Philip Serwer * and **Elena T. Wright**

Department of Biochemistry and Structural Biology, The University of Texas Health Science Center, San Antonio, TX 78229-3900, USA; wrighte@uthscsa.edu

* Correspondence: serwer@uthscsa.edu

Abstract: Studies of phage capsids have at least three potential interfaces with nanomedicine. First, investigation of phage capsid states potentially will provide therapies targeted to similar states of pathogenic viruses. Recently detected, altered radius-states of phage T3 capsids include those probably related to intermediate states of DNA injection and DNA packaging (dynamic states). We discuss and test the idea that some T3 dynamic states include extensive α-sheet in subunits of the capsid's shell. Second, dynamic states of pathogenic viral capsids are possible targets of innate immune systems. Specifically, α-sheet-rich innate immune proteins would interfere with dynamic viral states via inter-α-sheet co-assembly. A possible cause of neurodegenerative diseases is excessive activity of these innate immune proteins. Third, some phage capsids appear to have characteristics useful for improved drug delivery vehicles (DDVs). These characteristics include stability, uniformity and a gate-like sub-structure. Gating by DDVs is needed for (1) drug-loading only with gate opened; (2) closed gate-DDV migration through circulatory systems (no drug leakage-generated toxicity); and (3) drug release only at targets. A gate-like sub-structure is the connector ring of double-stranded DNA phage capsids. Targeting to tumors of phage capsid-DDVs can possibly be achieved via the enhanced permeability and retention effect.

Keywords: alpha-sheet; cancerous tumors; capsid dynamics; drug delivery vehicles; native gel electrophoresis; neurodegenerative disease; pathogenic viruses

1. Introduction

1.1. A Principle

The following principle (basics-focus principle) is proposed here as a foundation for uses of phages in nanomedicine. The curing of currently intractable, biochemically complex diseases requires understanding of the disease basics, but it does not require understanding of the disease details. Historical justifications for the basics-focus principle include (1) the development of smallpox [1] and rabies [2] vaccines before any details about virus composition and structure were known; (2) the discovery of bacterial cell wall-active [3,4] and ribosome-active [5] antibiotics before the composition and structure of bacterial cell walls and ribosomes were known and (3) the use of phage therapy for infectious disease in the absence of knowledge of the composition and structure of phages [6,7].

Applying the basics-focus principle does not mean neglecting the rest of the science. Louis Pasteur's basics-oriented work on improving wine and beer fermentation was a major part of the foundation for the fields of biochemistry and microbiology [2,8]. Antibiotics became major tools in investigating the composition, structure and dynamics of bacterial cell walls [4] and ribosomes [5,9].

A proposed corollary is the following. Basics-focus on practical aspects of neurodegenerative and other diseases will not compromise the remaining science. Indeed, as we will describe in Section 6, we think that the remaining science will also be promoted with this approach.

However, to get started, one has to know enough basics. We think likely that, at least in the case of neurodegenerative diseases, some basics will have to be assumed, without rigorous proof. In the discussion below, we will present both key assumptions and the phage-based evidence in the case of neurodegenerative disease. We will also present a basics-oriented strategy for malignant tumor therapy. This strategy includes use of a phage capsid-based drug delivery vehicle (DDV).

1.2. The Scientific Environment

Articulation of the basics-focus principle is motivated, in part, by a current environment in which progress is politely described as slow for the curing of both neurodegenerative diseases and metastatic cancer. Less politely, but probably more accurately, most (not all) research on neurodegenerative diseases appears to be mired in its focus on complex biochemical details, with only limited symptomatic relief achieved [10–13]. Focus on these details is the opposite of (1) what has historically been the most successful focus (previous section) and (2) what some fundamentals project to be the optimal focus for developing future therapies for pathogenic virus infections (Section 2.5), neurodegenerative diseases (Section 3.2) and cancerous tumors (Section 4.2). Lack of basics-focus is one possible explanation of why neurodegenerative diseases are incurable at this stage in history.

Two recent books appear to be warning signs that public patience is beginning to exhaust. A recent, cancer-oriented book comes very close to laying the slow-progress blame at the feet of science (really, scientists) [14]. A recent, polio vaccine-oriented book suggests the following. Without the intervention of the law partner of an American President, polio vaccine, as we know it, would not have existed as early as it did [15]. A logical rendition of the current state of phage therapy is likely to exhaust public patience to a new level because, in this case, successful application of the basics has already been achieved ([16], reviewed in [17,18]). However, apparently, in the US, one can receive phage therapy only on an ad hoc basis [16,19].

Our entire research history is in the area of phages, with a primary focus on phage assembly. Historically speaking, the Caltech phage group (the home of PS for four years) was, in its early years, supported by the foundation that previously supported the basics focus-oriented work on polio vaccines. This foundation, The March of Dimes (also called the National Foundation), also supported the Pauling-associated work on structure discussed below (see the credits in references [20–23]). That is to say, the basic philosophy in the current article appears to have an indirect linkage to the distant past.

1.3. Phage Assembly Basics

The composition and structure of phage T3 and its relative, T7, are illustrated in Figure 1d. The phage particle consists of (1) a DNA-encapsulating shell of the protein product of *gene 10*, called gp10; (2) an external structure (tail) adapted for specific binding to host cells and (3) an internal core stack [24–26]. The various phage proteins are labeled by gp, followed by the number of the encoding gene. A T3 gene is given the same number as its T7 counterpart [27].

The DNA-containing, protein capsid of all well-studied, double-stranded DNA phages begins its existence as a DNA-free capsid, called a procapsid. The T3/T7 procapsid (also called capsid I) is illustrated in Figure 1a [24]. The procapsid subsequently packages a DNA genome and, while so doing, changes its structure to form a more phage-like capsid, called capsid II for T3/T7 (Figure 2b). At the end of packaging, the T3/T7 tail is attached to a ring (called portal or connector) that is on a DNA-filled capsid (Figure 2c) called a head [26]. The connector (1) is the site of DNA entry; (2) occupies a 5-fold vertex of an icosahedral, DNA-containing gp10 shell and (3) forms the base of the core stack [24,25].

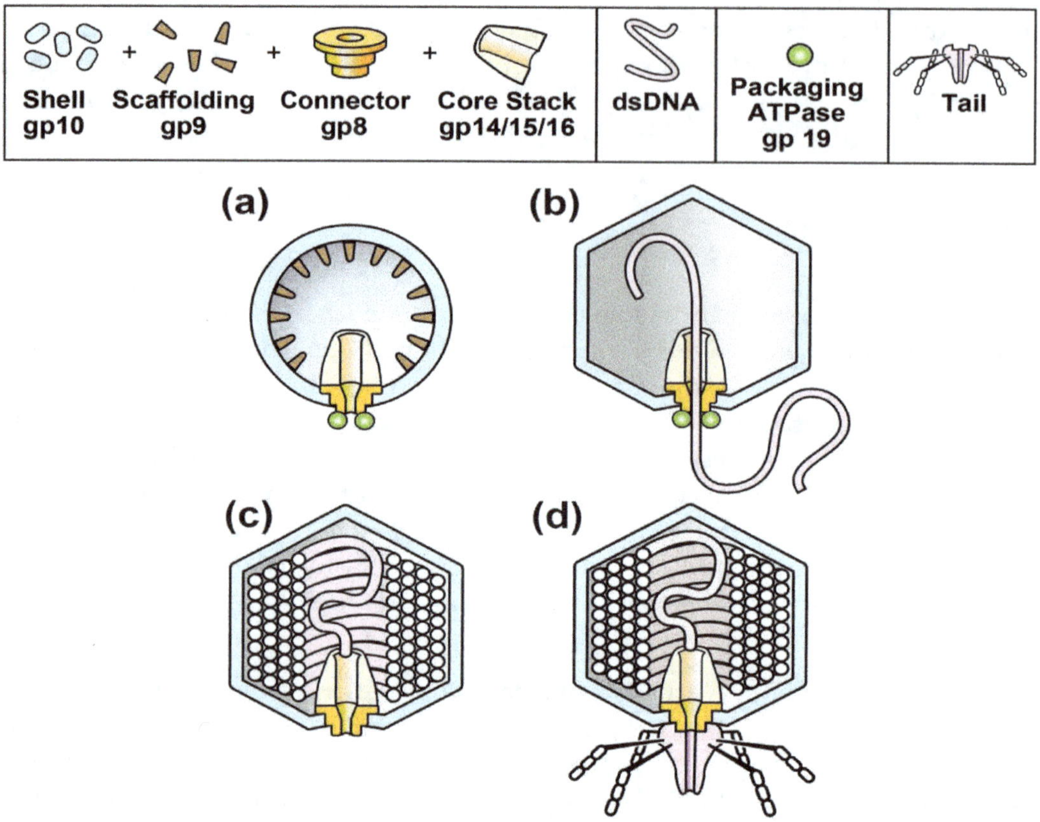

Figure 1. The progression of capsids during the in vivo assembly of phages T3 and T7. (**a**) capsid I; (**b**) capsid II packaging DNA; (**c**) head; (**d**) mature phage.

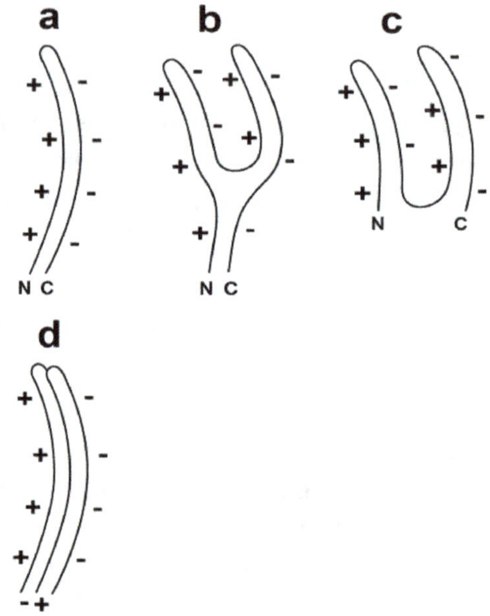

Figure 2. A line drawing of the proposed α-sheet-generating polypeptide backbone of the gp10 subunits of (**a**) hyper-expanded T3/T7 capsid II; (**b**) an intermediate converting from its state in hyper-expanded to its state in contracted capsid II and (**c**) contracted capsid II. N and C indicate the N- and C-terminals of gp10; (**d**) Assembly of two gp10 subunits is shown with the proposed radial staggering. The staggering improves electrical charge-charge-derived energetics. The + symbols indicate the + electrical charge of the α-amino edge; the − symbols indicate the − electrical charge of the α-carboxyl edge.

2. Phage Assembly and Dynamic States

2.1. *The Stability of the Icosahedral Shell of the Related Phages, T3 and T7*

One's first impression of the mature T3/T7 capsid is that the DNA-enclosing gp10 shell is extremely stable and unlikely to change states. The shell is resistant to both ionic detergent [28] and the proteases, trypsin and subtilisin [29]. In a mature phage T3/T7 capsid, the major shell protein has a conformation [25] in common among the various double-stranded DNA phages. This conformation is called the HK97 fold, named after the first phage found to have this fold [30].

The mature gp10 shell of T7 capsid II also has a surprisingly high stability to elevated temperature, as discussed in Section 4.3. This characteristic suggests resistance to damage during possible use as a capsid-drug delivery vehicle (DDV).

Nonetheless, dynamism of phage shells is suggested by cryo-electron microscopy (cryo-EM) analysis. After expulsion of DNA from phage HK97, the shell "bows out" more than it does before expulsion [31]. Similarly, T7 capsid II is found 1.4% larger than the mature capsid when purified [25] and in lysates [32]. If one makes the assumption that HK97-type shell proteins have incompressible components, without capacity for storing and releasing energy, the above result with HK97 is impossible. The reason is that DNA expulsion releases pressure from packaged DNA. This would cause contraction, not expansion, if one makes this assumption. Therefore, the shell of HK97 and probably T3/T7 can move internally to store and release energy.

Less direct evidence for shell dynamics arises from analysis of the leakage of DNA from (tail-free) phage T3 heads. The heads are obtained from a T3 mutant; almost no T3 and T7 heads are without a tail in a wild type infection [26]. The DNA leaks from heads in quantized amounts. This phenomenon is seen via the formation of sharp bands during agarose gel electrophoresis of the DNA remaining packaged after 1-hit restriction endonuclease digestion. The DNA remaining packaged is obtained by (1) DNase I-digestion of external DNA; and (2) expulsion from the capsid of DNA remaining packaged [26]. This leakage-quantization phenomenon is best explained by quantized gp10 shell contraction that evolved via selection for control of the rate of infection-initiating DNA injection [26,33].

In addition, some multi-site T3 mutants undergo enhanced in vivo production of the following gp10-shell variants of capsid II: hyper-expanded and contracted. Identification of hyper-expanded capsid II is made by both (1) electron microscopy and (2) calculation of hydration from the density during buoyant density centrifugation in Nycodenz density gradients of a Nycodenz-impermeable (low density, high hydration) capsid II [34,35]. A DNA-free version of these capsids was initially investigated [34].

Next, a DNA-containing version of hyper-expanded capsid II was isolated via its Nycodenz impermeability (sealing) and accompanying low density [35]. This particle underwent contraction in the presence of magnesium ATP, but not magnesium ADP. Binding of ATP to gp10 was proposed to be the source of energy [35]. The following was evidence that the sealing had been evolutionarily selected. For wild type capsid II, cryo-EM revealed the complexity of inter-gp10 subunit interactions to be so high [25] that accidental sealing was improbable. Thus, these capsids were proposed [35] to be in states selected for function during DNA packaging (details [36]). The hyper-expansion required shell thinning to the point that β-sheet was proposed as the likely dominant conformation of gp10 major shell protein [35].

2.2. *α-Sheet, Rather than β-Sheet, in Size-Altered Capsid II?*

α-sheets resemble parallel β-sheets in having a parallel and extended conformation. However, if the amino acids all have the same chirality (as they do in almost all current proteins), then amino acids alternate side chain positions. That is to say, the conformation is not a helical array of amino acids. If constituent amino acids either alternate in chirality or are all glycine, then an α-sheet can be a helical array of amino acids [21–23,37,38]. The above has suggested the possibility that α-sheets began existence abiotically, when proteins were glycine-rich and other amino acids involved were not chiral [37,38].

An α-sheet-like peptide of 3–6 amino acids is called a nest. Nests are typically glycine-rich. α-sheets and nests have α-amino groups segregated on one edge and α-carboxyl groups segregated on the opposing edge [21,37,38]. Nest-associated α-amino groups are known to bind anions, such as phosphate, via the α-amino group edge. P-loop ATP-binding sites typically have a phosphate-binding nest [37,38]. Thus, one projects that a more extensive α-sheet is also likely to bind phosphates, including those part of ATP.

α-sheets were discovered via model building in 1951 [21] and were originally called parallel pleated sheets before parallel β-pleated sheet was known to be the more frequent structure. However, extensive α-sheets were found, by the discoverers, to be unlikely for the real world of left-handed amino acids. The reason was "steric hindrance between adjacent side chains" [23] for left-handed, non-glycine amino acids. Bending of α-sheets can reduce steric hindrance enough to make extensive α-sheet possible [22].

Additional unfavorable energetics are expected from the charge separation of α-sheets in the absence of multiple sheet layers. The α-carboxyl edge is negatively charged; the α-amino edge is positively charged at physiological pH. Indeed, stable proteins have only a small percentage of α-sheet-like nests. Among the proteins stable enough to be characterized, α-sheet-like nests are found primarily in ATP binding sites and in the lining of transmembrane pores [37,38].

The expected increase of α-sheet content for abiotically generated peptides suggests that nests are imprints from times before the existence of living organisms. The idea is that this structure was not completely replaced when increased diversity and chirality arrived for biological amino acids [37,38].

In theory, evolutionary retention of α-sheet structure would be increased if (1) the α-sheet is curved; (2) cooperativity is symmetry-promoted by incorporation of the protein in a symmetrical structure and (3) the protein binds a nest-stabilizing ATP molecule, thereby initiating a cooperative transition to α-sheet structure. Thus, we propose the following hypothesis. During DNA packaging, the observed T3 capsid II shell dynamics (hyper-expansion and contraction) are caused by the adopting by gp10 of ATP-responsive, dynamic α-sheet conformations.

Specifically, single-layered α-sheet is the proposed structure for gp10 in the shell of the most hyper-expanded version of T3 capsid II (Figure 2a; orientation in the shell is discussed in the next paragraph). The thickness is 0.6–0.9 nm [21,37], which is thin enough to make possible covering of the entire surface of the observed hyper-expanded T3 capsid II [35]. In addition, the proposed structure for the contracted versions of T3 capsid II is multi-layer α-sheet (Figure 2c), generated by an event approximating the folding of the single-layered α-sheet (Figure 2b), without loss of gp10 subunits. This latter conversion would be assisted by favorable electrical charge-charge interactions (Figure 2b).

Assembly of multiple Figure 2a-like subunits will be inhibited by charge-charge interactions unless the radial positions of neighboring subunits vary so that the negatively and positively charged edges of neighboring subunits are juxtaposed (Figure 2d). This "staggering" of radial position will cause increase in apparent shell thickness when a shell is visualized in a two-dimensional projection of the three-dimensional structure.

Given the polar nature of the two edges of alpha-sheet, one edge is predicted to be at the outer surface of the capsid's shell; the other is predicted to be at the inner surface of the shell. Most likely, the negatively charged edge will be at the outer surface to minimize interaction with other intracellular proteins, most of which are negatively charged at neutral pH [39].

2.3. Test of a Prediction: Surface Charge

We have tested the prediction of a relatively high negative surface charge for the gp10 shell of hyper-expanded T3 capsid II. This was done for capsid II that had incompletely packaged DNA, abbreviated ipDNA; an ipDNA-containing capsid is called an ipDNA-capsid. The test was performed by native agarose gel electrophoresis in two dimensions (2d-AGE) (recent reference [26]):

0.30% agarose gel in the first dimension; 2.0% agarose gel in the second dimension. A band of hyper-expanded ipDNA-capsid II was seen (labeled HE-CII in Figure 3a) after GelStar staining, nucleic acid-specific. This band was also seen after Coomassie staining, protein-specific. The effective origin of electrophoresis is indicated by the letter o. For comparison, the position of wild type T3 capsid II was also determined by 2d-AGE (dot labeled WT-CII in Figure 3a). The latter 2d-AGE was performed in a separate first and second dimension gel embedded in the same agarose frame as the gel in Figure 3a. The position of capsid I was determined from a separate analysis (dot labeled CI).

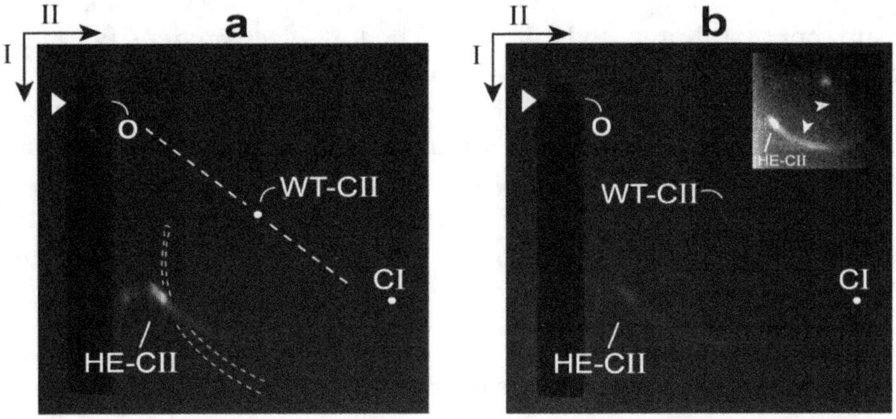

Figure 3. Analysis by 2d-AGE of hyper-expanded T3 ipDNA-capsid II. The ipDNA-capsid II is from the Nycodenz gradient-isolation in Figure 4b of reference [35]. Two fractions of the Nycodenz density gradient were analyzed by 2d-AGE, the (**a**) 1.073 g/mL and (**b**) 1.099 g/mL fractions. The procedure of 2d-AGE is described in reference [26]. The first dimension was run in a 0.30% agarose gel at 2.0 V/cm for 5.0 h. The second dimension was run in a surrounding 2.0% agarose gel at 1.8 V/cm V/cm for 16.0 h. The electrophoresis buffer was 0.09 M Tris-acetate, pH 8.3, 0.001 M $MgCl_2$. The temperature was 25 ± 0.3 °C. Seakem LE agarose was used (Lonza, Basel, Switzerland). The arrowheads indicate the leading edges of sample wells. The arrows indicate the directions of the first (I) and second (II) dimension electrophoresis. The curved dashed lines indicate the profile of variable length DNAs (no protein attached) from the DNA fraction of the same Nycodenz gradient. The DNA profile was obtained in a separate quadrant embedded in the same agarose frame as the gels of (**a**,**b**).

The effective radius of a particle (R_E), together with the radius of the gel's effective pore (P_E), uniquely determine the straight line drawn from the effective origin to the position of a particle in the gel. This line (dashed) is indicated for the WT-CII position in Figure 3a (R_E = 28.6 nm by small angle X-ray scattering [40]). The angle, θ, decreases as R_E increases [26]. Thus, the particles of the HE-CII band are confirmed in Figure 3a to be relatively large. The shape and orientation of HE-CII band indicate that these particles are also heterogeneous in R_E.

The average particle of the HE-CII band also had an average electrical surface charge density, σ, that was negative and was increased in absolute value. The reasoning is the following. The value of σ has, in general, been found to be proportional to the electrophoretic mobility in the absence of a gel. This mobility has been found to be independent of internal contents, such as ipDNA ([26] and included references). The ratio of σ value for HE-CII to σ value for WT-CII was 1.9. This ratio was determined from the ratio of average distances migrated in the first (low sieving) dimension, corrected for an estimated 5% greater effect of sieving on HE-CII in the first dimension. When CI was substituted for HE-CII, this ratio was also 1.9.

The source of the relatively high negative σ of hyper-expanded ipDNA-capsid II had to be either (1) the σ at the surface of shell-associated gp10 or (2) leakage from the capsid of a segment of (negatively charged) ipDNA, without dissociation of the ipDNA. We concluded that the former possibility was correct because electron micrographs did not reveal any leaked DNA [35]. Thus, qualitatively, the above prediction was confirmed.

Finally, the relatedness of HE-CII and WT-CII particles was confirmed in the 2d-AGE profile of a higher density fraction from the same Nycodenz density gradient. In this case, the HE-CII band was connected to a WT-CII-like band by an arc formed by capsid II particles with intermediate R_E and σ values (Figure 3b and inset). Contrast enhancement was used in the inset of Figure 3b to make the arc more easily seen. Presumably, the arc-forming particles were also intermediate in structure.

2.4. Electron Microscopy

In previous electron micrographs of specimens negatively stained with sodium phosphotungstate, hyper-expanded ipDNA-capsids appeared full [35]. This appearance was caused by impermeability to the negative stain, not by filling with DNA [35]. The low-electron density interior was occupied by dried buffer components. The appearance was similar after negative staining with uranyl acetate (Figure 4). However, the following feature was exaggerated in some particles of hyper-expanded NLD capsid II. A thin, dark layer of negative stain separated the light interior from the light gp10 shell (Figure 4). Apparently, after drying of most of the interior, negative stain leaked through the gp10 shell and then dried in a thin layer. In contrast, a contaminating wild type capsid II-like particle had the traditional negative stain-filled appearance (arrow in Figure 4). The shell of the latter capsids is 2 nm thick.

In the shell-revealing regions of hyper-expanded ipDNA-capsid II particles, the shell usually appears 3–5 nm thick. However, the thickness of the shell is likely to be significantly less than that (Section 2.2). This difference is enough so that a staining alone is unlikely to be the cause. Thus, the cause of the above difference resides in changes to the gp10 shell that occur in the latter stages of negative staining.

Details were suggested by the observation that, in some regions, the gp10 shell appeared doubled (arrowhead in Figure 4). The radial position staggering proposed in Figure 2d explained the observed doubling via an increase in the staggering distance at the latter stages of negative staining. By this explanation, embedding in the negative stain limited shell disruption to sub-observable levels. The regions of apparent shell thickening, without doubling, could be explained by smaller increase in the staggering distance and superposition of images from multiple planes perpendicular to the direction of observation. In summary, the electron microscopy produced images that were explained by a gp10 structure similar to the one in Figure 2d. Higher resolution, direct determination of structure is needed to test more directly for α-sheet and other structures.

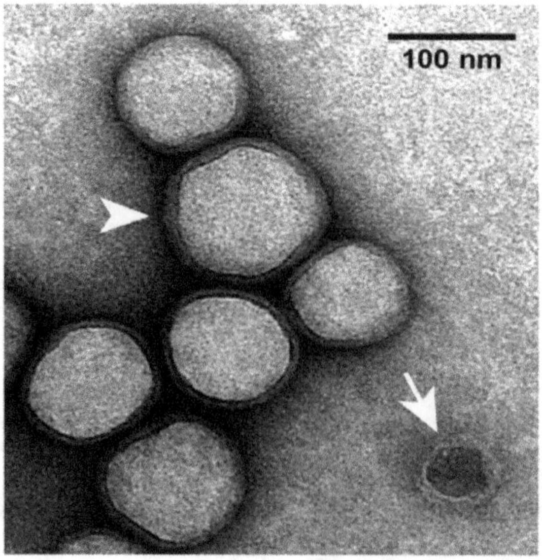

Figure 4. Electron microscopy of hyper-expanded NLD capsid II. The sample was the same as the sample used for Figure 2a. Particles in the sample were negatively stained with 1% uranyl acetate. The procedures of specimen preparation and EM were the same as used in reference [35].

2.5. *Possible Pathway to Anti-Viral Therapeutics*

The above analysis by 2d-AGE has been performed only for phage T3. This analysis is also a relatively simple, inexpensive, sensitive way to determine whether other viruses produce size-altered capsids like those of phage T3. For now, we make the following extrapolation. Pathogenic viruses do make extensive α-sheet-containing shell intermediates, at least in the case of the capsids of viruses with a double-stranded DNA genome, such as herpes viruses.

If, indeed, this is true, then the following is a possible strategy for anti-viral therapy. Find or design therapeutic compounds that target the backbone of extensive α-sheets. The low frequency of extensive α-sheets suggests low toxicity. This strategy has the projected advantage of being insensitive to bypass via single mutations. In addition, a relatively broad spectrum of viruses is likely to be susceptible. Two major factors in reducing the success of anti-viral drug therapies are evolution of resistant viral mutants and presence of narrow drug specificity (examples [41]).

3. Dynamic States and Neurodegenerative Disease

3.1. *Some Details*

Neurodegenerative diseases are all characterized by the presence of protein aggregates collectively called amyloid. This name is derived from starch-like texture. The aggregate-forming protein varies with the neurodegenerative disease (reviews: ALS [42], Alzheimer [43], Huntington [44], Parkinson [45], prion-associated [46]). When various proteins form amyloid, they convert from an original mixed-element structure to a predominantly β-sheet structure.

Although β-sheet is the predominant conformation of accumulating amyloid protein, computer simulations reveal that α-sheet is (1) accessible to amyloid-forming proteins [47,48] and (2) selectively accessible to them at low pH, especially in the protein region thought to initiate the transition to amyloid [48]. The proposal has been made that charge-charge interactions assist assembly to form an α-sheet intermediate that subsequently converts to β-sheet during the formation of amyloid. Molecular dynamics simulation has shown the possibility of conversion of α-sheet to β-sheet [47–49].

Given the association of nest-like structures with membrane pores, the theory is that extensive α-sheet-containing amyloid proteins, which are minority species, generate toxicity by making pores in cell membranes [37,38,47,48]. Mature amyloids are complex enough so that the amount of α-sheet associated with the β-sheet structure is, to our knowledge, not known.

Returning to the biology, the hypothesis has previously been proposed [50] (and expanded [51]) that one of the normal (non-toxic) functions of amyloid-forming proteins is innate immunity. This innate immunity occurs via neutralizing of products of both pathogen-generated and physical insults. By this hypothesis, neurodegenerative disease occurs when control of this insult-neutralizing activity is lost and, in some cases, the innate immune proteins start neutralizing themselves. A result is accumulation of toxic aggregates. Indeed, more recent studies have shown anti-bacterial and anti-viral effects of the amyloid for Alzheimer disease [52,53].

The expanded proposal [51] originally was that the neutralizing interaction occurred via the extension by innate immune proteins of β-sheet structure generated by the insult. However, the above considerations suggest substitution of α-sheet structure for β-sheet structure.

3.2. *A Proposal for Reduction to Practice*

We propose the following working assumptions to implement the basics-focus principle for neurodegenerative diseases. (1) Pathogenic viral infection and other insults trigger activation of amyloid-forming host innate immune proteins. These proteins convert to extensive α-sheet structure and, then, co-assemble with and inactivate extensive α-sheet-rich structures generated by the insults. These activities are currently obscure; (2) Extensive α-sheets are rare to non-existent in endogenous proteins of the virus host, other than activated innate immune system proteins; (3) Neurodegenerative diseases are caused by α-sheet-dependent damage caused by hyper-produced, activated innate immune proteins. These proteins eventually form β-sheet-rich amyloid.

Other virus-induced innate immunity activities (cytokine-induced neuroinflammation, for example) are known to be associated with the onset of neurodegenerative disease [54]. By the above hypothesis, these activities are secondary.

The proposal of (1)–(3) includes the fundamental idea that higher organisms have retained an immune system that has anciently derived targets and mode of action. This system is based on an abiotically derived protein structure. This system is difficult to analyze because proteins with more recently evolved composition and structure force both the innate immune system and its targets into intermittent status.

A way to reduce these ideas to practice is the following, as similarly suggested [51] before the introduction of the above α-sheet proposal. (1) Select for low-pathogenicity viruses that yield unstable, α-sheet-rich complexes when innate immune system proteins co-assemble with intracellular intermediates of this virus; (2) Infect patients with these viruses. To manage adaptive immunity, several such viruses would be used in succession. To isolate the needed viruses, one could begin by trying to find cases of spontaneous (presumably limited) disease-remission that have a correlation with a previous virus infection. The correlation would presumably not be dramatic and finding it would require determination. The reason is that most infections are expected to have the opposite effect, i.e., to trigger disease [54,55]. Therefore, the desired post-infection remissions are likely to be rare.

Of course, selecting for any immune system-avoiding virus has an obvious potential danger of ending with a problem of virus virulence. Presumably, virus virulence would be carefully monitored during selection. In any case, the selection would include screening for low virulence. Low virulence might, indeed, be necessarily co-selected during selection for an immune complex-disrupting phenotype.

4. Phage Assembly and Development of Gated, Targeted Drug Delivery Vehicles

4.1. *Avoiding Immune Systems*

Here, we change direction to discuss use of phage capsids as DDVs for the therapy of cancerous tumors. Administration of phage capsids is accompanied by the problem of phage capsid removal/neutralization by host immune systems. Both adaptive and innate immune systems will act to remove or neutralize foreign nanoparticles, including phages [56]. Optimization of a DDV-based strategy must include a response to this problem. One response to adaptive immunity is to serially change the phage-source of the DDV, in analogy to what is proposed above for low-virulence, therapeutic, eukaryotic viruses. In addition, phages propagate rapidly enough so that directed evolution can be used to reduce removal of a phage-derived DDV by any immune system.

However, probably the most significant potential advantage of a phage capsid-DDV is gating via the connector (formed by gp8 for T3/T7; Figure 1). The connector of phages T3 and T7 also exists for all other studied double-stranded DNA phages (reviews [57,58]). Thus, if connectors can, in general, be used as gates to capsid-DDVs, nature will provide many immunologically unrelated DDVs.

Before adaptive immunity becomes limiting (which takes 1–2 weeks in the case of phage K [56]), innate immune systems reduce effectiveness of any unprotected nanoparticle. Apparently the most active system is the mononuclear phagocyte system (MPS), formerly called the reticuloendothelial system. The MPS moves nanoparticles primarily to the liver and spleen [59,60]. Published tests of the efficiency of the MPS have been made with phages lambda, P22 and K. Phages lambda and P22 are lysogenic. When in a lysogenic state, lambda and P22 presumably could not have experienced and adapted to their environments. In contrast, phage K is lytic and had continuously experienced and adapted to its environment. Therefore, one expects that past genetic adaptation would cause phage K to have a longer lifetime in mouse circulation. This is the case. Complete removal occurs in over 24 hr for phage K [56] and 3–4 h for phages lambda and P22 [61,62].

That the above difference in lifetimes can be adaptive for phage lambda was shown by 10-cycles of the following: mutagenic growth, followed by mouse passage [61]. The adapted lambda was removed from mouse circulation more slowly than wild type lambda. Removal was by approximately a factor of 10 per day, several orders of magnitude more slowly than before adaptation [61]. A comparable removal rate was found for lytic phage T3 without any laboratory adaptation [63]. Apparently, the needed adaptation had occurred for T3 in the wild. Systematic attempts to make lytic phages more long-lived in circulation have not been made to our knowledge.

4.2. *Targeting Tumors*

Perhaps the most dramatic potential application of the basics-focus principle is to the therapy of cancerous tumors, including metastatic tumors. Focus on details (rather than basics) leads to targeting of either tumor-concentrated biochemistry or a patient's immune systems. However, the biochemistry of tumors overlaps the biochemistry of normal cells. Thus, essentially all drug and radiation therapies are toxic, which limits the extent of the therapy and makes patients sick [64–66]. Multi-faceted toxicity of immunotherapy is emerging as a problem [67]. In addition, when a cancer becomes metastatic, none of the above therapies can systematically block bypass of the therapy via evolution of cancer cells.

However, tumors do have one targetable, basic characteristic that they apparently cannot evolve to eliminate. This characteristic is the presence of 10–200 nm-sized pores in the blood vessels that feed tumors. Healthy blood vessels do not have these pores [68,69].

Thus, one should be able to achieve specific drug delivery to tumors with the following strategy: use of a DDV small enough to fit in these pores, but too large to pass through healthy blood vessels. In addition, tumors have relatively poor lymphatic drainage. Thus, when a DDV enters a tumor, it is slow to leave. The combined effect of the pores and the poor drainage is called the enhanced permeability and retention (EPR) effect [68,69].

Knowledge of the EPR effect is not new. The EPR effect has been the foundation for improved drug delivery via liposomal DDVs. Several products are FDA approved [70], the earliest being Doxil [71]. However, results have been disappointing. A recent review indicates the following limitations [72]: (1) "rapid loss of the drug cargo ... often immediately after ... systemic administration"; (2) "in many cases, less than 1% of the administered nanoparticle dose reaches the malignant tissue"; (3) "lack of release of drug in tumors"; (4) safety of the DDV; (5) safety of impurities (bacterial endotoxin which can generate a cytokine storm); (6) immunotoxicity (complement-generated, for example) and (7) standardization during manufacturing scale-up. A major uncertainty is the extent of the EPR effect in early metastases.

In evaluating the use of a phage capsid-DDV, we initially note that phages have never been found to be toxic to humans. Phages T3 and T7 were presumably isolated from sewage [73] and have passed through humans many times with the associated likelihood that they and their capsids already have evolved to avoid human immune systems. Furthermore, phage capsids with DDV potential are assembled in vivo and easily purified with structure uniform enough to obtain a 3–4 Å cryo-EM structure of the shell in the case of T7 [25]. If gated, a phage capsid is well on its way to removing the above limitations.

Indeed, one form of T3 capsid II is sufficiently gated so that it does not allow Nycodenz (molecular weight = 821) in its internal cavity until the temperature is raised. That is to say, the elevated temperature opens a gate. The gate is closed by lowering temperature, with the loaded Nycodenz not leaking detectably. The connector is the likely location of the gate [74].

4.3. *Adequate Loading of a Phage Capsid-DDV*

Nonetheless, a major barrier still exists to implementation of a gated T3 phage capsid-DDV. When we used elevated temperature to increase permeability (open the gate) in the presence of 10 mg/mL doxorubicin, the T3 gp10 shell was damaged. Native gel electrophoresis suggested disassembly to small aggregates, possibly monomers of gp10 (unpublished data). This limitation was possibly caused by detergent characteristics of doxorubicin, which has a positively charged and a hydrophobic region. These two regions are also present in most anti-cancer drugs.

Although this limitation has not yet been bypassed, we have found that the T7 counterpart of the loadable T3 capsid II is much more stable (80–83 °C). Future work will focus on the T7 version of this capsid II.

5. Prospects for the Future

Hypothesis-derived theory and practice can differ, possibly because of variables not in the theory. Also, hypotheses can be incorrect. The above proposals for treating neurodegenerative disease and cancerous tumors are not exempt from this pattern. Nonetheless, when basic considerations point in a simple direction, attempts to go in that direction should, it seems to us, be made. However, in the absence of a change in priorities, the probability is very close to zero that the above strategies will be tested experimentally within the next few years.

As support for this pessimism, exhibit A is the current status of phage therapy of infectious disease. For phage therapy, the basics point in a clear direction (details [17]). In addition, going in this direction is supported by (1) historical [6,7,73] and present-day [16] examples of dramatic success of doing that and (2) recent development of technologies that should dramatically improve results, if deployed. These include technologies of computerized database use/phage storage/phage retrieval, in addition to updated procedures of phage isolation and characterization. Yet, few, if any, systematic attempts at implementation are being made. Phage therapy is performed on an ad hoc basis [16,75], which slows implementation, sometimes to the point that it is too late.

6. Relationship to Scientific Details

In Section 1.1, we gave examples of scientific output triggered by de facto past clinically oriented use of the basics-focus principle. The finding of a curative, low-virulence virus, as proposed above, should do the same in the case of neurodegenerative diseases. Existence of such a virus would likely trigger experiments that use the virus to determine the interaction of virus metabolism and assembly with cellular events. Cellular events would be better understood. One finding might be that the above assumptions need to be modified or abandoned. However, the overall direction embodied in the assumptions could still be accurate enough to obtain a cure.

A gated, uniform-size, uniform-structure DDV can also be used to track the progression of tumors via the loading of the DDV with a trackable compound. Nycodenz, for example, is x-ray opaque and can be tracked by use of micro-CT. Such tracking would be used both clinically and also to analyze pathways of receptor mediated endocytosis and lysosome targeting of endocytic vesicles.

In other words, the division between basic science and clinical practice is not sharp.

Author Contributions: P.S. wrote the manuscript, designed the experiment for Figure 3 and performed the electron microscopy; E.T.W. contributed procedural suggestions, performed the 2d-AGE and prepared specimens for electron microscopy.

Acknowledgments: We thank Martin Adamo and Richard Ludueña for comments on drafts of this manuscript. This work was supported by the Welch Foundation (AQ-764) and the San Antonio Area Foundation.

References

1. Riedel, S. Edward Jenner and the history of smallpox and vaccination. *Proc. (Baylor Univ. Med. Cent.)* **2005**, *18*, 21–25. Available online: https://www.ncbi.nlm.nih.gov/pmc/articles/PMC1200696/ (accessed on 10 March 2018). [CrossRef] [PubMed]
2. Schwartz, M. The life and works of Louis Pasteur. *J. Appl. Microbiol.* **2001**, *91*, 597–601. [CrossRef] [PubMed]
3. Tan, S.Y.; Tatsumura, Y. Alexander Fleming (1881–1955): Discoverer of penicillin. *Singap. Med. J.* **2015**, *56*, 366–367. [CrossRef] [PubMed]
4. Kong, K.-F.; Schneper, L.; Mathee, K. Beta-lactam antibiotics: From antibiosis to resistance and bacteriology. *APMIS* **2010**, *118*, 1–36. [CrossRef] [PubMed]

5. Jelić, D.; Antolović, R. From erythromycin to azithromycin and new potential ribosome-binding antimicrobials. *Antibiotics* **2016**, *5*, 29. [CrossRef] [PubMed]
6. Summers, W.C. *Felix d'Herelle and the Origins of Molecular Biology*; Yale Univ. Press: New Haven, CT, USA, 1999.
7. Kutter, E.; De Vos, D.; Gvasalia, G.; Alavidze, Z.; Gogokhia, L.; Kuhl, S.; Abedon, S.T. Phage therapy in clinical practice: Treatment of human infections. *Curr. Pharm. Biotechnol.* **2010**, *11*, 69–86. [CrossRef] [PubMed]
8. Dubos, R.J. *Louis Pasteur: Free Lance of Science*; Scribner: New York, NY, USA, 1976.
9. Yonath, A. Antibiotics targeting ribosomes: Resistance, selectivity, synergism, and cellular regulation. *Annu. Rev. Biochem.* **2005**, *74*, 649–679. [CrossRef] [PubMed]
10. Graham, W.V.; Bonito-Oliva, A.; Sakmar, T.P. Update on Alzheimer's disease therapy and prevention strategies. *Annu. Rev. Med.* **2017**, *68*, 413–430. [CrossRef] [PubMed]
11. Piemontese, L. New approaches for prevention and treatment of Alzheimer's disease: A fascinating challenge. *Neural. Regen. Res.* **2017**, *12*, 405–406. [CrossRef] [PubMed]
12. Gitler, A.D.; Dhillon, P.; Shorter, J. Neurodegenerative disease: Models, mechanisms, and a new hope. *Dis. Models Mech.* **2017**, *10*, 499–502. [CrossRef] [PubMed]
13. Santiago, J.A.; Potashkin, J.A. A network approach to clinical intervention in neurodegenerative diseases. *Trends Mol. Med.* **2014**, *20*, 694–703. [CrossRef] [PubMed]
14. Leaf, C. *The Truth in Small Doses*; Simon and Schuster: New York, NY, USA, 2013.
15. Rose, D.W. *Friends and Partners: The Legacy of Franklin D. Roosevelt and Basil O'Connor in the History of Polio*; Elsevier: Amsterdam, The Netherlands, 2016.
16. Schooley, R.T.; Biswas, B.; Gill, J.J.; Hernandez-Morales, A.; Lancaster, J.; Lessor, L.; Barr, J.J.; Reed, S.L.; Rohwer, F.; Benler, S.; et al. Development and use of personalized bacteriophage-based therapeutic cocktails to treat a patient with a disseminated resistant *Acinetobacter baumannii* infection. *Antimicrob. Agents Chemother.* **2017**, *61*, e00954-17. [CrossRef] [PubMed]
17. Serwer, P. Restoring logic and data to phage-cures for infectious disease. *AIMS Microbiol.* **2017**, *3*, 706–712. [CrossRef]
18. Sankar, A.; Merril, C.R.; Biswas, B. Therapeutic and prophylactic applications of bacteriophage components in modern medicine. *Cold Spring Harb. Perspect. Med.* **2014**, *4*, a012518. [CrossRef]
19. Weber, L. Sewage Saved This Man's Life. Someday It Could Save Yours. Bacteriophages—Viruses Found in Soil, Water and Human Waste—May Be the Cure in a Post—Antibiotic World. HuffPost, US Edition. 2017. Available online: http://www.huffingtonpost.com/entry/antibioticresistant-superbugs-phage-therapy_us_5913414de4b05e1ca203f7d4 (accessed on 10 March 2018).
20. Edgar, R.S.; Feynman, R.P.; Klein, S.; Lielausis, I.; Steinberg, C.M. Mapping Experiments with R Mutants of Bacteriophage T4D. *Genetics* **1962**, *47*, 179–186. Available online: https://www.ncbi.nlm.nih.gov/pmc/articles/PMC1210321/ (accessed on 12 March 2018). [CrossRef]
21. Pauling, L.; Corey, R.B. The pleated sheet, a new layer configuration of polypeptide chains. *Proc. Natl. Acad. Sci. USA* **1951**, *37*, 251–256. [CrossRef] [PubMed]
22. Pauling, L.; Corey, R.B. The structure of feather rachis keratin. *Proc. Natl. Acad. Sci. USA* **1951**, *37*, 256–261. [CrossRef] [PubMed]
23. Pauling, L.; Corey, R.B. Configurations of polypeptide chains with favored orientations around single bonds: Two new pleated sheets. *Proc. Natl. Acad. Sci. USA* **1951**, *37*, 729–740. [CrossRef] [PubMed]
24. Guo, F.; Liu, Z.; Vago, F.; Ren, Y.; Wu, W.; Wright, E.T.; Serwer, P.; Jiang, W. Visualization of uncorrelated, tandem symmetry mismatches in the internal genome packaging apparatus of bacteriophage T7. *Proc. Natl. Acad. Sci. USA* **2013**, *110*, 6811–6816. [CrossRef] [PubMed]
25. Guo, F.; Liu, Z.; Fang, P.-A.; Zhang, Q.; Wright, E.T.; Wu, W.; Zhang, C.; Vago, F.; Ren, Y.; Jakana, J.; et al. Capsid expansion mechanism of bacteriophage T7 revealed by multistate atomic models derived from cryo-EM reconstructions. *Proc. Natl. Acad. Sci. USA* **2014**, *111*, E4606–E4614. [CrossRef] [PubMed]
26. Serwer, P.; Wright, E.T.; Liu, Z.; Jiang, W. Length quantization of DNA partially expelled from heads of a bacteriophage T3 mutant. *Virology* **2014**, *456–457*, 157–170. [CrossRef] [PubMed]
27. Pajunen, M.I.; Elizondo, M.R.; Skurnik, M.; Kieleczawa, J.; Molineux, I.J. Complete nucleotide sequence and likely recombinatorial origin of bacteriophage T3. *J. Mol. Biol.* **2002**, *319*, 1115–1132. [CrossRef]
28. Serwer, P.; Pichler, M.E. Electrophoresis of bacteriophage T7 and T7 capsids in agarose gels. *J. Virol.* **1978**, *28*, 917–928. [PubMed]

29. Serwer, P.; Hayes, S.J.; Watson, R.H. The structure of a bacteriophage T7 procapsid and its *in vivo* conversion product probed by digestion with trypsin. *Virology* **1982**, *122*, 392–401. [CrossRef]
30. Wikoff, W.R.; Liljas, L.; Duda, R.L.; Tsuruta, H.; Hendrix, R.W.; Johnson, J.E. Topologically linked protein rings in the bacteriophage HK97 capsid. *Science* **2000**, *289*, 2129–2133. [CrossRef] [PubMed]
31. Duda, R.L.; Ross, P.D.; Cheng, N.; Firek, B.A.; Hendrix, R.W.; Conway, J.F.; Steven, A.C. Structure and energetics of encapsidated DNA in bacteriophage HK97 studied by scanning calorimetry and cryoelectron microscopy. *J. Mol. Biol.* **2009**, *391*, 471–483. [CrossRef] [PubMed]
32. Yu, G.; Vago, F.; Zhang, D.; Snyder, J.E.; Yan, R.; Zhang, C.; Benjamin, C.; Jiang, X.; Kuhn, R.J.; Serwer, P.; et al. Single-step antibody-based affinity cryo-electron microscopy for imaging and structural analysis of macromolecular assemblies. *J. Struct. Biol.* **2014**, *187*, 1–9. [CrossRef] [PubMed]
33. Serwer, P.; Wright, E.T.; Demeler, B.; Jiang, W. States of phage T3/T7 capsids: Buoyant density centrifugation and cryo-EM. *Biophys. Rev.* **2018**, in press. [CrossRef] [PubMed]
34. Serwer, P.; Wright, E. Testing a proposed paradigm shift in analysis of phage DNA packaging. *Bacteriophage* **2016**, *6*, e1268664. [CrossRef] [PubMed]
35. Serwer, P.; Wright, E.T. ATP-driven contraction of phage T3 capsids with DNA incompletely packaged *in vivo*. *Viruses* **2017**, *9*, 119. [CrossRef] [PubMed]
36. Serwer, P. Proposed ancestors of phage nucleic acid packaging motors (and cells). *Viruses* **2011**, *3*, 1249–1280. [CrossRef] [PubMed]
37. Milner-White, E.J.; Russell, M.J. Predicting the conformations of peptides and proteins in early evolution. A review article submitted to Biology Direct. *Biol. Direct* **2008**, *3*, 3. [CrossRef] [PubMed]
38. Milner-White, E.J.; Russell, M.J. Functional capabilities of the earliest peptides and the emergence of life. *Genes* **2011**, *2*, 671–678. [CrossRef] [PubMed]
39. Naryzhny, S. Towards the full realization of 2DE power. *Proteomes* **2016**, *4*, 33. [CrossRef] [PubMed]
40. Stroud, R.M.; Serwer, P.; Ross, M.J. Assembly of bacteriophage t7. Dimensions of the bacteriophage and its capsids. *Biophys. J.* **1981**, *36*, 743–757. [CrossRef]
41. Van der Linden, L.; Wolthers, K.C.; van Kuppeveld, F.J.M. Replication and inhibitors of enteroviruses and parechoviruses. *Viruses* **2015**, *7*, 4529–4562. [CrossRef] [PubMed]
42. Taylor, P.J.; Brown, R.H., Jr.; Cleveland, D.W. Decoding ALS: From genes to mechanism. *Nature* **2016**, *539*, 197–206. [CrossRef] [PubMed]
43. Van Dam, D.; Vermeiren, Y.; Dekker, A.D.; Naudé, P.J.W.; De Deyn, P.P. Neuropsychiatric disturbances in Alzheimer's disease: What have we learned from neuropathological studies? *Curr. Alzheimer Res.* **2016**, *13*, 1145–1164. [CrossRef] [PubMed]
44. Bates, G.P.; Dorsey, R.; Gusella, J.F.; Hayden, M.R.; Kay, C.; Leavitt, B.R.; Nance, M.; Ross, C.A.; Scahill, R.I.; Wetzel, R.; et al. Huntington Disease. *Nat. Rev. Dis. Primers* **2015**, *1*, 15005. Available online: https://www.nature.com/articles/nrdp20155 (accessed on 13 March 2018). [CrossRef] [PubMed]
45. Kalia, L.V.; Lang, A.E. Parkinson disease in 2015: Evolving basic, pathological and clinical concepts in PD. *Nat. Rev. Neurol.* **2016**, *12*, 65–66. [CrossRef] [PubMed]
46. Collinge, J. Mammalian prions and their wider relevance in neurodegenerative diseases. *Nature* **2016**, *539*, 217–226. [CrossRef] [PubMed]
47. Daggett, V. Alpha-sheet: The toxic conformer in amyloid diseases? *Acc. Chem. Res.* **2006**, *39*, 594–602. [CrossRef] [PubMed]
48. Armen, R.S.; Alonso, D.O.V.; Daggett, V. Anatomy of an amyloidogenic intermediate: Conversion of β-sheet to α-sheet in transthyretin. *Structure* **2004**, *12*, 1847–1863. [CrossRef] [PubMed]
49. Kellock, J.; Hopping, G.; Caughey, B.; Daggett, V. Peptides composed of alternating L- and D-amino acids inhibit amyloidogenesis in three distinct amyloid systems independent of sequence. *J. Mol. Biol.* **2016**, *428*, 2317–2328. [CrossRef] [PubMed]
50. Bandea, C.I. Aβ, tau, α-synuclein, huntingtin, TDP-43, PrP and AA are members of the innate immune system: A unifying hypothesis on the etiology of AD, PD, HD, ALS, CJD and RSA as innate immunity disorders. *bioRχiv* **2013**. [CrossRef]
51. Serwer, P. Hypothesis for the cause and therapy of neurodegenerative diseases. *Med. Hypotheses* **2018**, *110*, 60–63. [CrossRef] [PubMed]
52. Vijaya Kumar, D.K.; Choi, S.H.; Washicosky, K.J.; Eimer, W.A.; Tucker, S.; Ghofrani, J.; Lefkowitz, A.;

McColl, G.; Goldstein, L.E.; Tanzi, R.E.; et al. Amyloid-β peptide protects against microbial infection in mouse and worm models of Alzheimer's disease. *Sci. Transl. Med.* **2016**, *8*, 340ra72. [CrossRef] [PubMed]

53. White, M.R.; Kandel, R.; Hsieh, I.-N.; De Luna, X.; Hartshorn, K.L. Critical role of C-terminal residues of the Alzheimer's associated β-amyloid protein in mediating antiviral activity and modulating viral and bacterial interactions with neutrophils. *PLoS ONE* **2018**, *13*, e0194001. [CrossRef] [PubMed]
54. Deleidi, M.; Isacson, O. Viral and inflammatory triggers of neurodegenerative diseases. *Sci. Transl. Med.* **2012**, *4*, 121ps3. [CrossRef] [PubMed]
55. Itzhaki, R.F.; Lathe, R.; Balin, B.J.; Ball, M.J.; Bearer, E.L.; Braak, H.; Bullido, M.J.; Carter, C.; Clerici, M.; Cosby, S.L.; et al. Microbes and Alzheimer's disease. *J. Alzheimers Dis.* **2016**, *51*, 979–984. [CrossRef] [PubMed]
56. Hodyra-Stefaniak, K.; Miernikiewicz, P.; Drapała, J.; Drab, M.; Jończyk-Matysiak, E.; Lecion, D.; Kaźmierczak, Z.; Beta, W.; Majewska, J.; Harhala, M.; et al. Mammalian host-versus-phage immune response determines phage fate *in vivo*. *Sci. Rep.* **2015**, *5*, 14802. [CrossRef] [PubMed]
57. Aksyuk, A.A.; Rossmann, M.G. Bacteriophage assembly. *Viruses* **2011**, *3*, 172–203. [CrossRef] [PubMed]
58. Fokine, A.; Rossmann, R.G. Molecular architecture of tailed double-stranded DNA phages. *Bacteriophage* **2014**, *4*, e28281. [CrossRef] [PubMed]
59. Hume, D.A. The mononuclear phagocyte system. *Curr. Opin. Immunol.* **2006**, *18*, 49–53. [CrossRef] [PubMed]
60. Yona, S.; Gordon, S. From the reticuloendothelial to mononuclear phagocyte system—The unaccounted years. *Front. Immunol.* **2015**, *6*, 328. [CrossRef] [PubMed]
61. Merril, C.R.; Biswas, B.; Carlton, R.; Jensen, N.C.; Creed, G.J.; Zullo, S.; Adhya, S. Long-circulating bacteriophage as antibacterial agents. *Proc. Natl. Acad. Sci. USA* **1996**, *93*, 3188–3192. [CrossRef] [PubMed]
62. Merril, C.R.; Scholl, D.; Adhya, S.L. The prospect for bacteriophage therapy in Western medicine. *Nat. Rev. Drug Discov.* **2003**, *2*, 489–497. [CrossRef] [PubMed]
63. Serwer, P.; Wright, E.T.; Williams, T.L.; Demeler, B.; Lee, J.C. Phage-based therapies: What is the "real thing"? In Proceedings of the XXV Biennial Conference on Phage/Virus Assembly, Ellicott City, MD, USA, 20–25 August 2017.
64. Schaue, D.; McBride, W.H. Opportunities and challenges of radiotherapy for treating cancer. *Nat. Rev. Clin. Oncol.* **2015**, *12*, 527–540. [CrossRef] [PubMed]
65. Chabner, B.A.; Roberts, T.G. Chemotherapy and the war on cancer. *Nat. Rev. Cancer* **2005**, *5*, 65–72. [CrossRef] [PubMed]
66. Mehlen, P.; Puisieux, A. Metastasis: A question of life or death. *Nat. Rev. Cancer* **2006**, *6*, 449–458. [CrossRef] [PubMed]
67. Kottschade, L.A. Incidence and management of immune-related adverse events in patients undergoing treatment with immune checkpoint inhibitors. *Curr. Oncol. Rep.* **2018**, *20*, 24. [CrossRef] [PubMed]
68. Kobayashi, H.; Watanabe, R.; Choyke, P.L. Improving conventional enhanced permeability and retention (EPR) effects; what is the appropriate target? *Theranostics* **2014**, *4*, 81–89. [CrossRef] [PubMed]
69. Nakamura, H.; Fang, J.; Maeda, H. Development of next-generation macromolecular drugs based on the EPR effect: Challenges and pitfalls. *Expert Opin. Drug Deliv.* **2015**, *12*, 53–64. [CrossRef] [PubMed]
70. Weissig, V.; Pettinger, T.K.; Murdock, N. Nanopharmaceuticals (Part 1): Products on the market. *Int. J. Nanomed.* **2014**, *9*, 4357–4373. [CrossRef] [PubMed]
71. Barenholz, Y. Doxil®—The first FDA-approved nano-drug: Lessons learned. *J. Control. Release* **2012**, *160*, 117–134. [CrossRef] [PubMed]
72. Anchordoquy, T.J.; Barenholz, Y.; Boraschi, D.; Chorny, M.; Decuzzi, P.; Dobrovolskaia, M.; Farhangrazi, Z.S.; Farrell, D.; Gabizon, A.; Ghandehari, H.; et al. Mechanisms and barriers in cancer nanomedicine: Addressing challenges, looking for solutions. *ACS Nano* **2017**, *11*, 12–18. [CrossRef] [PubMed]
73. Summers, W.C. Bacteriophage therapy. *Annu. Rev. Microbiol.* **2001**, *55*, 437–451. [CrossRef] [PubMed]
74. Serwer, P.; Wright, E.T.; Chang, J.T.; Liu, X. Enhancing and initiating phage-based therapies. *Bacteriophage* **2014**, *4*, e961869. [CrossRef] [PubMed]
75. Boodman, E. To Save a Young Woman Besieged by Superbugs, Scientists Hunt a Killer Virus. Available online: https://www.pbs.org/newshour/health/to-save-a-young-woman-besieged-by-superbugs-scientists-hunt-a-killer-virus (accessed on 13 November 2017).

16

Novel T7 Phage Display Library Detects Classifiers for Active Mycobacterium Tuberculosis Infection

Harvinder Talwar [1], Samer Najeeb Hanoudi [2], Sorin Draghici [2] and Lobelia Samavati [1,3,*]

1 Department of Medicine, Division of Pulmonary, Critical Care and Sleep Medicine, Wayne State University School of Medicine and Detroit Medical Center, Detroit, MI 48201, USA; ar8673@wayne.edu

2 Department of Computer Science, Wayne State University, Detroit, MI 48202, USA; ei1875@wayne.edu (S.N.H.); sorin@wayne.edu (S.D.)

3 Center for Molecular Medicine and Genetics, Wayne State University School of Medicine, 540 E. Canfield, Detroit, MI 48201, USA

* Correspondence: lsamavat@med.wayne.edu or ay6003@wayne.edu

Abstract: Tuberculosis (TB) is caused by *Mycobacterium tuberculosis* (MTB) and transmitted through inhalation of aerosolized droplets. Eighty-five percent of new TB cases occur in resource-limited countries in Asia and Africa and fewer than 40% of TB cases are diagnosed due to the lack of accurate and easy-to-use diagnostic assays. Currently, diagnosis relies on the demonstration of the bacterium in clinical specimens by serial sputum smear microscopy and culture. These methods lack sensitivity, are time consuming, expensive, and require trained personnel. An alternative approach is to develop an efficient immunoassay to detect antibodies reactive to MTB antigens in bodily fluids, such as serum. Sarcoidosis and TB have clinical and pathological similarities and sarcoidosis tissue has yielded MTB components. Using sarcoidosis tissue, we developed a T7 phage cDNA library and constructed a microarray platform. We immunoscreened our microarray platform with sera from healthy ($n = 45$), smear positive TB ($n = 24$), and sarcoidosis ($n = 107$) subjects. Using a student t-test, we identified 192 clones significantly differentially expressed between the three groups at a False Discovery Rate (FDR) <0.01. Among those clones, we selected the top ten most significant clones and validated them on independent test set. The area under receiver operating characteristics (ROC) for the top 10 significant clones was 1 with a sensitivity of 1 and a specificity of 1. Sequence analyses of informative phage inserts recognized as antigens by active TB sera may identify immunogenic antigens that could be used to develop therapeutic or prophylactic vaccines, as well as identify molecular targets for therapy.

Keywords: T7phage library; sarcoidosis; tuberculosis; microarray; immunoscreening

---

1. Introduction

Tuberculosis (TB) remains a serious global health threat with 10 million new cases and 1.7 million deaths each year [1,2]. Currently, we have limited tools available to diagnose active TB, predict treatment efficacy and the cure of tuberculosis, or to detect the reactivation of a latent tuberculosis infection, and assay the induction of protective immune responses through vaccination. A major obstacle to global control of TB remains inadequate case detection [3]. Efforts during the past decade to consistently diagnose and treat most infectious cases have slowed the TB incidence rate, but have not yielded substantial progress [3]. The existing TB diagnostic pipeline still does not have a simple, rapid, inexpensive point-of-care test [3]. Qualified tuberculosis biomarkers are most urgently needed as predictors of reactivation and cure, and indicators of vaccine-induced protection [3].

Pulmonary tuberculosis has clinical and pathological similarities with sarcoidosis. Sarcoidosis is a systemic granulomatous disease of unknown etiology with predominant involvement of the lungs, among other organs [4–7]. Several studies have suggested that the cellular and humoral responses associated with granuloma formation in sarcoidosis are the consequence of an exaggerated immune response to specific *Mycobacterium tuberculosis* (MTB) antigens [4,8]. Sarcoidosis tissue has yielded MTB components including, ESAT6 and catalase—peroxidase (mKatG) [9]. Despite the presence of specific TB antigens in sarcoidosis lung tissues [8,10–12], patients with sarcoidosis negatively respond to the tuberculin skin test and are considered to be anergic [13]. Additionally, sarcoidosis subjects rarely ever develop tuberculosis. Lungs are highly involved both in sarcoidosis and TB. Resident alveolar macrophages (AMs) play an important role in the pathogenesis and host defense of both diseases [4,14–16]. It has been shown that AMs provide a reservoir for MTB and other slow growing organisms [11,14,15,17]. Additionally, AMs play an integral role in autoimmunity and the initiation of fibrosis [14]. Based on this knowledge, we hypothesized that bronchioalveolar cells (BALs) of sarcoidosis subjects may harbor degradation products of specific pathogen(s), including MTB. We constructed four T7 phage display cDNA libraries, two of which originate from sarcoidosis BAL cells and white blood cells (WBCs), and two others derived from cultured human embryonic fibroblasts and splenic monocytes, and combined all four libraries into a complex library [18,19]. We randomly selected 1070 clones through biopanning and constructed a microarray platform with the selected clones. Previously, upon immunoscreening of this platform with sera from healthy controls, sarcoidosis and culture positive TB patients, we showed that we can detect highly sensitive and specific biomarkers for TB in the sera of subjects with culture positive MTB [18,20]. In that study, the TB patients were smear negative but culture positive and at the time of sera collection, they were on treatment with anti-tuberculosis agents [18,20]. To investigate whether our display library also detects specific biomarkers in sera from smear positive MTB patients and if these biomarkers differ from those of smear negative but culture positive TB, we immunoscreened T7 phage display libraries with sera of smear-positive TB patients. The objective of the present study was to identify the specific diagnostic biomarkers from the sera of TB patients who had active TB. We discovered reactive clones that distinguished sera from active TB patients from sarcoidosis patients and uninfected control sera with a high sensitivity and specificity.

2. Materials and Methods

2.1. Chemicals

All chemicals were purchased from Sigma-Aldrich (St. Louis, MO, USA) unless specified otherwise. LeukoLOCK filters and RNAlater were purchased from Life Technologies (Grand Island, NY, USA). The RNeasy Midi kit was obtained from Qiagen, (Valencia, CA, USA). The T7 mouse monoclonal antibody was purchased from Novagen (San Diego, CA, USA). Alexa Fluor 647 goat anti-human IgG and Alex Fluor goat anti-mouse IgG antibodies were purchased from Life Technologies (Grand Island, NY, USA).

2.2. Patient Selection

This study was approved by the institutional review board at Wayne State University, and the Detroit Medical Center. Sera were collected from 3 groups: (1) healthy volunteers; (2) sarcoidosis subjects; and (3) smear positive pulmonary TB patients. All study subjects signed a written informed consent. All methods were performed in accordance with the human investigation guidelines and regulations by the IRB (protocol No = 055208MP4E) at Wayne State University. All sarcoidosis subjects were ambulatory patients. Sera from patients with tuberculosis were obtained from the Foundation for Innovative New Diagnostics (FIND, Geneva, Switzerland). All TB patients had smear positive sputum.

2.3. Serum Collection

Using standardized phlebotomy procedures blood samples were collected and stored at −80 °C [18].

2.4. Construction and Biopanning of T7 Phage Display cDNA Libraries

We have used the same T7 phage display libraries as before [18,19]. Briefly, T7 phage display libraries from BALs, WBCs, EL-1 and MRC5 were made to generate a complex sarcoid library (CSL) [18,19]. Differential biopanning for negative selection was the performed using sera from healthy controls to remove the non-specific IgG, and sarcoidosis sera for positive enrichment as described previously [18,19].

2.5. Microarray Construction and Immunoscreening

A total of 1070 individually picked phage clones from the biopannings 3 and 4 for microarray construction were same as used in previous studies [18,19]. The phage lysates were arrayed in quintuplicates onto nitrocellulose FAST slides (Grace Biolabs, OR, USA) using the ProSys 5510TL robot (Cartesian Technologies, CA, USA). The nitrocellulose slides were hybridized with sera and processed as described previously [18].

2.6. Sequencing of Phage cDNA Clones

Individual phage clones were PCR amplified using T7 phage forward primer 5′ GTTCTAT CCGCAACGTTATGG 3′ and reverse primer 5′ GGAGGAAAGTCGTTTTTTGGGG 3′ and sequenced by Genwiz (South Plainfield, NJ, USA), using T7 phage sequence primer TGCTAAGGACAACG TTATCGG. cDNA sequences of T7 phage clones obtained from Genwiz were translated into peptide/protein sequences using ExPASy *translate tool*. The length of each peptide clone is determined after the last amino acid of linker sequence (GDPNSS) inserted in frame of T7 phage till the stop codon of the sequence. Using NCBI protein BLAST site each identified sequence was used for further BLAST. For each peptide, we performed three BLASTs. First, the identified sequences were randomly blasted to the sequence data without indication of specific species. Second, we used random BLAST to the human genome and thirdly to the mycobacterium genome. We selected the proteins with highest homology with our peptide sequence.

2.7. Data Acquisition and Pre-Processing

Following the immunoreaction, the microarrays were scanned in an Axon Laboratories 4100 scanner (Palo Alto, CA, USA) using 532 and 647 nm lasers to produce a red (Alexa Fluor 647) and green (Alexa Fluor 532) composite image. Cy5 (red dye) labeled anti-human antibody was used to detect IgGs in human serum that were reactive to peptide clones, and a Cy3 (green dye) labeled antibody was used to detect the phage capsid protein [18]. Using the ImaGene 6.0 (Biodiscovery) image analysis software, the binding intensity of each peptide with IgGs in sera was expressed as log_2 (red/green) fluorescent intensities. These data were pre-processed using the limma package in the R language environment [19,21,22] and the normexp method was applied to correct the background [19,23]. Within array normalization was performed using the *LOESS* method [18,23,24]. The scale method was applied to normalize between arrays [23,24]. Intensity ratio of a clone in active TB samples divided by the same clone intensity ratio from healthy control samples were calculated to determine the fold change of a clone.

2.8. Statistical Analyses

To detect differentially expressed antigens for TB, we applied a two-tailed *t*-test. In order to correct for multiple comparisons, we applied the false discovery rate (FDR) algorithm with a threshold of 0.01 FDR [25]. We identified 192 significant clones at 0.01 FDR. All significant clones were sorted in an increasing order. The top ten highly significant clones were considered as "classifier clones". We randomly split the TB and healthy controls samples into: (i) training; (ii) test sets. Out of the 24 TB samples, 12 samples were randomly assigned to training set and 12 samples to testing set. The training and testing sets for the 45 healthy controls were randomly selected to 23 training and 22 test sets. A *t*-test was applied between TB-training samples versus healthy controls training samples. All 107 sarcoidosis samples were assigned to the testing set. To assess the performance of classifiers clones, we applied principal component analysis (PCA), agglomerative hierarchal clustering (HC), heatmap, and linear discriminant analysis (LDA). The LDA model was built on the training samples to predict TB samples from others (healthy controls and sarcoidosis) samples, and tested the classification model on the testing set (samples not used in the training set). We performed the classification on the classifiers clones. We applied principal component analysis (PCA), agglomerative hierarchal clustering (HC), and heatmap with all samples (training and testing) twice. Those analyses were first applied to all clones (1070 clones) and then with the highly significant 10 classifier clones.

3. Results

3.1. Complex Sarcoidosis (CSL) Library Detects Unique Antigens in the Sera of Active Tuberculosis Patients

A panel of potential antigens was randomly selected from two highly enriched pools of T7 phage cDNA libraries through biopanning of the CSL library [18,19]. The constructed microarray platform was immunoscreened with 176 sera (45 healthy controls, 24 smear-positive TB patients, and 107 sarcoidosis patients). The demographics of the study subjects are shown in Table 1. Following immunoreaction, the microarray data were pre-processed and then analyzed. First, we performed an unsupervised PCA using all 1070 clones with data from 176 study subjects. As shown in Figure 1A, several healthy controls and sarcoidosis patients clustered together with TB subjects. We also performed unsupervised hierarchical clustering with all 1070 clones on these 176 samples. We observed the magenta cluster has a mix of samples and lacks specific sub-clusters of TB samples (Figure 1B). Next, we applied a two-tailed *t*-test and identified 192 clones that were differentially expressed in sera of smear-positive TB as compared to sarcoidosis patients and healthy controls at the FDR < 0.01. To determine whether the selected 192 significant clones can improve the class separation of TB samples from healthy controls and sarcoidosis patients, we constructed a PCA plot. As shown in Figure 1C, there is a good separation of TB samples from sarcoidosis and healthy controls, in which twenty six percent of variance was along the PCA1. Similarly, when clustering algorithm was performed using 192 TB clones on all subjects, we observed a distinct hierarchical linkage clearly separating TB samples from healthy controls and sarcoidosis patients (Figure 1D). Furthermore, we constructed a PCA plot using 10 classifier clones that can differentiate TB patients from healthy controls and sarcoidosis patients. The result in Figure 1E shows a clear separation of TB samples from healthy controls and sarcoidosis patients. Fifty four percent of variance was explained along the PCA 1. Similarly, when the clustering algorithm was performed using 10 TB classifier clones, we observed a distinct hierarchical linkage separating the TB patients from others (Figure 1F).

Table 1. Subjects demographics.

Characteristic	Control Subjects	Sarcoidosis Subjects	TB Subjects
Age (Mean ± SEM)	40.3 ± 7.5	30.6 ± 11.8	40.5 ± 8.5
Gender, N (%)			
Male	11 (25)	27 (25)	18 (64)
Female	34 (75)	80 (75)	10 (36)
Race, N (%)			
African American	31 (69)	95 (89)	
African	-		4 (25)

Table 1. *Cont.*

Characteristic	Control Subjects	Sarcoidosis Subjects	TB Subjects
Caucasian	-	12 (11)	
Asians	14 (31)		20 (75)
BMI (Mean ± SEM)	27 ± 3.8	28 ± 10.5	28 ± 6.9
Organ involvement			
Neuro-ophthalmologic	NA	31 (29)	-
Lung	NA	101 (94)	24 (100)
Skin	NA	46 (43)	-
Multiorgan	NA	65 (61)	-
PPD [a]	NA	Negative	-
TB smear [b]	NA	Negative	Positive

NA = not applicable; [a] PPD = Mantoux test (purified protein derivative); [b] TB Smear obtained from sputum.

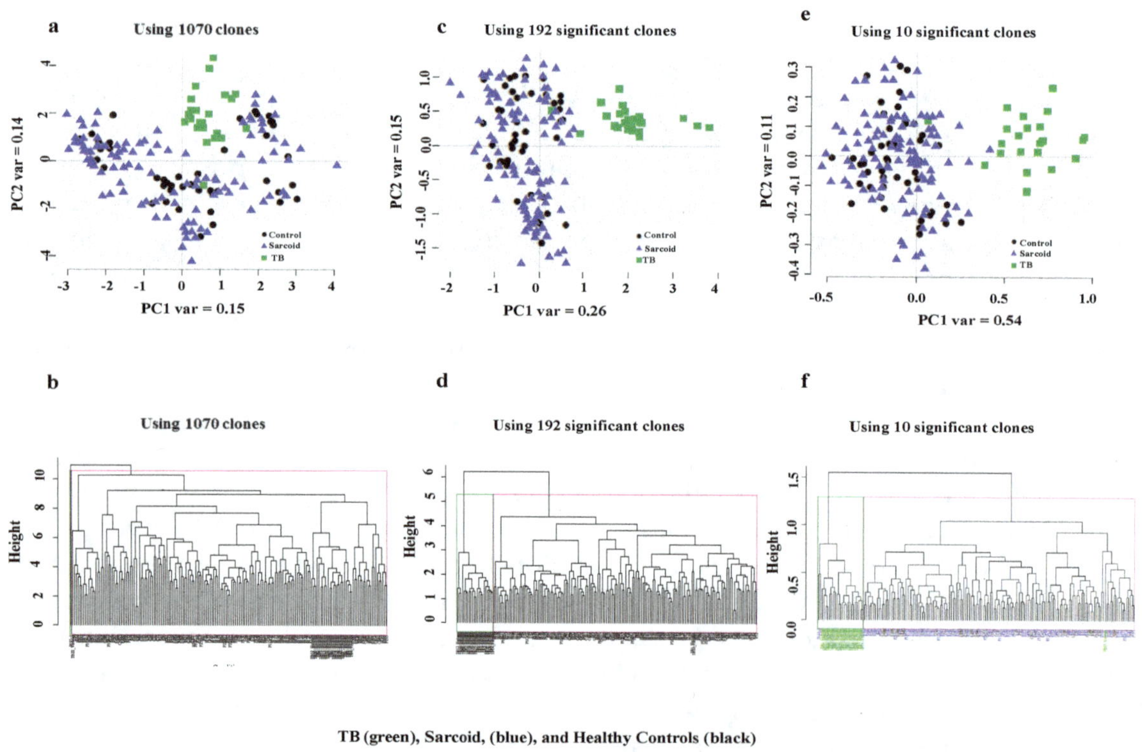

Figure 1. PCA and Hierarchal clustering. (**a**) PCA score plots along PC1 and PC2 generated with 1070 clones of three groups: (1) healthy control samples (black circles); (2) TB samples (green squares) and; (3) Sarcoidosis samples (blue triangle). Biomarker clusters along the PCA1 explain a variance of only 0.15, while the variance along PC2 was about 0.14. (**b**) The hierarchal clustering was applied on the healthy controls (black labels), TB patients (green labels) and sarcoidosis (blue labels) with 1070 clones. (**c**) PCA score plots along the PC1 and 2 results when applied on 192 TB clones. The PC1 explained 0.26 of variance, whereas PC2 explained 0.15 of variance. As shown, the TB samples are well separated from the healthy controls and sarcoidosis samples. (**d**) Hierarchal clustering using only the highly significant 192 TB clones. The blue and black clusters include sarcoidosis and healthy controls, the green cluster includes all the TB samples except one. (**e**) PCA score plots along PC1 and 2 generated with top 10 highly significant clones. The PC1 explained 0.54 of variance, whereas PC2 explained 0.11 of variance. (**f**) Hierarchal clustering using 10 top significant TB clones. This figure demonstrates better clustering with 192 TB clones and the highly significant 10 TB clones (panels c, d, e, and f) when compared the clustering using all clones (panels a and b).

Figure 2A displays a heatmap plot of the distinct expression features of 192 TB clones among the study subjects. The heatmap using ten significant TB clones (classifiers) among study subjects is highlighted as a plot in Figure 2B.

Furthermore, we applied the classifier model and calculated the AUC values using 192 TB clones on testing sets. As shown in Figure 3A, the AUC under the ROC using 192 clones was one with no false positive and no false negative prediction. Next, we applied the classifier model on the test set (12 TB patients, 107 sarcoidosis patients, and 22 healthy controls) using 10 classifier clones. Figure 3B shows that despite reduction of clones to 10, the AUC under the ROC remained one, again with no false positive or false negative class labeling. These suggest robust classifier performance.

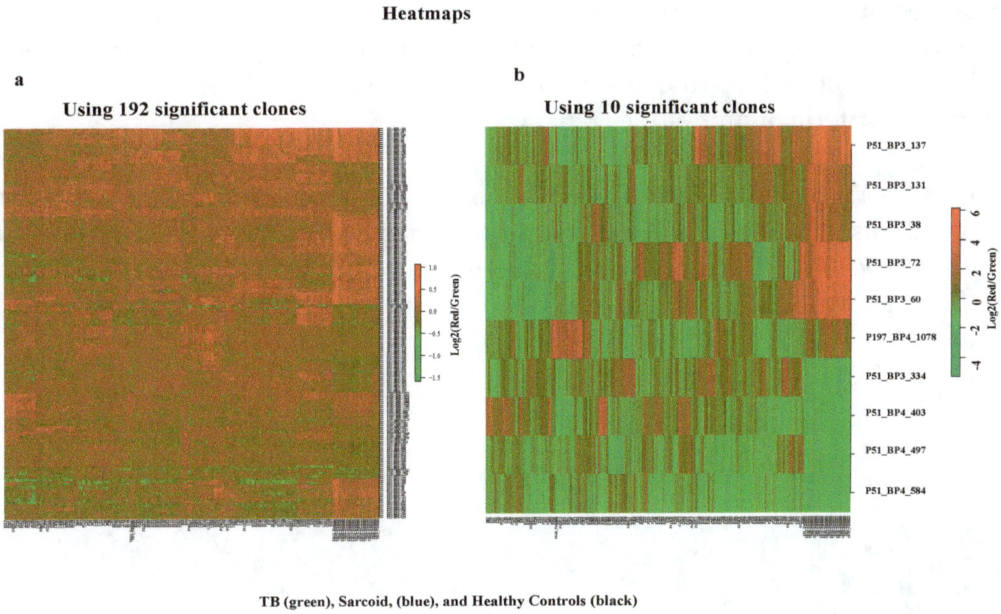

Figure 2. Heatmaps generated based on 192 clones and the 10 highly significant clones from the data of 176 study subjects (45 healthy controls, 24 with TB, and 107 with sarcoidosis) (**a**,**b**). Each row represents a clone, while each column represents a study subject. As shown in Figure 2, most of TB samples clustered to the right side of heatmap plots, while sarcoidosis samples and healthy controls clustered on the left side of the plot, indicating different expression profiles.

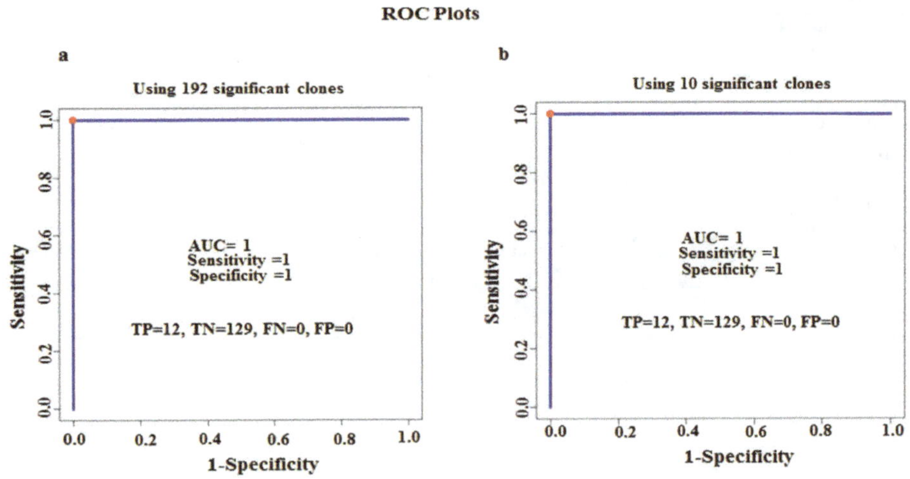

Figure 3. Classifiers to predict TB from healthy controls and sarcoidosis patients. (**a**) Performance of 192 clones on test set. (**b**) Performance of the top 10 classifier clones on test set. The ROC curves demonstrate excellent classification performance with AUC of 1 with sensitivity of 1 and specificity of 1.

3.2. Characterization of Ten TB Classifiers

Based on the results of training and test sets, we characterized the top 10 highly performing active TB clones through sequencing. After obtaining the sequences of clones, the Expasy program was used to translate the cDNA sequences to peptide/protein sequences [18,19].

Protein blast using algorithms of the BLAST program were applied to identify the highest homology to identified peptides [18,19]. The identified clones were blasted with human and MTB genomes and then selected those specific peptide sequences with the highest homology of amino acids sequence. All top 10 clones have the highest homology with TB sequences. Additionally, we compared these results with corresponding nucleotide sequences using nucleotide BLAST and determined the predicted amino acids in frame with T7 phage 10B gene capsid proteins. All of the 10 classifier clones are coded by the inserted gene fragments leading to out-of-frame peptides, therefore meeting the criteria of mimotopes [26] (Table 2).

As sera of active TB patients reacted with these out-of-frame peptides, it is likely that these TB clones are produced as a result of altered reading frames or alternative splicing, as described in previous studies [18,19,26]. Full length of peptides and genes of the ten classifiers clones are shown in Table S1. Table 2 shows the 10 most significant TB antigens, gene names, sensitivity, specificity, and FDR adjusted *p*-values. Figure 4 shows the ROC curves for six clones that are increased in TB, while Figure 5 shows ROC curves for four clones decreased in TB.

Table 2. 10 Top Significant TB Clones.

Clone	Increased in Tuberculosis (TB)	Gene Name	*p* Value	FDR Corrected *p* Value	AUC 95% CI	Sensitivity, 95% CI	Specificity, 95% CI
P51_BP3_38	Polyketide synthase	Pks13 Rv3800c	4.7×10^{-7}	2.79×10^{-5}	0.98	1	0.97
P51_BP3_60	Hydrolase	Rv1723	1.62×10^{-8}	3.46×10^{-6}	0.95	0.92	0.95
P51_BP3_72	Ferredoxin	fdxA Rv2007c	1.36×10^{-9}	7.27×10^{-7}	0.92	0.91	0.89
P51_BP3_131	Dihydroxy acid dehydratase	ilvD Rv0189c	2.15×10^{-8}	3.84×10^{-6}	0.95	0.92	0.98
P51_BP3_137	Transketolase	TKT Rv1449c	7.14×10^{-8}	9.72×10^{-6}	0.95	1	0.81
P197_BP4_1078	Signal peptidase	lepB Rv2903	1.44×10^{-7}	1.36×10^{-5}	0.78	0.92	0.64
	Decreased in Tuberculosis (TB)						
P51_BP3_334	TetR family transcriptional regulator	MRA_2532	4.02×10^{-10}	4.3×10^{-7}	0.98	1	0.91
P51_BP4_403	Menaquinone biosynthesis protein	menD Rv0555	7.27×10^{-8}	9.71×10^{-6}	0.95	1	0.87
P51_BP4_497	Cobalamin biosynthesis protein	CobN Rv2062c	1.11×10^{-8}	2.96×10^{-6}	0.88	0.92	0.78
P51_BP4_584	5-oxoprolinase	OplA Rv0266c	5.82×10^{-9}	2.10×10^{-6}	0.94	1	0.83

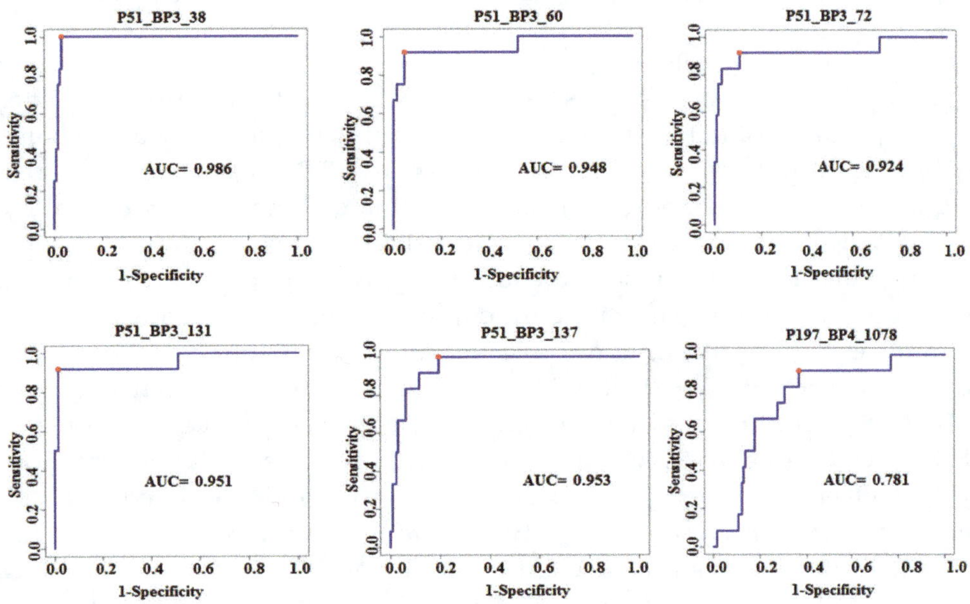

Figure 4. ROCs for top 6 significant clones that are increased in TB sera compared to healthy controls and sarcoidosis.

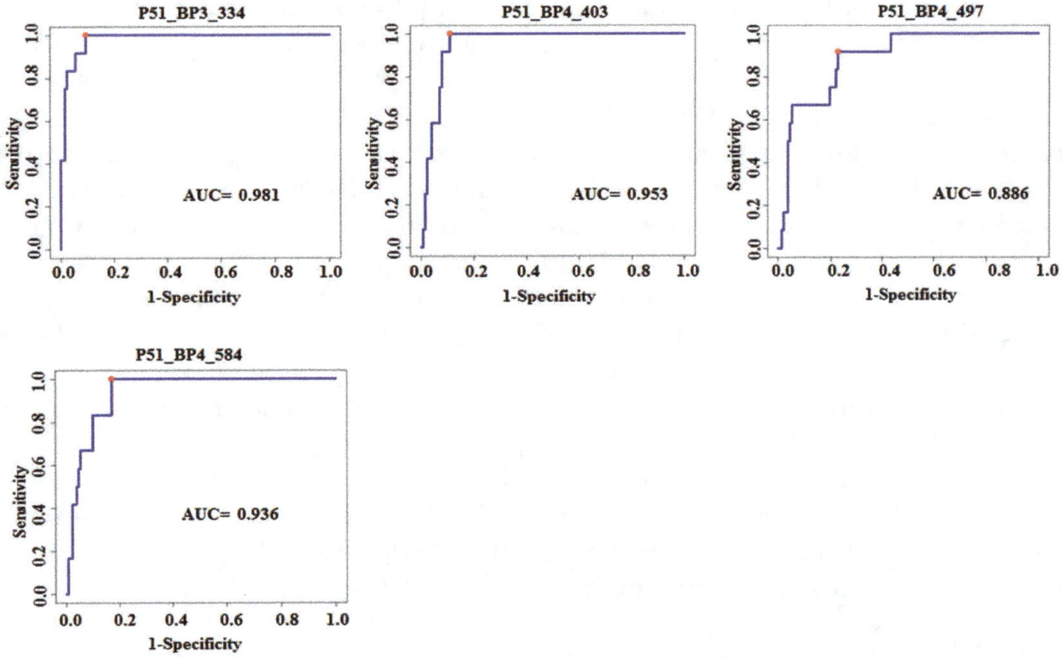

Figure 5. ROC for the top 4 significant clones that are decreased in TB sera compared to healthy controls and sarcoidosis. This figure demonstrates reasonable classification performance when classification was applied to one clone.

4. Discussion

Standard methods to diagnose TB and to monitor response to treatment rely on sputum microscopy and culture. The current CDC/NIH roadmap emphasizes the need for development of new TB biomarkers as alternative methods [2]. Recently, a tremendous effort has been put forward elucidating the antibody responses to MTB antigens, which has implications for the development of new antigens to diagnose and monitor successful treatment, as well as to develop effective vaccination [27]. Most other studies searching for TB antigens have identified unspecific markers primarily involving host response such as C-reactive protein, serum amyloid A and other non-specific markers [28,29].

In view of this background, we hypothesized complex library derived from sarcoidosis tissue may harbor degradation products of MTB antigens and these antigens can be used as a bait to specifically and selectively bind to antibodies present in sera from active TB subjects. Our microarray platform identified 10 highly significant TB clones that can discriminate TB patients from healthy controls and sarcoidosis patients. All of these clones are TB specific and related to bacterial growth of *M. Tuberculosis* and its metabolism (Table 2). We sequenced the top 10 highly significant clones for TB and identified homologies in a public database. The range in length of identified peptides for TB antigens was between 6–23 amino acids (AA). Among the 10 TB specific phage peptides, six out-of-frame peptides were increased in sera of active TB patients (Table 2). One of the highly sensitive and specific peptide antigens (P51_BP3_38) identified in sera from active TB subjects is polyketide synthase (PKS). There are about 24 PKS encoding genes in *M. Tuberculosis*. This is an essential enzyme for mycolic acid formation [30]. The cell envelope of *M. Tuberculosis* is distinctive and associated with its pathogenicity and resistance. Mycolic acid is a long chain fatty acid found in the cell wall of *M. Tuberculosis* and this compound constitutes major strategic elements of the protective coat surrounding the tubercle bacillus [30]. Moreover, the cyclopropane ring of mycolic acid protects the bacteria from oxidative stress [31]. Another identified peptide antigen (P51_BP3_60) highly reactive to sera of MTB patients was hydrolase. *M. Tuberculosis* secretes hydrolases that have lipase activity and catalyzes lipid hydrolysis. They are responsible for the degradation of host lipid material [31]. It has become clear that in vivo MTB prefers to consume fatty acids and lipids over carbohydrates. Tubercle bacillus utilizes the host derived lipids/fatty acids as nutrients for prolonged persistence in a hypoxic environment [31].

Ferredoxin is another antigen (P51_BP3_72) significantly increased in sera of MTB patients. Ferredoxins are acidic, soluble iron–sulfur proteins. They act as redox partner for the cytochrome P450 enzyme (CYP51B). The *M. Tuberculosis* genome contains 20 CYPs. They are involved in metabolic processes such as epoxidation, sulfoxidation, and hydroxylation. *M. Tuberculosis*'s CYPs and their redox partners such as ferredoxin are essential for pathogen viability [32]. Another important MTB specific peptide antigen (P197_BP4_1078) belongs to the signal peptidase I (SPase I) enzyme. This is a membrane-bound endopeptidase responsible for cleavage of signal peptides of secreted proteins [33]. SPase I is an attractive target for the development of novel anti-tuberculosis treatments because first, it is essential for survival in all bacterial species; and secondly, bacterial SPase I is distinctively different from eukaryotic SPase I. Similarly, peptide antigen (P51_BP3_131) dihydroxyacid dehydratase (DHAD), which is involved in the growth of *Mycobacterium* is significantly increased. It is a key enzyme involved in branched-chain amino acid synthesis and also catalyzes the synthesis of 2-ketoacids from dihydroxyacids. It has been shown that the downregulation of this enzyme inhibits the growth of *M. Tuberculosis* in vitro and in mice model of TB infection [34].

The peptide antigen (P51_BP3_137) increased in MTB belongs to transketolase (Tkt) enzyme. This enzyme catalyzes the synthesis of ribose-5-phosphate (R5P) from the intermediates of the oxidative pentose phosphate pathway. Studies have shown that the depletion of Tkt using RNA silencing and protein degradation systems arrested the growth of *M. Tuberculosis* in vitro. The studies further demonstrated, using an ex vivo model of TB transfection in THP-1 cells, that Tkt-depleted bacteria showed less virulence as compared to wild type bacilli, confirming the essentiality of this enzyme in intracellular growth [35]. The three peptide antigens (transketolase, ferredoxin, and dihydroxy acid dehydratase) identified with the present study were also identified in our previous published study using sera from culture positive but smear negative patients [18]. These results clearly demonstrate the importance of these peptide antigens in TB. Among ten mimotopes, we found four with decreased expression in TB patients (Table 2). Interestingly, one of these four peptides with higher sensitivity and specificity (P51_BP3_334), belongs to repressor transcriptional regulators such as TetR [36]. TetR is involved in the regulation of antibiotic resistance and controls the expression of membrane-associated proteins involved in antibiotic resistance [37,38].

In this study, we have identified 10 highly significant clones from the sera of smear positive TB patients. These identified clones are mostly involved in the growth and virulence of *M. Tuberculosis*. Most of these clones have high specificity and sensitivity. Previous studies using a combination of ESAT-6 and CFP10 antigens, which are two *Mycobacterium tuberculosis*-specific antigens, to diagnose TB provided a sensitivity of 73% and 93% of specificity [39,40]. While studies in countries with higher TB prevalence has shown even lower sensitivity and specificity using various antigens including ESAT-6 and CFP10 [41]. Interestingly, Drake and et.al showed that higher percentage of sarcoidosis subjects (16/26) exhibit immunoreactivity to ESAT-6 and katG [9]. Our results appear to have a higher sensitivity and specificity as compared with those studies. One limitation of our study is that we did not include infected subjects with non-tuberculous mycobacteria. Although among the control group, 16 Asian subjects had BCG vaccination and 6 had positive quantiferon gold tests, we did not have enough power to detect possible differences between subjects with latent TB and active TB infection. Larger studies using sera from diverse populations including, subjects with non-tuberculous mycobacterial infection, latent TB infection and after BCG vaccination need to further validate the sensitivity and specificity of our classifiers.

We detected these novel antigens using a heterologous library derived from sarcoidosis subjects. Lungs are highly exposed to numerous bacteria and our library is predominantly derived from sarcoidosis BAL cells and WBCs containing diverse immune cells, including macrophages that were exposed to various pathogens. We postulate that the CSL represents a segment of the lung microbe containing diverse antigens for TB, sarcoidosis, and cystic fibrosis [18–20].

There are various applications of a phage display. In the current work, we used a phage display for the discovery of TB biomarkers. The same system can be applied to identify novel markers for multi-drug resistance in TB, which is becoming a major issue in TB treatment. Additionally, phage displays can be used for the development of specific targeted therapies [42]. The phage display technology and immunoscreening has utilities not only in identifying diagnostic biomarkers, but also may enable us to develop a novel targeted therapy utilizing the peptide sequences (mimotopes) as vehicles to deliver specific drugs. The identified sequences can be used to develop peptide/protein-coated magnetic nanoparticles for clinical testing or for applications in drug delivery [43]. Additionally, this technology might enable us to discover unknown epitopes targeting specific bacterial antigens leading to immunogenicity and antibody production in TB subjects, as well as providing us with a better understanding of host immune defenses in TB subjects. For instance, TB sera were less reactive to some of the identified clones (TetR, menD, CobN, and OplA), these clones are less likely to be used for diagnostic purposes. However, these clones can be used to develop new vaccine and to boost the immunity against TB infection. Furthermore, this microarray platform can be hybridized to detect IgA in sputum of TB patients that may have clinical values. Moreover, antibody detection in the sera of patients has a potential value in clinical practice, as it is non-invasive and requires a minimal amount of blood or other bodily fluids.

The lack of sensitivity and specificity and cross-reactivity of biomarkers with other diseases dampened the enthusiasm in TB biomarker discovery studies. However, our study shows excellent sensitivity and specificity, not only as compared to healthy controls but also to another granulomatous disease. Other studies using gene expression profiling between TB and sarcoidosis found 94% similarities [44,45]. Our system has the advantage of detecting TB clones with high sensitivity and specificity and is based on an immune reaction rather than gene expression. The detection of this immune reaction, in form of antibodies, relies on a complex interaction between antigen presenting cells, T cells and B cells that leads to a specific antibody production in response to a TB infection. Highly specific biomarkers may have a potential role as candidate antigens in the development of novel vaccination for TB or for multidrug resistant bacterial infections.

Author Contributions: H.T. contributed to the methodology, sample processing and conducted the analysis. S.H. performed the preprocessing, validation of data, and statistical analysis using computer software. S.D. contributed to formal data analysis. L.S. contributed to conceptualization and designing of the study. L.S. and H.T. contributed to Writing-Original draft preparation, writing, reviewing and editing the data. L.S. supervise the project, designing the project and funding acquisition of this study. Wayne State University Department of Internal Medicine, Pulmonary division and Center for Medical Medicine and Genetics are involved in data curation.

Acknowledgments: We thank all patients and healthy volunteers for their participation in this study. This project was funded by NIH grant R21HL104481-01A1 awarded to L.S. and with the support of the Department of Medicine, Wayne State University. We would like to thank Foundation for Innovative New Diagnostics (FIND, Geneva, Switzerland) for providing TB samples. This work has been partially supported by the following grants: NIH R01 DK089167, NIH STTR R42GM087013, NSF DBI-0965741 (to Sorin Draghici), and by the Robert J. Sokol Endowment in Systems Biology. The technology of this study has a national and international technology Transfer ID: WO2016141347.

References

1. Lawn, S.D.; Zumla, A.I. Tuberculosis. *Lancet* **2011**, *378*, 57–72. [CrossRef]
2. Nahid, P.; Saukkonen, J.; Mac Kenzie, W.R.; Johnson, J.L.; Phillips, P.P.; Andersen, J.; Bliven-Sizemore, E.; Belisle, J.T.; Boom, W.H.; Luetkemeyer, A.; et al. CDC/NIH Workshop. Tuberculosis biomarker and surrogate endpoint research roadmap. *Am. J. Respir. Crit. Care Med.* **2011**, *184*, 972–979. [CrossRef] [PubMed]
3. Wallis, R.S.; Pai, M.; Menzies, D.; Doherty, T.M.; Walzl, G.; Perkins, M.D.; Zumla, A. Biomarkers and diagnostics for tuberculosis: Progress, needs, and translation into practice. *Lancet* **2010**, *375*, 1920–1937. [CrossRef]
4. Dubaniewicz, A.; Dubaniewicz-Wybieralska, M.; Sternau, A.; Zwolska, Z.; Izycka-Swieszewska, E.; Augustynowicz-Kopec, E.; Skokowski, J.; Singh, M.; Zimnoch, L. Mycobacterium tuberculosis complex and mycobacterial heat shock proteins in lymph node tissue from patients with pulmonary sarcoidosis. *J. Clin. Microbiol.* **2006**, *44*, 3448–3451. [CrossRef] [PubMed]
5. Moller, D.R. Potential etiologic agents in sarcoidosis. *Proc. Am. Thorac. Soc.* **2007**, *4*, 465–468. [CrossRef] [PubMed]
6. Song, Z.; Marzilli, L.; Greenlee, B.M.; Chen, E.S.; Silver, R.F.; Askin, F.B.; Teirstein, A.S.; Zhang, Y.; Cotter, R.J.; Moller, D.R. Mycobacterial catalase-peroxidase is a tissue antigen and target of the adaptive immune response in systemic sarcoidosis. *J. Exp. Med.* **2005**, *201*, 755–767. [CrossRef] [PubMed]
7. Rastogi, R.; Du, W.; Ju, D.; Pirockinaite, G.; Liu, Y.; Nunez, G.; Samavati, L. Dysregulation of p38 and MKP-1 in response to NOD1/TLR4 stimulation in sarcoid bronchoalveolar cells. *Am. J. Respir. Crit. Care Med.* **2011**, *183*, 500–510. [CrossRef] [PubMed]
8. Carlisle, J.; Evans, W.; Hajizadeh, R.; Nadaf, M.; Shepherd, B.; Ott, R.D.; Richter, K.; Drake, W. Multiple Mycobacterium antigens induce interferon-gamma production from sarcoidosis peripheral blood mononuclear cells. *Clin. Exp. Immunol.* **2007**, *150*, 460–468. [CrossRef] [PubMed]
9. Drake, W.P.; Dhason, M.S.; Nadaf, M.; Shepherd, B.E.; Vadivelu, S.; Hajizadeh, R.; Newman, L.S.; Kalams, S.A. Cellular recognition of Mycobacterium tuberculosis ESAT-6 and KatG peptides in systemic sarcoidosis. *Infect. Immun.* **2007**, *75*, 527–530. [CrossRef] [PubMed]
10. Hajizadeh, R.; Sato, H.; Carlisle, J.; Nadaf, M.T.; Evans, W.; Shepherd, B.E.; Miller, R.F.; Kalams, S.A.; Drake, W.P. Mycobacterium tuberculosis Antigen 85A induces Th-1 immune responses in systemic sarcoidosis. *J. Clin. Immunol.* **2007**, *27*, 445–454. [CrossRef] [PubMed]
11. Oswald-Richter, K.; Sato, H.; Hajizadeh, R.; Shepherd, B.E.; Sidney, J.; Sette, A.; Newman, L.S.; Drake, W.P. Mycobacterial ESAT-6 and katG are recognized by sarcoidosis CD4+ T cells when presented by the American sarcoidosis susceptibility allele, DRB1*. *J. Clin. Immunol.* **2010**, *30*, 157–166. [CrossRef] [PubMed]
12. Richmond, B.W.; Ploetze, K.; Isom, J.; Chambers-Harris, I.; Braun, N.A.; Taylor, T.; Abraham, S.; Mageto, Y.; Culver, D.A.; Oswald-Richter, K.A.; et al. Sarcoidosis Th17 cells are ESAT-6 antigen specific but demonstrate reduced IFN-gamma expression. *J. Clin. Immunol.* **2013**, *33*, 446–455. [CrossRef] [PubMed]
13. Smith-Rohrberg, D.; Sharma, S. Tuberculin skin test among pulmonary sarcoidosis patients with and without tuberculosis: Its utility for the screening of the two conditions in tuberculosis-endemic regions. *Sarcoidosis Vasc. Diffuse Lung Dis.* **2006**, *23*, 130–134. [PubMed]

14. Murray, P.J.; Wynn, T.A. Protective and pathogenic functions of macrophage subsets. *Nat. Rev. Immunol.* **2011**, *11*, 723–737. [CrossRef] [PubMed]
15. Peyron, P.; Vaubourgeix, J.; Poquet, Y.; Levillain, F.; Botanch, C.; Bardou, F.; Daffé, M.; Emile, J.-F.; Marchou, B.; Cardona, P.-J.; et al. Foamy macrophages from tuberculous patients' granulomas constitute a nutrient-rich reservoir for M. tuberculosis persistence. *PLoS Pathog.* **2008**, *4*, e1000204. [CrossRef] [PubMed]
16. Talreja, J.; Talwar, H.; Ahmad, N.; Rastogi, R.; Samavati, L. Dual Inhibition of Rip2 and IRAK1/4 Regulates IL-1beta and IL-6 in Sarcoidosis Alveolar Macrophages and Peripheral Blood Mononuclear Cells. *J. Immunol.* **2016**, *197*, 1368–1378. [CrossRef] [PubMed]
17. Murphy, J.; Summer, R.; Wilson, A.A.; Kotton, D.N.; Fine, A. The prolonged life-span of alveolar macrophages. *Am. J. Respir. Cell Mol. Biol.* **2008**, *38*, 380–385. [CrossRef] [PubMed]
18. Talwar, H.; Rosati, R.; Li, J.; Kissner, D.; Ghosh, S.; Madrid, F.F.; Samavati, L. Development of a T7 Phage Display Library to Detect Sarcoidosis and Tuberculosis by a Panel of Novel Antigens. *EBioMedicine* **2015**, *2*, 341–350. [CrossRef] [PubMed]
19. Talwar, H.; Hanoudi, S.N.; Geamanu, A.; Kissner, D.; Draghici, S.; Samavati, L. Detection of Cystic Fibrosis Serological Biomarkers Using a T7 Phage Display Library. *Sci. Rep.* **2017**, *7*, 17745. [CrossRef] [PubMed]
20. Talwar, H.; Talreja, J.; Samavati, L. T7 Phage Display Library a Promising Strategy to Detect Tuberculosis Specific Biomarkers. *Mycobact. Dis.* **2016**, *6*, 214. [CrossRef] [PubMed]
21. Ritchie, M.E.; Phipson, B.; Wu, D.; Hu, Y.; Law, C.W.; Shi, W.; Smyth, G.K. limma powers differential expression analyses for RNA-sequencing and microarray studies. *Nucleic Acids Res.* **2015**, *43*, e47. [CrossRef] [PubMed]
22. *R: A Language and Environment for Statistical Computing*; R Foundation for Statistical Computing: Vienna, Austria, 2015.
23. Ritchie, M.E.; Silver, J.; Oshlack, A.; Holmes, M.; Diyagama, D.; Holloway, A.; Smyth, G.K. A comparison of background correction methods for two-colour microarrays. *Bioinformatics* **2007**, *23*, 2700–2707. [CrossRef] [PubMed]
24. Yang, Y.H.; Dudoit, S.; Luu, P.; Lin, D.M.; Peng, V.; Ngai, J.; Speed, T.P. Normalization for cDNA microarray data: A robust composite method addressing single and multiple slide systematic variation. *Nucleic Acids Res.* **2002**, *30*, e15. [CrossRef] [PubMed]
25. Benjamini, Y.; Hochberg, Y. Controlling the false discovery rate: A practical and powerful approach to multiple testing. *J. R. Stat. Soc. Ser. B (Methodol.)* **1995**, 289–300.
26. Wang, X.; Yu, J.; Sreekumar, A.; Varambally, S.; Shen, R.; Giacherio, D.; Mehra, R.; Montie, J.E.; Pienta, K.J.; Sanda, M.G.; et al. Autoantibody signatures in prostate cancer. *N. Engl. J. Med.* **2005**, *353*, 1224–1235. [CrossRef] [PubMed]
27. Kunnath-Velayudhan, S.; Salamon, H.; Wang, H.Y.; Davidow, A.L.; Molina, D.M.; Huynh, V.T.; Cirillo, D.M.; Michel, G.; Talbot, E.A.; Perkins, M.D.; et al. Dynamic antibody responses to the Mycobacterium tuberculosis proteome. *Proc. Natl. Acad. Sci. USA* **2010**, *107*, 14703–14708. [CrossRef] [PubMed]
28. De Groote, M.A.; Nahid, P.; Jarlsberg, L.; Johnson, J.L.; Weiner, M.; Muzanyi, G.; Janjic, N.; Sterling, D.G.; Ochsner, U.A. Elucidating novel serum biomarkers associated with pulmonary tuberculosis treatment. *PLoS ONE* **2013**, *8*, e61002. [CrossRef] [PubMed]
29. Agranoff, D.; Fernandez-Reyes, D.; Papadopoulos, M.C.; Rojas, S.A.; Herbster, M.; Loosemore, A.; Tarelli, E.; Sheldon, J.; Schwenk, A.; Pollok, R. Identification of diagnostic markers for tuberculosis by proteomic fingerprinting of serum. *Lancet* **2006**, *368*, 1012–1021. [CrossRef]
30. Gavalda, S.; Bardou, F.; Laval, F.; Bon, C.; Malaga, W.; Chalut, C.; Guilhot, C.; Mourey, L.; Daffé, M.; Quémard, A. The polyketide synthase Pks13 catalyzes a novel mechanism of lipid transfer in mycobacteria. *Chem. Biol.* **2014**, *21*, 1660–1669. [CrossRef] [PubMed]
31. Aggarwal, A.; Parai, M.K.; Shetty, N.; Wallis, D.; Woolhiser, L.; Hastings, C.; Dutta, N.K.; Galaviz, S.; Dhakal, R.C.; Shrestha, R.; et al. Development of a novel lead that targets M. tuberculosis polyketide synthase. *Cell* **2017**, *170*, 249–259. [CrossRef] [PubMed]
32. Ouellet, H.; Johnston, J.B.; de Montellano, P.R.O. The Mycobacterium tuberculosis cytochrome P450 system. *Arch. Biochem. Biophys.* **2010**, *493*, 82–95. [CrossRef] [PubMed]
33. Ollinger, J.; O'Malley, T.; Ahn, J.; Odingo, J.; Parish, T. Inhibition of the sole type I signal peptidase of Mycobacterium tuberculosis is bactericidal under replicating and nonreplicating conditions. *J. Bacteriol.* **2012**, *194*, 2614–2619. [CrossRef] [PubMed]

34. Singh, V.; Chandra, D.; Srivastava, B.S.; Srivastava, R. Downregulation of Rv0189c, encoding a dihydroxyacid dehydratase, affects growth of Mycobacterium tuberculosis in vitro and in mice. *Microbiology* **2011**, *157*, 38–46. [CrossRef] [PubMed]
35. Kolly, G.S.; Sala, C.; Vocat, A.; Cole, S.T. Assessing essentiality of transketolase in Mycobacterium tuberculosis using an inducible protein degradation system. *FEMS Microbiol. Lett.* **2014**, *358*, 30–35. [CrossRef] [PubMed]
36. Balhana, R.J.; Singla, A.; Sikder, M.H.; Withers, M.; Kendall, S.L. Global analyses of TetR family transcriptional regulators in mycobacteria indicates conservation across species and diversity in regulated functions. *BMC Genom.* **2015**, *16*, 479. [CrossRef] [PubMed]
37. Liu, H.; Yang, M.; He, Z.-G. Novel TetR family transcriptional factor regulates expression of multiple transport-related genes and affects rifampicin resistance in Mycobacterium smegmatis. *Sci. Rep.* **2016**, *6*, 27489. [CrossRef] [PubMed]
38. Cuthbertson, L.; Nodwell, J.R. The TetR family of regulators. *Microbiol. Mol. Biol. Rev.* **2013**, *77*, 440–475. [CrossRef] [PubMed]
39. Van Pinxteren, L.A.; Ravn, P.; Agger, E.M.; Pollock, J.; Andersen, P. Diagnosis of tuberculosis based on the two specific antigens ESAT-6 and CFP. *Clin. Diagn. Lab. Immunol.* **2000**, *7*, 155–160. [PubMed]
40. Xu, J.-N.; Chen, J.-P.; Chen, D.-L. Serodiagnosis efficacy and immunogenicity of the fusion protein of Mycobacterium tuberculosis composed of the 10-kilodalton culture filtrate protein, ESAT-6, and the extracellular domain fragment of PPE. *Clin. Vaccine Immunol.* **2012**, *19*, 536–544. [CrossRef] [PubMed]
41. Kumar, G.; Dagur, P.K.; Singh, P.K.; Shankar, H.; Yadav, V.S.; Katoch, V.M.; Bajaj, B.; Gupta, R.; Sengupta, U.; Joshi, B. Serodiagnostic efficacy of Mycobacterium tuberculosis 30/32-kDa mycolyl transferase complex, ESAT-6, and CFP-10 in patients with active tuberculosis. *Arch. Immunol. Ther. Exp.* **2010**, *58*, 57–65. [CrossRef] [PubMed]
42. Wu, C.-H.; Liu, I.-J.; Lu, R.-M.; Wu, H.-C. Advancement and applications of peptide phage display technology in biomedical science. *J. Biomed. Sci.* **2016**, *23*, 8. [CrossRef] [PubMed]
43. Rana, S.; Bajaj, A.; Mout, R.; Rotello, V.M. Monolayer coated gold nanoparticles for delivery applications. *Adv. Drug Deliv. Rev.* **2012**, *64*, 200–216. [CrossRef] [PubMed]
44. Koth, L.L.; Solberg, O.D.; Peng, J.C.; Bhakta, N.R.; Nguyen, C.P.; Woodruff, P.G. Sarcoidosis blood transcriptome reflects lung inflammation and overlaps with tuberculosis. *Am. J. Respir. Crit. Care Med.* **2011**, *184*, 1153–1163. [CrossRef] [PubMed]
45. Maertzdorf, J.; Weiner, J.; Mollenkopf, H.J.; Network, T.B.; Bauer, T.; Prasse, A.; Muller-Quernheim, J.; Kaufmann, S.H. Common patterns and disease-related signatures in tuberculosis and sarcoidosis. *Proc. Natl. Acad. Sci. USA* **2012**, *109*, 7853–7858. [CrossRef] [PubMed]

Permissions

All chapters in this book were first published in MDPI; hereby published with permission under the Creative Commons Attribution License or equivalent. Every chapter published in this book has been scrutinized by our experts. Their significance has been extensively debated. The topics covered herein carry significant findings which will fuel the growth of the discipline. They may even be implemented as practical applications or may be referred to as a beginning point for another development.

The contributors of this book come from diverse backgrounds, making this book a truly international effort. This book will bring forth new frontiers with its revolutionizing research information and detailed analysis of the nascent developments around the world.

We would like to thank all the contributing authors for lending their expertise to make the book truly unique. They have played a crucial role in the development of this book. Without their invaluable contributions this book wouldn't have been possible. They have made vital efforts to compile up to date information on the varied aspects of this subject to make this book a valuable addition to the collection of many professionals and students.

This book was conceptualized with the vision of imparting up-to-date information and advanced data in this field. To ensure the same, a matchless editorial board was set up. Every individual on the board went through rigorous rounds of assessment to prove their worth. After which they invested a large part of their time researching and compiling the most relevant data for our readers.

The editorial board has been involved in producing this book since its inception. They have spent rigorous hours researching and exploring the diverse topics which have resulted in the successful publishing of this book. They have passed on their knowledge of decades through this book. To expedite this challenging task, the publisher supported the team at every step. A small team of assistant editors was also appointed to further simplify the editing procedure and attain best results for the readers.

Apart from the editorial board, the designing team has also invested a significant amount of their time in understanding the subject and creating the most relevant covers. They scrutinized every image to scout for the most suitable representation of the subject and create an appropriate cover for the book.

The publishing team has been an ardent support to the editorial, designing and production team. Their endless efforts to recruit the best for this project, has resulted in the accomplishment of this book. They are a veteran in the field of academics and their pool of knowledge is as vast as their experience in printing. Their expertise and guidance has proved useful at every step. Their uncompromising quality standards have made this book an exceptional effort. Their encouragement from time to time has been an inspiration for everyone.

The publisher and the editorial board hope that this book will prove to be a valuable piece of knowledge for researchers, students, practitioners and scholars across the globe.

List of Contributors

Piotr Jarocki, Elwira Komoń-Janczara, Oleksandr Kholiavskyi, Monika Pytka and Monika Kordowska-Wiater
Department of Biotechnology, Microbiology and Human Nutrition, University of Life Sciences in Lublin, 8 Skromna St., 20-704 Lublin, Poland

Marcin Podleśny
Process and Development Department, Al. Tysiąclecia Państwa Polskiego 13, Grupa Azoty Zakłady Azotowe "Puławy" S.A, 24-110 Puławy, Poland

Nikolay V. Volozhantsev, Alexander I. Borzilov, Ekaterina V. Komisarova, Valentina M. Krasilnikova and Vladimir V. Verevkin
State Research Center for Applied Microbiology and Biotechnology, 142279 Obolensk, Moscow Region, Russia

Anna M. Shpirt, Alexander S. Shashkov and Yuriy A. Knirel
N.D. Zelinsky Institute of Organic Chemistry, Russian Academy of Sciences, Leninsky Prospekt 47, 119991 Moscow, Russia

Sergey A. Buth and Petr G. Leiman
Institute of Physics of Biologic Systems, École Polytechnique Fédérale de Lausanne (EPFL), BSP-415, 1015 Lausanne, Switzerland

Mikhail M. Shneider
Institute of Physics of Biologic Systems, École Polytechnique Fédérale de Lausanne (EPFL), BSP-415, 1015 Lausanne, Switzerland
Shemyakin Ovchinnikov Institute of Bioorganic Chemistry, 16/10 Mikluho Maklaya Str., Moscow 117997, Russia

Dean Scholl
Pylum Biosciences, 385 Oyster Point Blvd., Suite 6A, South San Francisco, CA 94080, USA

Eunsu Ha, Bokyung Son and Sangryeol Ryu
Department of Food and Animal Biotechnology, Seoul National University, Seoul 08826, Korea
Department of Agricultural Biotechnology, and Research Institute of Agriculture and Life Sciences, Seoul National University, Seoul 08826, Korea

Eugenijus Šimoliūnas, Lidija Truncaitė, Rasa Rutkienė, Simona Povilonienė and Rolandas Meškys
Department of Molecular Microbiology and Biotechnology, Institute of Biochemistry, Life Sciences Centre, Vilnius University, Saulėtekio av. 7, LT-10257 Vilnius, Lithuania

Karolis Goda
Sector of Microtechnologies, Institute of Biotechnology, Life Sciences Centre, Vilnius University, Saulėtekio av. 7, LT-10257 Vilnius, Lithuania

Algirdas Kaupinis and Mindaugas Valius
Proteomics Centre, Institute of Biochemistry, Life Sciences Centre, Vilnius University, Saulėtekio av. 7, LT-10257 Vilnius, Lithuania

Valery A. Petrenko
Department of Pathobiology, Auburn University, Auburn, AL 36849-5519, USA

Jan W. Kretzer, Mathias Schmelcher and Martin J. Loessner
Institute of Food, Nutrition and Health, ETH Zurich, Schmelzbergstrasse 7, 8092 Zurich, Switzerland

Eva González-Menéndez, Lucía Fernández, Diana Gutiérrez, Beatriz Martínez, Ana Rodríguez and Pilar García
Instituto de Productos Lácteos de Asturias (IPLA-CSIC), Paseo Río Linares s/n, 33300 Villaviciosa, Spain

Daniel Pando
Nanovex Biotechnologies S.L., Parque Tecnológico de Asturias, CEEI, 33428 Llanera, Spain

Éva Viola Surányi, Kinga Nyíri, Bianka Kőhegyi and Beáta G. Vértessy
Department of Applied Biotechnology and Food Sciences, Budapest University of Technology and Economics, H-1111 Budapest, Hungary
Institute of Enzymology, RCNS, Hungarian Academy of Sciences, H-1111 Budapest, Hungary

Rita Hírmondó, Szilvia Tarjányi and Judit Tóth
Institute of Enzymology, RCNS, Hungarian Academy of Sciences, H-1111 Budapest, Hungary

Nataliia P. Antonova and Vladimir A. Gushchin
N.F. Gamaleya Federal Research Centre for Epidemiology and Microbiology, Ministry of Health of the Russian Federation, 123098 Moscow, Russia
Lomonosov Moscow State University, 119991 Moscow, Russia

Daria V. Vasina, Evgeny V. Usachev, Alexander L. Gintsburg and Artem P. Tkachuk
N.F. Gamaleya Federal Research Centre for Epidemiology and Microbiology, Ministry of Health of the Russian Federation, 123098 Moscow, Russia

Anastasiya M. Lendel
Lomonosov Moscow State University, 119991 Moscow, Russia

Valentine V. Makarov
Center for Strategic Planning of the Ministry of Health of the Russian Federation, 119435 Moscow, Russia

Jumpei Uchiyama, Hironobu Murakami, Naoki Watanabe, Tadahiro Nasukawa, Keijiro Mizukami, Masaya Ogata and Masahiro Sakaguchi
School of Veterinary Medicine, Azabu University, Kanagawa 252-5201, Japan

Hidehito Matsui and Hideaki Hanaki
Kitasato Institute for Life Sciences, Kitasato University, Tokyo 108-8641, Japan

Shin-ichiro Kato
Research Institute of Molecular Genetics, Kochi University, Kochi 783-8502, Japan

Shigenobu Matsuzaki
Kochi Medical School, Kochi University, Kochi 783-8505, Japan

Ben H. Burrowes
Department of Immunology and Molecular Microbiology, Texas Tech University Health Sciences Center, 3601 4th Street, Lubbock, TX 79430, USA
Roche Molecular Systems, 983 University Avenue B200, Los Gatos, CA 95032, USA

Ian J. Molineux
Center for Infectious Disease, Department of Molecular Biosciences, The University of Texas at Austin, 1 University Station A5000, Austin, TX 78712, USA

Joe A. Fralick
Department of Immunology and Molecular Microbiology, Texas Tech University Health Sciences Center, 3601 4th Street, Lubbock, TX 79430, USA

Vasanthan Devaraj
Research Center for Energy Convergence and Technology Division, Pusan National University, Busan 46241, Korea

Jiye Han and Chuntae Kim
Department of Nano Fusion Technology, Pusan National University, Busan 46241, Korea
BK21 Plus Nanoconvergence Technology Division, Pusan National University, Busan 46241, Korea

Yong-Cheol Kang
Department of Chemistry, Pukyong National University, Busan 48513, Korea

Jin-Woo Oh
Research Center for Energy Convergence and Technology Division, Pusan National University, Busan 46241, Korea
Department of Nano Fusion Technology, Pusan National University, Busan 46241, Korea
BK21 Plus Nanoconvergence Technology Division, Pusan National University, Busan 46241, Korea
Department of Chemistry, Pukyong National University, Busan 48513, Korea
Department of Nanoenergy Engineering, Pusan National University, Busan 46241, Korea

Marek Harhala, Paulina Miernikiewicz, Beata Brzezicka, Joanna Majewska, Aleksander Szymczak, Dorota Lecion and Karolina Lahutta
Bacteriophage Laboratory, Institute of Immunology and Experimental Therapy, Polish Academy of Sciences, Rudolfa Weigla Street 12, 53-114 Wroclaw, Poland

Daniel C. Nelson, Ryan D. Heselpoth, Sara B. Linden and Xiaoran Shang
Institute for Bioscience and Biotechnology Research, University of Maryland, Rockville, MD 20850, USA

Karolina Marek-Bukowiec, Marlena Kłak and Andrzej Konieczny
Research and Development Center, Regional Specialized Hospital, 51-124 Wrocław, Poland

Bartosz Wojciechowicz
Perlan Technologies Sp. z o. o., 02-785 Warsaw, Poland

Krystyna Dąbrowska
Bacteriophage Laboratory, Institute of Immunology and Experimental Therapy, Polish Academy of Sciences, Rudolfa Weigla Street 12, 53-114 Wroclaw, Poland
Research and Development Center, Regional Specialized Hospital, 51-124 Wrocław, Poland

Philip Serwer and Elena T. Wright
Department of Biochemistry and Structural Biology, The University of Texas Health Science Center, San Antonio, TX 78229-3900, USA

Harvinder Talwar
Department of Medicine, Division of Pulmonary, Critical Care and Sleep Medicine, Wayne State University School of Medicine and Detroit Medical Center, Detroit, MI 48201, USA

Samer Najeeb Hanoudi and Sorin Draghici
Department of Computer Science, Wayne State University, Detroit, MI 48202, USA

Lobelia Samavati
Department of Medicine, Division of Pulmonary, Critical Care and Sleep Medicine, Wayne State University School of Medicine and Detroit Medical Center, Detroit, MI 48201, USA
Center for Molecular Medicine and Genetics, Wayne State University School of Medicine, 540 E. Canfield, Detroit, MI 48201, USA

Index

P

R

S

T

V

www.ingramcontent.com/pod-product-compliance
Lightning Source LLC
LaVergne TN
LVHW081024210926
848485LV00004B/941